V. C. SRIVASTAVA

ACHIEVING SUSTAINABILITY IN AGRICULTURE

Issues, Challenges and Opportunities

ACHIEVING SUSTAINABILITY IN AGRICULTURE

Issues, Challenges and Opportunities

V. C. Srivastava
Ph.D., FISA
Ex- University Professor & Chairman
Department of Agronomy
and
Former Dean, Faculty of Agriculture
Birsa Agricultural University, Ranchi (India)

AGROBIOS (INDIA)

Published by:

AGROBIOS (INDIA)

Agro House, Behind Nasrani Cinema
Chopasani Road, Jodhpur 342 002
Phone: 91-0291-2642319, Fax: 2643993
E. mail: agrobios@sify.com

AGROBIOS (INIDA)

9 788177 544145

IS Rs. 2500.00 / US$ .00 4-5

P Published by: AGROBIOS (INDIA)

Published by: Dr. Updesh Purohit for Agrobios (India), Jodhpur
Laser Typeset at: Yashee Computers and Printers, Jodhpur
Cover Design by: Reena
Printed at: Babloo Offset, Jodhpur

Dedicated to
My Parents
Smt. Chandrawati
&
Sri. Kailash Prasad

PREFACE

Global population was only one billion in 1800 AD, doubled by 1930 and reached three billion mark by 1960. However, it took only 39 years to double itself again by 1999, when it reached to six billion mark. It is estimated that by 2100 it will again double itself to twelve billion mark. Most of the future increase in population will be in developing countries of Asia, Africa and Latin America, where more and more area under agricultural production systems have been developed at the cost of natural eco-systems, which gradually deplete the soils of organic matter, nutrients and water, making them unsuitable in the long-run leading to the question on their sustainability. Sustainability is generally seen as human oriented, long–term and involving interaction with natural systems. A sustainable agricultural production system must, therefore, produce economically the needed food, fibre, feed and fuel without depleting the natural resources and without deterrent to environment. It must discourage regional imbalance and encourage gender equality.

Through a large amount of information has been generated on various aspects of sustainable agriculture, they are scattered in literatures. Thus, a compilation of available information has been strongly felt keeping in view the requirement of students, teachers, research scientists, administrators and those interested in the subject. The book contains chapters on sustainable agriculture, farming systems, agro-forestry, organic farming and crop and environment. The chapters have been arranged systematically, simple to understand and comprehend in an integrated manner. There may still be some lacuna in the book, which will be corrected in the subsequent editions; if these are brought to my knowledge. Any suggestion from readers shall be welcomed for further improvement of the book in future. In writing the book, works of authors quoted and materials incorporated freely from various sources are gratefully acknowledged. Finally, I wish to place on record my appreciation for the perseverance, support and encouragement provided by my wife Kiran, sons Rajnish and Manish and daughters-in-law Monica and Shweta, which made this book a reality.

2010

V.C. Srivastava

CONTENTS

I SUSTAINABLE AGRICULTURE 1
1.1 Introduction 1
1.1.1 Concept and Goal 2
1.1.2 Definition 3
1.2 Classification of Relative Sustainability 5
1.3 Essential Components of a Sustainable Agriculture 5
1.4 Low Input Sustainable Agriculture (LISA) 6
1.5 Criteria of Sustainability 8
1.6 Evaluation of Sustainability 9
1.6.1 Assessment of Agricultural System for Evaluation 9
1.6.2 Criteria for Evaluating Sustainability 10
1.6.3 Qualitative Evaluation of a Sustainable System 11
1.7 Approach to Achieve Sustainability 12
1.8. Soil and water Resource Management 15
1.8.1 Soil Resource Conservation 15
1.8.1.1 Soil Degradation 15
1.8.1.2 Soil Conservative Measures 17
1.8.2 Water Resource Conservation 20
1.8.2.1 Increasing Efficiency of Rain Water 21
1.8.2.2 Water Harvesting and Recycling 21
1.8.2.3 Increased Water Use Efficiency (WUE) 23
1.8.2.4 Sustainable Utilization of Water Resources 23
1.8.2.5 Water Conservation in Irrigated Agriculture 24

1.9 Conservation Tillage 24
1.9.1 Types of Tillage Systems 25
1.10 Soil Fertility Management 29
1.10.1 Concept of Soil Fertility 29
1.10.2 Role of Organic Matter and Bio-fertilizers 29
1.10.3 Adverse Effect of Nutrient Accumulation 30
1.10.4 Impact of Fertilizer Use on Environment 31
1.10.5 Nutrient Imbalance 32
1.10.6 Integrated Nutrients Management – Key for Sustainable Production 33
1.10.6.1 Concept of Integrated Nutrient Management System (INMS) 34
1.10.6.2 Nutrient Management for Crops and Cropping Systems 36
1.10.6.3 Integrated Nutrient Management Practices 37
1.11 Efficient Cropping Systems 37
1.11.1 Concept of Cropping Systems 37
1.11.2 Eco-friendly Management of Cropping System 38
1.11.3 Conservation Cropping Pattern for Sustained Production 40
1.11.3.1 Cropping Pattern Suited to Land Capabilities 40
1.11.3.2 Crop Management for Rainfed Condition 40
1.11.3.3 Cropping through a Watershed Management approach 44
1.11.3.4 Cropping in Acid Soil 45
1.11.3.5 Cropping in Alkali and Saline Soils 46
1.12 Integrated Pest Management and Strategies 47
1.12 Strategies for Sustainable Agricultural Development and Environment Conservation 51
References 53

II FARMING SYSTEMS 59
2.1 Introduction 59
2.1.1 Concept and Definition 59
2.1.2 Scope and Objectives 62
2.1.2.1 Scope 62
2.1.2.2 Objectives 63
2.2 Plant Interactions 63

2.3 Components of Farming Systems 66
2.3.1 Management of Animal Components 66
2.3.1.1 Cattle / Dairy Management 66
2.3.1.2 Sheep and Goat Rearing 68
2.3.1.3 Poultry and Duck Rearing 69
2.3.1.4 Piggery 71
2.3.1.5 Rabbit Rearing 72
2.3.2 Management of Other Allied Components 72
2.3.2.1 Aquaculture (Fish culture) 72
2.3.2.2 Apiculture 73
2.3.2.3 Sericulture 74
2.3.2.4 Mushroom cultivation 75
2.3.2.5 Biogas 76
2.3.3 Terms and Definitions 76
2.3.3.1 Crop Components 76
2.3.3.2 Tree Based Components 79
2.3.3.3 Animal Based Component 80
2.4 Integrated Farming System (IFS) 81
2.4.1 Concept and Definition 81
2.4.2 Classification of Farming System Interactions. 81
2.3.3 Advantages of Integrated Farming System (IFS) 82
2.4.4 Aims 83
2.4.5 Farming Situations 83
2.4.6 Integrated Aquaculture - Agriculture Components 84
2.4.6.1 Cropping with Aquaculture 84
2.4.6.2 Rice-fish Farming 84
2.4.6.3 Rice-fish-azolla Farming 85
2.4.6.4 Rice-fish-vegetable Farming 85
2.4.6.5 Horticulture with Fish Farming 85
2.4.6.6 Fish-cum-sericulture 86
2.4.7 Integrated Agriculture and Livestock Components 86
2.4.7.1 Cropping with Dairy 86
2.4.7.2 Cropping with Goat Rearing 86
2.4.7.3 Cropping with Animal and Biogas System 86
2.4.8 Integrated Aquaculture and Animal Components 88

2.4.8.1 Fish-cum-pig-farming 88
2.4.8.2 Fish-cum-poultry /Duck Farming 88
2.4.9 Integrated Agriculture, Animal and Tree-Components 88
2.4.9.1 Cropping with Goat and Silvi-Pasture in Rainfed Areas 88
2.4.9.2 Cropping with Dairy, Biogas and Silviculture in Irrigated Areas 89
2.4.10 Crops and Cropping /Farming Systems and Prospects of Diversification 89
2.4.10.1 Western Himalayan Region 90
2.4.10.2 Eastern Himalayan Region 90
2.4.10.3 Indo-gangetic (Lower, Middle, Upper and Trans Gangetic) Plains Region 90
2.4.10.4 Plateau (Eastern, Western, Southern and Central) Region 91
2.4.10.5 Coastal (Eastern and Western) Region 99
2.4.10.6 Western Dry Region 99
2.4.10.7 Island Region 99
2.4.10.8 Diversification of Area for Vegetable, Aromatic and Medicinal Plants and Floriculture 100
2.5 Harmonic Farming System 100
2.6 Determinants of Farming System 101
2.7 Technology 105
2.8 Recommendation Domain 105
2.9 Farming System Research (FSR) 107
2.9.1 Introduction 107
2.9.2 Why the Technologies Generated Become Inappropriate? 107
2.9.3 Limitations of Present Research Thrust 108
2.9.4 Concept, Definition and Approach of FSR 109
2.9.5 Farming System Research versus Conventional Agricultural Research 110
2.9.6 Goal of FSR 111
2.9.7 Principles involved in FSR 111
2.9.8 Scope of FSR 111
2.9.9 Objectives of FSR 112

2.9.10 Classification of Farming System Research (FSR) 112
2.9.11 Characteristics of Farming System Research (FSR) 114
2.9.12 Methodology 116
2.9.13 Steps followed for Implementation of a Project 120
2.9.14 Strategy 121
2.10 Shift in Future Strategies of Research 122
References 124

III AGROFORESTRY 128

3.1 Introduction 128
3.1.1 Historical Background 129
3.1.2 Concept of Agroforestry 131
3.1.3 Definitions 132
3.1.4 Other Related Terms 135
3.1.5 Scope and Need 136
3.1.6 Benefits of Agroforestry 138
A. Economic 138
B. Enviornmental 139
C. Social 141
3.1.7 Limitation of Agroforestry 142
3.2 Classification of Agroforestry Systems 144
3.2.1 Criteria of Classification 144
3.2.2 Structural Classification 144
3.2.3 Functional Classification of Agroforestry Systems 152
3.2.4 Socioeconomic Classification 152
3.2.5 Ecological Basis of Classification 153
3.3 Agroforestry System 154
3.3.1 Traditional Agroforestory Systems in India 161
3.3.1.1 Forest-based Agroforestry Systems or Swidden Cultivation 161
3.3.1.2 Farm- based Agrorestry Practices 164
3.4 Tree- Crop Interactions 179
3.4.1 Nature of Interactions 180
3.4.2 Factors Affecting Interactions 181
3.4.3 Types of Interactions 182
3.4.4 Methods for Quantifying Interactions 184
3.5 Management of Agroforestry System 187

3.5.1 Characteristics of Tree Species for Agroforestry 187
3.5.2 Suitable Areas for Agroforestry 188
3.5.3 Management Practices in Agroforestry 188
3.5.3.1 Management of Component Species 188
3.5.3.2 Nursery Management 190
3.5.3.3 Planting and After Care 191
3.5.3.4 Management of Trees 192
3.6 Design and Analysis of Agroforestry Experiments 195
3.6.1 Criteria for Good Agroforestry Design 195
3.6.2 Important Parameters of Agroforestry Research 195
3.6.3 Features of Agroforestry System and Measurment 196
3.6.4 Design Requirements 196
3.6.5 Design for Agroforestry 197
3.7 Strategy for Agroforestry Developments 199
3.7.1 Future Thrust 199
3.7.2 Research Needs and Priority Areas 201
References 204
IV ORGANIC FARMING 208
4.1 Introduction 208
4.1.1 Use of Chemical Inputs and their Adverse Effects 210
4.1.1.1 Pesticides 210
4.1.1.2 Insecticides and Fungicides 210
4.1.1.3 Herbicides 212
4.1.1.4 Fertilizers 212
4.1.2 Need for Organic Farming 213
4.2 Concept Definitions, Characteristics and Advantages 213
4.2.1 Basic Concept 213
4.2.2 Definitions 214
4.2.3 Organic Versus Natural Farming 215
4.2.4 Characteristics 215
4.2.5 Options 216
4.2.6 Objectives 216
4.2.7 Components 217
4.2.8 Organic Vs Conventional Farming 217
4.2.9 Advantages 218
4.2.10 Limitations 219

4.3 Principles and Practices of Organic Farming 220
4.3.1 Principles of Organic Farming 220
4.3.2 Practices of Organic Farming 221
4.4 Nutrient Management in Organic Farming 221
4.4.1 Green Manuring (GM) 222
4.4.1.1 Background and Scope 222
4.4.1.2 Definition 222
4.4.1.3 Types of GM 222
4.4.1.4 Subsidiary Objectives of GM 223
4.4.1.5 Advantages of GM 223
4.4.1.6 Disadvantages of GM 224
4.4.1.7 Classification of Green Manures 225
4.4.1.8 Plant Suitable for GM 225
4.4.1.9 The Methods of Growing GM Crops *in situ* 226
4.4.1.10 Selection of GM Crops and their Agronomy for Growing *in situ* 227
4.4.1.11 Agronomy of Certain Green Leaf Manure Shrubs/Trees 228
4.4.1.12 Technique of Green Manuring *in situ* 230
4.4.1.13 Green Manure in Cropping Systems 230
4.4.1.14 Decomposition of Green Manure 231
4.4.2 Inclusion of Legumes in Cropping Systems 234
4.4.3 Crop Residue Management 236
4.4.3.1 Introduction 236
4.4.3.2 Crop Residues 237
4.4.3.3 Crop Residue Potential 238
4.4.3.4 Residue Management Practices 240
4.4.3.5 Crop Residues and Physical Properties of Soil 242
4.4.3.6 Crop Residues and Chemical Properties of Soil 243
4.4.3.7 Crop Residues and Biological Properties of Soil 245
4.4.4 Farmyard Manure (FYM) 245
4.4.5 Biofertilizer and Organic Fertilizer/Native Phosphate 247
4.4.5.1 Biofertilizers 247

4.4.5.2 Types of Biofertilizers-factors affecting use and Methods of Application 248
4.4.6 Waste Recycling 256
4.5 Weed Management 257
4.6 Insect, Pest and Disease Management 259
4.7 Global Scenario and Prospects of Organic Farming 260
4.7.1 Organic Farming in USA 261
4.7.2 Organic Farming in other Countries 262
4.7.3 Organic Farming in India 263
References 266

V AGRICULTURE AND ENVIRONMENT 275

5.1 Introduction 275
5.2 Environment and Plant Growth 278
5.2.1 Definition and Scope 278
5.2.2 Physical Environment 280
5.2.3 Energy Environment 281
5.2.4 Microclimate of Crops 282
5.3 Climatic Factors and Crops 282
5.3.1 Weather and Climate 283
5.3.2 Solar Energy 283
5.3.2.1 Radiant Energy 283
5.3.2.2 Measurement of Solar Energy Use 284
5.3.2.3 Solar Radiation Effect on Crops 285
5.3.3 Temperature and Plant Growth 288
5.3.3.1 Role of Temperature 289
5.3.3.2 Temperature Requirement of Certain Crops 291
5.3.4 Atmospheric Humidity 293
5.3.5 Precipitation 293
5.3.6 Wind Velocity 295
5.3.7 Composition of the Atmosphere 297
5.4 Edaphic Factors and Crops 298
5.4.1 Soil Air 298
5.4.2 Soil Temperature 299
5.4.3 Soil Moisture 301
5.4.4 Soil Reaction 302
5.4.5 Mineral Nutrient Supply 303
5.4.6 Biotic Factors 304
5.5 Environmental Issues in Agriculture 306

5.5.1 Air Pollution 306
5.5.1.1 Depletion of Ozone Layer 307
5.5.1.2 Greenhouse effect and Global Warming 309
5.5.1.3 Climate Change 311
5.5.1.4 Acid Rain 314
5.5.2 Water Crises and Pollution 315
5.5.3 Soil Degradation and Pollution 317
5.5.4 Pesticide Residues 318
5.5.5 Loss of Biodiversity 319
5.6 Mitigating Pollution in Crop Environment 321
5.6.1 Reducing Greenhouse Gases from Agricultural Source 321
5.6.1.1 Nitrous Oxide 321
5.6.1.2 Methane 322
5.6.1.3 Carbon dioxide 322
5.6.1.4 Conventional Fossil Fuel 322
5.6.1.5 Degradation of Heavy Metals by Microbes and Plants 323
5.6.1.6 Degradation of Hazardous Chemical by Microorganism 323
5.6.2 Measures to Reduce Air Pollution 324
5.6.2.1 Farming Systems Approach 324
5.6.2.2 Tree Plantation 324
5.6.2.3 Conservation of Energy 324
5.6.3 Control of Water Pollution 325
5.6.4 Control of Soil Pollution 326
5.6.5 Conservation of Biodiversity 326
5.7 Protected Agriculture 327
5.7.1 Drought Proofing 328
5.7.2 Irrigation and Rainwater Management 328
5.7.3 Response Farming 328
5.7.4 Manipulation of Micro-climate 329
References 331

Subject Index 336

CHAPTER I

SUSTAINABLE AGRICULTURE

1.1 INTRODUCTION

According to World Commission on Environment and Development "the next few decades present a greater challenge to the world food system than they may ever face again. The effort needed to increase production in pace with unprecedented increase in demand, while retaining the essential ecological integrity of food system, is colossal both in its magnitude and complexity. Given the obstacles to be overcome, most of them man-made, it can fail more easily than it can succeed."

Global populations are increasing at a rate of 1.6 to 1.7 percent resulting in almost 90 million more consumers of agricultural products annually. More than 90 percent of this growth is occurring in developing world, many parts of which are already suffering from inadequate food. Demand for food is influenced by a number of forces including population growth and movements, income levels and economic growth, human resource development, lifestyle and preferences. People's access for food depends on income. Currently, more than 1.3 billion people are absolutely poor with incomes of a dollar a day or less per person, while onother 2 billion people are only marginally better off (World Bank, 1997). Although the absolute population increase during the next quarter century will be highest in Asia, the relative increase will be greatest in Sub-Sahara Africa where the population is expected to almost double. The World Bank estimates that about one third of the developing world population does not have enough calories for an active working life. Moreover, approximately one-fourth of the population of both Africa and South Asia do not receive enough calories to prevent malnutrition and serious health risks, according to United Nations agencies.

Growth in population not only substantially increases the demand for food but also contributes to environment and natural resource degradation problems in ways that threaten the continued productivity of many areas. For example, growing needs have forced developing world farmers to clear and cultivate forests, hill-sides and other fragile lands with the intensification of traditional cropping cycles. This has resulted in greater accelerated rates of erosion and loss of soil productivity. The forage needs of livestock in this part of world now exceeds the carrying capacity of grassland leading to severe deterioration and often to desertification. Current and future food productions in virtually all developing countries suffer from seasonal or chronic water shortage. Thirty five countries have annual internal renewable water resources that are below the threshold required for socioeconomic and environmental quality (Engelman and LeRoy, 1993; Reserigrant, 1997) and could experience water scarcity by 2020. The water-related challenges on the horizon include growing competition for water between countries increasing the potential for water wars; conflict between farmers, industrialists and urban residents; and rising costs of developing new water resources. Water resources are being depleted and air and soil water are being contaminated by waste products of industry, urban development and some agricultural practices. Thus, the increasing demand for food, fibre, fuel and fodder on one hand and the deterioration of environment due to indiscriminate use of natural resources on the other have caused great concern world over for sustaining. It must be remembered that reduction in poverty is an essential element of sustainable development since poor people cannot afford conservation efforts that deprive them of their livelihood.

1.1.1 Concept and Goal

'Sustainable agriculture' according to Lockeretz (1988) encompasses a range of strategies for addressing many of the problems that afflict agriculture. These problems include loss of soil productivity from intensive cultivation / excessive erosion and associated plant nutrient losses / depletion; surface and ground water pollution from pesticides, fertilizers and sediments; impending shortage of non-renewable resources, and low farm income due to low commodity prices and high production costs.

There are five important aspects of sustainable agriculture and achievement of all these is a formidable task. These are:

(i) Meeting the changing needs of today and tomorrow,

(ii) Economic viability of enhanced productivity level,

(iii) Successful management of resources, internal, external, renewable or non-renewable,

(iv) Maintenance, preferably enhancement of quality environment, and

(v) Conservation of natural resources particularly soil and water, which form the pivot of agriculture.

A System can be considered sustainable if it ensures that today economic development is not at the expense of tomorrow's development prospects (World Commission on Environment and Development, 1987). Traditional agricultural systems that met the test of sustainability have not been able to respond adequately to modern rate of growth in demand for agricultural commodities as demanded by the current population pressure of man and animal and rapidly declining resources of arable good quality lands and water resources. For example, the shifting cultivation system of long years of rotation supporting small population of tribal societies of north-east India, which was sustainable and good in the past, is not sustainable today to meet the needs of future and the present day population adequately. Hence, an alternative and more productive but sustainable system are needed. Thus, the need of the time is sustainability of agriculture at enhancing rate of productivity.

1.1.2 Definition

Sustainable agriculture has been defined in many different ways. The definitions which appear to be appropriate are given below:

- According to definition proposed by a group constituted by American Society of Agronomy (1988), "sustainable agriculture is one that over the long term (i) enhances environmental quality and the resource base on which agriculture depends, (ii) provides for basic human food and fibre needs, (iii) is economically viable, and (iv) enhances the quality of life for farmers and society as a whole."
- "An agriculture that can evolve indefinitely towards greater human utility, greater efficiency of resource use and a balance

with the environment that is favourable to human and most other species."

- "The successful management of resources for agriculture to satisfy changing human needs while maintaining or enhancing the natural resource base and avoiding environmental degradation."
- The subgroup on maintainability and economic productivity of the standing policy planning committee of Indian Council of Agricultural Research accepted the definition that "a sustainable system is the one with a non-negative trend in a measured output. The system is sustainable over a defined period if output does not decrease when inputs are not increased and benefits derived from year to year and in principle from a system should be maintained in practice from generation to generation."
- "Sustainable agriculture is a management system for renewable agriculture resources including soil, wild life, forests, crops, fish, livestock, plant genetic resources and ecosystems to provide food, income and livelihood for current and future generations and that maintains or improves the economic productivity and ecosystem services of these resources" (USAID)
- "Sustainable agriculture is the successful management of resources for agricultural production to satisfy the changing human needs, while maintaining or enhancing the quality of its environment and conserving natural resources." (TAC, 1988)
- "A sustainable agriculture is committed to maintain and preserve the agricultural base of soil, water, and atmosphere ensuring future generations the capacity to feed themselves with adequate supply of safe and wholesome food. It neither allows surrender of modern technology nor employs it recklessly; the former is prescription of starvation and later one is prescription for catastrophe. Hence, we must choose the course in between that is to use technology wisely, much as we use fire ensuring that it warms but does not burn." (Gracey, 1990)
- "Sustainable agriculture system is one that ensures over time secure nutrition and minimum living comforts to the growing human population with environment preservation and producer's satisfaction." In brief "sustainable agriculture

means increased agriculture production in tune with efficient, economic and environmentally sound production principles."

1.2 CLASSIFICATION OF RELATIVE SUSTAINABILITY

The conceptual elegance of the term "Sustainable" is that it has a time dimension in practice. Based on the length of time expected to elapse before the continued use of a particular system would become unacceptable, Smith and Dumnaski (1993) proposed a classification of relative sustainability (Table 1). The classes are a measure of evaluator's confidence in the stability of factors affecting each system.

TABLE 1 Classes of sustainability proposed (Smith and Dumanski, 1995)

Class		Confidence Limits (Years)
Sustainable	1. Sustainable in Long term	>25
	2. Sustainable in medium term	15 – 25
	3. Sustainable in short term	7- 15
Unsustainable	1. Slightly unstable	5 – 7
	2. Moderately unstable	2 –7
	3. Highly unstable	< 2

Most of the low-input farming systems and high input systems, sooner or later would probably fail because they are neither economically nor environmentally sustainable over the long term. To achieve "sustainable agriculture" settled in as the ultimate goal depends on creative and innovative conservation / restoration and production practices that provide farmers with economically viable and environmentally sound alternatives or options in their farming systems (Parr *et al.*, 1990)

1.3 ESSENTIAL COMPONENTS OF A SUSTAINABLE AGRICULTURE

Essential components of a sustainable agriculture (Kumar Swamy, 1994) are listed below:

I. Related to production targets

   a) Stable supply of food and living comforts to the growing human population.

b) Adequate supply of industrial raw materials to encourage employment in non-farm sectors.

c) Regular growth, both quantitatively and qualitatively, to keep pace with growing needs of the mankind.

d) Overcoming natural calamities like drought, flood, etc.

e) Paying for the production inputs and for constant upgrading of the farm resource base.

II. Related to production inputs

a) Production with minimum exploitation of physical and biological environment.

b) Conservation of natural resources for continued production in future also.

c) Prevention of environmental degradation and maintenance of environmental quality.

d) Less dependence on non- renewable source of energy and other inputs from outside the system.

e) Less dependence on harmful chemicals that cause environmental imbalance.

III. Related to producers satisfaction

a) Economic viability of the system at the farm level with a favorable cost: benefit ratio.

b) Minimum risk of failure due to natural factors, management factors or social-economic factors.

c) Less dependence on outside input and easy availability of required inputs.

d) Adequate additional net income to the farmer to meet cost living with constant upgradation of living standards.

e) Minimum physical drudgery and mental botheration to the farmer and the labourers.

1.4 LOW INPUT SUSTAINABLE AGRICULTURE (LISA)

Interest in LISA, an acronym for “low- input sustainable agriculture” was stimulated in the mid-1980 in USA by the emergence of some major public concern about agriculture, particularly water quality, soil conservation and other environmental issues. The LISA issue clearly revealed the need for a major infusion of systems for production and marketing systems research. A system approach is required to investigate the complex interactions that occur within and among agricultural systems and to predict the effects of these interactions. LISA concept stimulated for finding more productive, cast-efficient, environmentally safe and resource conserving methods of producing crops.

In LISA, the level of input usage is tied directly to the sustainability issue. Many believe it is serious mistake to link low inputs directly with sustainability. The "low inputs" is a very imprecise term. How low? Low in relation to what? Whar inputs? Systems involving reduced level of some inputs, such as herbicides, may actually require higher levels of other inputs, such as labour. It should be recognised that there is no adequate substitute for the use of commercial fertilizers if agricultural systems, world wide, are to satisfy the food needs of steadily increasing numbers of people at prices which most can afford to pay. Where fertilizers are contributing to environmental problems (such as nitrate pollution of ground water) research should be accelerated to develop means of overcoming the problems. Chemical fertilizers are essential for plant growth while some of these nutrients can be supplied by soils as nutrient levels in soil are usually inadequate for optimum crop performance, Some nutrients particularly N may be provided by plant and animal residues. However, such residues are usually not available to most farmers in amounts needed for optimum crop performance. It should be noted that less than one percent (1%) US food is estimated to be produced organically - that is without the use of chemical inputs. However, the cost of such food is usually substantially higher because of lower production levels and at times greater cost of other inputs such as labour (York, 1989).

Low input systems are least appropriate for operations with high fixed costs. A lower price does not result in a proportionately lower cost of the optimum combination of inputs. If low-Input systems are those that involve lower levels of inputs than the level leading to minimum cost per unit of output, low input systems cannot be self-sustaining in a mature agricultural industry (Holt, 1989).

Inspite of merits and many benefits offered by animal manures, the most controversial issues pertaining to agricultural sustainability relates to use of inputs, especially pesticides and fertilizers. Some believe that high levels of such inputs threaten sustainability, others suggest that without greater input use, sustainability objectives cannot be achieved. For example, without adequate inputs, many believe, it would be impossible to meet the food needs of increasing population unless additional land that is less sustainable is brought into cultivation, thereby damaging or destroying the natural ecosystems. People advocate organic farming and use of no commercial chemicals. This is sharply in contrast to the type of intensive farming found throughout world including those self sufficient in food.

1.5 CRITERIA OF SUSTAINABILITY

(i) The sustainability of agricultural systems must:

- be productive and economically viable both in the short-term and long-term perspectives;
- be dynamic and adaptable to changing needs and conditions, i.e. implementable with reasonable efforts;
- be ecologically sound to protect environment and to give priority attention for providing genetic and ecological integrity of the renewable natural resource base; and
- be human and socially just and to enhance health and safety.

(ii) In the design of sustainable agricultural systems, requirement of future generation must be considered.

(iii) In the system, natural resources also perform eco-system services beyond the production of food, fibre, fuel, and income, e.g.

- detoxification of noxious chemicals within soils;
- purification of waters;
- control of local climate; and
- regulation of hydrological process within watersheds.

(iv) It must prevent degradation and excessive loss of soil by erosion and excessive loss of rain water by runoff.

(v) It should replenish nutrients removed by harvesting of crops and grazing of livestock through a judicious combination of recycling of nutrients contained in crop residues, animal wastes and human wastes, those mobilized by natural biological process, and those applied through fertilizers and manures.

(vi) It should control weeds, pests and diseases through integrated control mechanism and cultural practices that minimize off-site cost and inputs and protect beneficial organisms.

1.6 EVALUATION OF SUSTAINABILITY

1.6.1 Assessment of Agricultural System for Evaluation

For evaluating the sustainability of an agricultural system (Virmani & Eswaran, 1990), the assessment consists of:

(i) **An assessment of risk:** The farmer's goal is to be able to produce at a threshold level which is a function of his socio-economic status. His assessment of risk is a function of the production in relation to his inputs.

(ii) **An assessment of Production:** Increase in production, which may be yield, quality, durability of the products, etc. may be achieved through better fertilization, resistance to pest damage or any other components of the technology.

(iii) **An assessment of maintenance of production over a time-frame-stability of the system:** In this case technology is evaluated in the severest stress situation whether abiotic or biotic (e.g., drought, flood, pest damage.)

(iv) **An assessment of the impact of the farming system-Degradation**: Different soils behave differently. Changes in productivity as a result of erosion on vertisols which may be imperceptible in the initial stages and only become evident when the effective soil volume is reduced to a critical depth. An alfisol on the other hand behave differently; loss of a few centimeters of top soil may show marked differences in productivity.

(v) **An assessment of the economics of the system from the farmers' perspective- Profitability**: Profitability must also be considered over a time frame. The basic question is, can the farmer survive during the bad years when crops fail and the production levels are low and uneconomical?

(vi) **An assessment of the environmental soundness:** The current need is quantitative approach.

(vii) **Other factors:** The socio-economic and institutional factors such as distance to market, land tenure, and improvement of the livelihood of the farmer are as important as the technological factors in consideration of sustainability.

1.6.2 Criteria for Evaluating Sustainability

The basic criteria for evaluating sustainability may be obtained by answers to the following questions based on the principles of agro-technology transfer (Silva and Uehara, 1985):

(i) Is it technologically feasible?

(ii) Is it economically viable?

(iii) Is it politically desirable?

(iv) Is it administratively manageable?

(v) Is it socially acceptable?

(vi) Is it environment sound?

Technical Feasibility (T): The real test of the system is the response during adverse weather or adverse soil conditions. Although enhancing productivity is the goal, the behavior of the system during bad years ensures acceptance.

Economic Viability (E): Marking facilities, distance to markets, etc., are the variables apart from the value of the product, which determines economic feasibility.

Political Desirability (P): The system must be in tune with the political aspirations and strategies of the country. For example, crop diversification is a common goal of many countries and this must be respected.

Administrative Manageability (A): Improved technologies frequently require greater inputs including labour. New systems must be introduced gradually and with good training and field demonstrations. Each farmer has a ceiling of performance and exceeding this may prove detrimental.

Social Acceptability (S): For example, if labour requirement coincides with important festive occasions, the system is in jeopardy.

Environment Soundness (N): It is largely a function of soil qualities and methodologies to assess impact on environment which are not sufficiently well conceived. There is an urgent need to establish indicators of change which must necessarily be at several levels.

An assessment of sustainability of agricultural practices can be made using the following relationships:

$$ASI = (T \times E \times P \times A \times S \times N) / 1000$$

Where, T, E, P, A, S and N are ranked from O to 10

and,
ASI = Agriculture Sustainability Index
T = Technological Feasibility
E = Economic Viability
P = Political Desirability
A = Administrative Manageability.
S = Social Acceptability
N = Environmental Soundness

1.6.3 Qualitative Evaluation of a Sustainable System

Qualitative evaluation of a sustainable system is imperative to accept or to reject and tailor it to needs if to commensurate with production priorities under local environments. The important indices of sustainability proposed so far are:

1. **Coefficient of Sustainability** (Lal, 1991)

   It is a measure of change in soil properties in relation to production under specific management system.

$$C_s = F(O_i, O_d.O_m)\ t$$

Where,
C_s = Coefficient of sustainability.
Oi = output per unit that maximises per capita productivity or profit.
O_d = output per unit decline in the most limiting or non-renewable resource.
O_m = minimum assured output, and
t = time, the time scale is important and must be carefully selected.

2. **Total Factor Productivity** (Herdt, 1993)

   It is defined as productivity per unit cost of all factors involved.

$$TFP = \frac{P}{\sum_{i=1}^{n} (R_i \times C_i)}$$

Where,
P = Total production

R_i = Resource used

C_i = Cost of the resource

n = Number of resources used in achieving total production.

3. **Index of Sustainability** (Lal, 1993; Lal and Miller,1993)

It is a measure of sustainability relating productivity to change in soil and environmental characteristics.

$$I_s = f\,(P_i.S_i.W_i.C_i)\,t$$

Where, I_s = Index of sustainability.

S_i = Alteration in soil properties

W_i = Change in water resources and quality.

C_i = Modification in climatic factors

t = time

4. **Agricultural Sustainability** (Lal, 1993)

It is a broad index based on several parameters associated with agricultural production.

$$A_s = d\,(P_t.\,S_p.\,W_t.\,C_t.)\,d_t$$

Where, A_s = Agricultural sustainability

P_t = Productivity per unit input of the limited or non- renewable sources

S_p = Critical soil property of rooting depth, soil organic matter content

W_t = Available water capacity including water quality

C_t = Climatic factor, such as gaseous flux from agricultural activities

t = Time

1.7 APPROACH TO ACHIEVE SUSTAINABILITY

Various approaches are being advocated to achieve sustainable agricultural systems. These include those proposing "alternative", "regenerative" or "organic" farming systems. True approaches to sustainable agricultural systems emphasize a number of practices including soil and water resource conservation, conservation tillage, soil fertility management, efficient crops and cropping systems, integrated disease and pest managements and other agronomic manipulations.

The means of achieving sustainability are low input methods and skilled management which seek to optimize the management and use of internal production input (i.e. on farm resources) in ways that provide acceptable levels of sustainable crop yields and livestock production and result in economically profitable returns (Fig.1). This approach emphasizes such cultural and management practices as crop rotation, recycling of animal manures, and conservation tillage to maintain or enhance soil productivity (Parr *et al.*, 1989,1990; Parr and Harnick, 1990). The key to achieve the sustainability of intensive farming systems is an outcome of interplay between the process of soil degradation and conservation. An important relationship of soil degradation processes and soil conservation practices is ahown in Fig. 2 for improving the sustainability of dryland farming system.

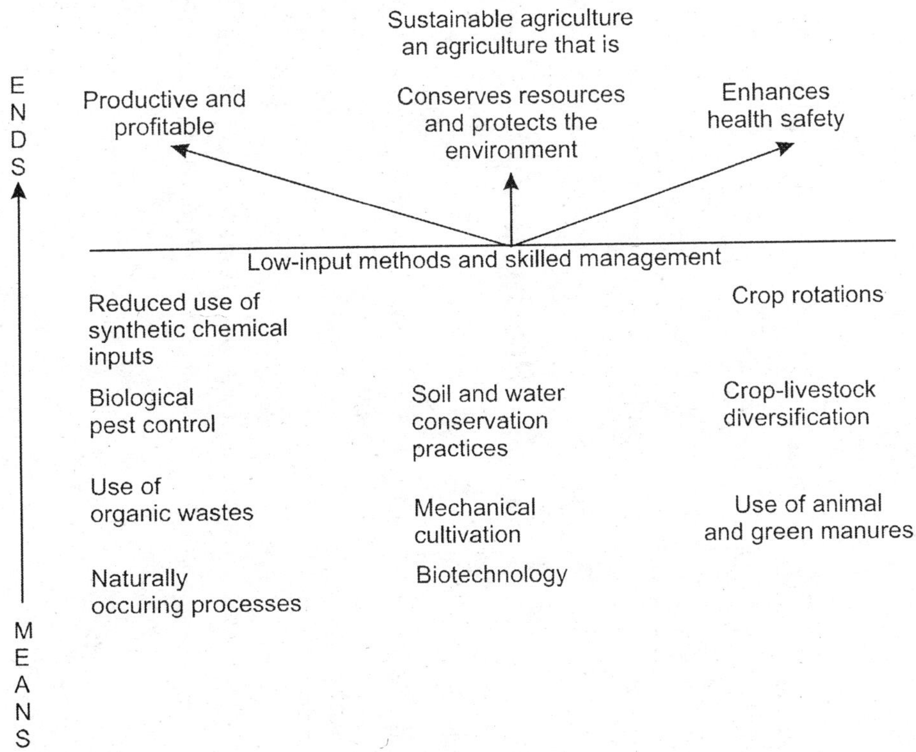

FIG. 1: A current concept of sustainable agriculture (source- Parr et.al., 1990)

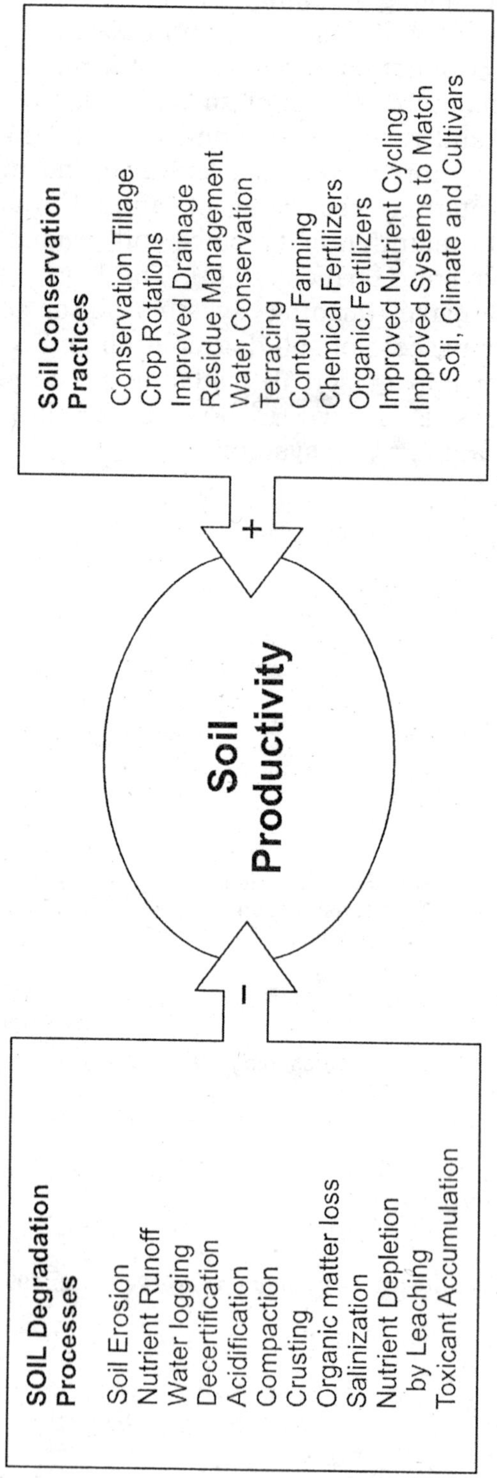

FIG. 2: Relationships of soil productivity to soil degradative process and soil conservasation practices (Stewart et. Al., 1090 as adapted from Hornick and Parr. 1987).

As soil degradation processes proceed and intensify, soil productivity decreases concomitantly. Conversely, soil conservation practices tend to slow down these degrading processes and increase soil productivity. Generally, under intensive cultivation the most serious degradation practices are depletion of nutrient resources of soils through crop removal and leaching as well as of soil organic matters. The key to improving the sustainability of intensive farming system is to halt any further deterioration of the natural resource base. This can be achieved largely by developing and / or implementing sound soil nutrient and water management practices.

1.8. SOIL AND WATER RESOURCE MANAGEMENT

1.8.1 Soil Resource Conservation

The soil is a natural resource which is not renewable in a short time. Under most conditions soil is formed at a rate of approximately 1 cm every 100 to 400 years. The soil is very expensive to reclaim or to improve once it is eroded by water or wind, physically degraded and chemically depleted (Oldmen, 2000).

1.8.1.1 Soil Degradation

Soil degradation refers to decrease in soil productivity including its present and potential capabilities through deterioration of physical, chemical and biological features. Human induced degradation can be either displacement of soil material mainly by water and wind, or soil deterioration resulting from salt accumulation, loss of nutrients or through physical processes including water-logging.

According to Eswaran and Virmani (1990), improper soil use and management, in many instances, is the cause of soil degradation. Absence of appropriate management techniques is often related to poverty, field and farm size, under development, and other socio-economic constraints. Degradation is driven by demographic and land-use trends and aggravated by various obstacles to agricultural progress and sustainable land-use. These include legal, macroeconomic, policy and institutional impediments to resource conservative land use.

Erosion is the important degrading process manifesting itself both at site and off site as given below (El Swaify *et. al.*, 1984):

- Reduction in agricultural land and declining land value.
- Sedimentation of reservoirs, rivers and drainage system

- Increased frequency and severity of flooding
- Loss of water resources
- Reduction in fish stocks and breeding grounds
- Effect on power generation
- Effect on health and the quality of life

The lateritic soils suffer severely from soil erosion losing almost 40 t per ha per year in absence of adequate soil conservation measures (Velayuthum, 1998). Due to water erosion, the silting of irrigations dams (reservoirs) in India decreases their storage capacity for water by 1-2 percent annually. Wind erosion is serious in arid and dry semi-arid regions of India particularly Gujrat, Haryana and Punjab. Wind erosion as also prevalent in coastal areas where sandy soils are predominant and in the cold desert regions, such as Ladakh and Lahaulspitti areas. The problem is accentuated by deforestation and overgrazing (Prasad and Biswas, 1998). The total area degraded in India with different degree of severity is given in Table 2.

TABLE 2: Extent of soil degradation in India

Degradation type		Area affected (million ha)	
		Total	**% of geographical area**
Water erosion		148.9	45.3
	Loss of top soil	132.5	40.3
	Terrain deformation	16.4	5.0
Wind erosion		13.5	4.1
	Loss of top soil	6.2	1.9
	Loss of top soil / terrain deformation	4.6	1.4
Terrain deformation / over-blowing		2.7	0.8
Chemical deterioration		13.8	4.2
	Salination	10.1	1.1
Loss of nutrient		3.7	3.1
Physical deterioration		11.6	3.5
	Water logging	11.6	3.5
	Total effected area	**187.7**	**57.1**

Source: Paroda (1998)

The problem of salinization and alkalization is increasing at an alarming rate in the canal irrigated areas through increase in water table, poor water management and lack of drainage. The rate is fast in black soil regions of heavy texture with serious drainage problem.

Soil degradation causes reduced crop yield (FAO, 1988) by:

(i) leaching and washing out of plant nutrients and fertilizers,

(ii) deterioration of soil structure and texture and exposure of subsoil,

(iii) reduction of soil rooting depth,

(iv) poor aeration of soil, and

(v) chemical imbalance.

1.8.1.2 Soil conservative measures:

Using the Hornick and Parr (1987) concept, a sustainable system is any system where the benefits from the soil conservation practices are equal to or greater than the negative effects of soil degrading processes. Sustainable crop production aims at preventing and reducing environmental pollution while emphasizing reduction in chemical and energy intensive-industrial inputs. It is based on the use of innovative soil and crop management techniques to obtain satisfactory returns, optimize resource use and presence of a healthy balance of soil, forest, people and environment. The research efforts of the past 3 to 4 decades have identified a number of resource conservation techniques, which reduced the risk of soil's productive potential, decrease the level of inputs required, and sustain crop productivity over the long run. These measures include (a) Land shaping (b) Agronomic manipulation (c) Biological barriers, and (d) Runoff harvesting and recycling. (Samra, *et al.*, 1998).

(a) Land shaping involves mechanical measures such as bunding and terracing of sloping agricultural lands by dividing a long slope into several short ones so as to reduce the velocity of runoff, and thus, prevent erosion. These measures increase the time of concentration by intercepting the runoff and thereby providing enhanced opportunity of infiltration of water.

(i) **Contour bunds** are narrow trapezoidal or parabolic shaped bunds which are recommended on gently slopes (0.5 – 6, 0%) where annual rainfall is low (750 – 1000 mm) and soils are relatively permeable. In North Eastern Hill regions of India, the parabolic channels are constructed on steep slopes upto 33 per cent and dug out soil is kept in the form of bund at the lower edge of the channel. In due course, the inter-bunded area gets leveled-up due to silt deposition and take the shape of terrace in 4-8 years. The technique is economical and efficient in controlling 50- 60 percent runoff and soil loss (Singh, 1998).

(ii) **Graded bunds** are recommended upto 10 percent slopes, where annual rainfall exceeds 800 mm and soils are clayey with poor drainage. The surface water from the field is made to flow at non-erosive velocities through a graded channel constructed on the higher side of the bund and finally disposed off into the grass waterway.

(iii) **Bench terracing** is the most common conservation measures in hill areas on slopes upto 33 percent. This practice consists of construction of step-like fields or benches along the contour by cut and fill method, thus reducing the length as well as degree of slope and impounding the rain-water for rice cultivation or channelising it for safe disposal.

(iv) **Conservation bench terraces,** also called zing terraces, are designed to control runoff and soil erosion and stabilize yield of rainfed crops by soil moisture storage and inter-plot water harvesting, The runoff contributing area is either kept fallow or under a well drained crop like maize, whereas the lower reaches of the field are leveled to grow crop like rice in high rainfall regions.

(b) Agronomic manipulations such as contour farming, strip cropping, diversification of crops and cropping systems, optimum date and rate of seeding and crop geometry are recommended on mild slopes but on steep slopes, these are required to be adopted in conjunction with land shaping measures.

(i) **Contour farming**: The high intensity rains occurring over short periods cause excessive runoff and soil loss from sloping agricultural lands, carrying out farm operations, such as ploughing, seeding, planting and inter-culturing along the contour lines results in creation of furrows which act as miniature reservoirs to hold the excess runoff, dissipate the energy of flow and provide increased opportunity time for runoff absorption. Thus, the erosion velocity of runoff is considerably reduced, and soils and nutrients are conserved. This practice is effective on moderate slopes (2-6%) and on deep permeable soils which are not prone to surface sealing.

(ii) **Strip cropping**: Strip cropping on contours shortens the slopes, checks the movement of runoff, helps to arrest the soil loss by providing a biological barrier / filter and increases absorption of rainwater by the soil. Growing maize and cluster-bean in alternate strips each 9 m long was affective in reducing runoff and soil loss, while giving 15 percent higher income over pure maize in Shivalik foothills (Mittal *et al.*, 1988), Further strip cropping of maize and cowpea in 2:1 ratio reduced the runoff loses from 43 to 37 percent and soil loss from 43 to 37 percent and soil loss 21 to 11 t/ha, and was as effective as graded contour bundings (Bhardwaj, 1994).

(iii) **Diversification of crop and cropping systems**: It is a necessary strategy for maintaining soil fertility and ensuring suitable crop production, An ideal cropping system involves sequential cropping of cereal followed by a legume, shallow-rooted by deep-rooted crops, fertility depleting by fertility restoring crops, soil degrading by soil regenerating crops and the crops demanding heavy inputs by those that can survive on low inputs.

The cover crops, such as greengram, blackgram groundnut, soybean sunhemp, and sesbania restore soil fertility, control weeds, conserve rainwater and reduce energy costs, beside reducing soil erosion and improving soil physical conditions.

Cropping systems have undergone considerable change in different regions due to availability of high yielding varieties and modern technologies along with other infrastructure development. In the Nilgiris of southern India, the farmers have switched over to tea and other plantation crops by diminishing terraces originally meant for vegetable crops. This has resulted in reduction in the area under potatoes with a increase in the area under tea plantation, thus reducing soil loss sharply over the years (Tripathi and Samraj, 1994).

(iv) **Optimum sowing date and seed rate**: Date and rate of seeding ensure a good stand and ground cover by the time intense rainfall takes place. Studies at Dehra Dun revealed that maize sown on 25th June gave higher canopy at peak rainy season and thereby less splash (25.9 g) as compared to crop sown on 1st July (28.3 g) and 7th July (39.3 g) (Prasad *et al.*, 1980). Similarly, the canopy coverage and crop yields were higher under early sown dates in maize + cowpea intercropping system. (Khola *et al.*, 1999)

(v) **Crop geometry**: Manipulation of crop geometry in order to achieve closer spacing with the rows across the slope provides greater resistance to overland flow resulting in reduction in runoff and soil loss. Although dense plant population are better from soil and water conservation points of view, the excessive plant population may lead to severe moisture, nutrient and light stress and consequently reduced crop yields.

1.8.2 Water Resource Conservation

Water is a limiting factor for crop production and water management is key to development of sustainable agriculture for both irrigated as well as rainfed areas. Though there is a considerable scope for increasing area under irrigation, the prohibitive cost of irrigation development including high costs of reclamation of soils damaged through poor water management and its environmental consequences are seriously limiting the extension of irrigation. Moreover, it will face serious competition from industry and other users of water, hence for sustainability of irrigated farming the consideration of productivity, economics and environmental safety becomes of paramount importance. Drainage is an essential component of irrigation system which has to be designed taking into consideration the soil and climatic factors. In future more gains will come from improving efficiency per unit of water than from extending irrigation to newer area at high costs.

More gains are expected from development of minor irrigation and *in situ* water harvesting and recycling then from building gigantic storage dams. The modern irrigation techniques, such as drip irrigation, which are remunerative for high value crops, hi-tech and optimum inputs are very much in demand and call for integrated use of soil, crop, fertilizer, pesticide management technologies and agronomic management system. Drip method of irrigation saves nearly 50 percent of pond water compared to modified sprinkler or surface method.

1.8.2.1 *Increasing efficiency of rain water*

For drylands, increased efficiency of rain water use is essential and it can be achieved in the following ways:

(i) To retain precipitation *in situ* and minimize the run–off to reduce erosion of soil and incidence of flash flood.

(ii) To reduce evaporation in relation to transpiration.

(iii) To use drought tolerant crops that fit in the rainfall pattern.

(iv) To recycle the runoff water after water harvesting and to reduce stagnation of rainwater in the cropped area by providing drainage.

(v) To use watershed concept for maximization of rain water *in situ* and to store inevitable runoff within the catchment.

In fact, it is a combination of all these that ensures best result with minimum degradation of environment. It hardly needs to be emphasized that the most critical factors of rainfed zone are the moisture stress and nutrient deficiency stress and an integrated technology is needed for removing these stresses to secure sustainable agriculture.

1.8.2.2 *Water harvesting and recycling*

A critical appraisal of existing rain water availability in India (Paroda, 1998) shows that

- India receives 400 million ha m of rain water annually.
- Nearly 160 million ha m falls on agricultural land.
- Around 24 million ha m equivalent rainwater is available for harvesting in small scale water harvesting structures.
- Nearly 186 million ha m goes to rivers as runoff.
- About one fourth of the total annual rainfall is received before or after the cropping season.

Therefore, future success to attain the sustainability goals will depend upon the effective rain water harvesting and management. Besides *in situ* water conservation, water harvesting in natural and man made structures is a potential means of providing water for crop production in rainfed areas. The main aim of water harvesting is to make crop production economically feasible in regions where rainfall is normally insufficient for cropping even with the best water conservation measures. There are two basic components of a water harvesting system: One is the catchment of surface area over which runoff is maximized and the other is the storage facility or use area. The water harvesting systems have been classified in three types:

(i) System with an external catchment area for collection of runoff or flood water from small watersheds,

(ii) Systems for storage and agricultural use of flood water,

(iii) Systems with "a within field" catchment area.

One additional benefit of some water harvesting system is "nutrient harvesting" through runoff water.

Runoff farming is basically water harvesting system specifically designed to provide supplemental or life saving irrigation to a crop or crops, especially during the period of moisture stress.

To get the maximum runoff for storage, catchment area treatments are of these categories:

(1) **Topography modifications** involve low initial costs but they may have low runoff efficiency.

(2) **Soil modifications** involving application of chemicals to the soil surface by spraying or mixing to reduce or stop water infiltration. The treatments provide large quantities of water at low cost. Bitumen or asphalt, best suited for use in sandy soils, have been widely tested. Salt treatment is cheapest soil modification technique consisting of mixing about 11 t per ha of water soluble sodium based salt (NaCl) into the top 2 cm of soil. The area is then wetted and compacted to a firm smooth surface.

(3) **Impermeable covering** involves use of water proof sheeting or membrane as catchment covering. Most of these coverings are susceptible to mechanical damage. A membrane asphalt saturated fabric is relatively resistant to damage by wind, animals and weathering processes.

Water storage techniques for holding the water collected from a catchment area are of two types: (1) The soil profile or monolith and (2) tanks or ponds. In designing soil monolith storage factors like (i) depth of soil profile (ii) water holding capacity of the soil, and (iii) the infiltration rate of the soil surface are considered.

Supply water through water harvesting is a costly proposition. It is, therefore, incumbent to control evaporation losses an integral part of all water storage facilities.

1.8.2.3 *Increased water use efficiency (WUE)*

The agronomic definition of water use efficiency consists of two major terms: (i) Biological component (commonly called the transpiration efficiency) that specifies the amount of dry matter produced per unit of water transpired, and (ii) a management component that specifies the fraction of total water supply use for transpiration (Gregory, 1988). The overall water use efficiency is considerably influenced by the management of crop and soils. Some agronomic practices used to improve water use efficiency are listed in table 3.

TABLE 3: Some Agronomic practices used to improve water Use Efficiency (W U E)

Reducing evaporation from the soil surface	Increasing the water supply to plants
Mulch	Rain harvesting
Modifying plant population and spacing	Supplemental irrigation
Selecting varieties with rapid early growth	Cultivation to improve infiltration and reduced runoff
Early sowing	Weed Control, fallowing
Application of fertilizers	Application of fertilizers, multiple / relay cropping, selecting varieties with deeper root system

1.8.2.4 *Sustainable utilization of water resources*

The present practices of water utilization are not sustainable because they largely depend upon the luxurious use of water. At present water use, as much as 3000 tons of water is used to produce 1 ton of husked rice. Similarly, 1000 tons of water is used to produce 1 ton of wheat. On the quality front, the present extravagant irrigation practices have led to declining water table, increasing N content in the ground water and pollution of surface water resources.

The decline in water tables has resulted in decrease in well yield, failure of wells/ tube wells, increased pumping costs and higher consumption of energy and ingress of sea water in coastal areas. The decline in water table can be arrested by either restricting the groundwater draft or increasing the groundwater recharge (Khepar e*t al*., 1996). The groundwater resources can be augmented through artificial recharge. The success of ground water recharge depends upon the availability of good quality water, suitability of site and appropriate recharge techniques. The water resources which can be used for artificial groundwater recharge include surplus monsoon runoff, canal water during rainy season and treated sewage water.

1.8.2.5 Water conservation in irrigated agriculture

Water conservation in irrigated agriculture can be achieved by (1) reducing conveyance losses, (ii) efficient canal water management, (iii) rainfall conservation, (iv) reducing water demand, and (v) reuse of waste water.

Water is lost during conveyance through seepage which accounts for 40 to 50 percent of the water delivered into a canal. In order to reduce these losses, lining of canal network should be done relatively based on economic consideration. During rainy days or a lean crop water requirement, the surplus canal water in excess of irrigation requirement goes waste. This water could be stored by providing auxiliary storage of water course level during peak supply period for releasing the stored water during peak demand period.

The efficiency of irrigation water can be increased by its judicious use on the farm. The on–farm water management including improving the conveyance efficiency of irrigation channels / canal water courses, application efficiency, scheduling of irrigation and precision land leveling increases the water use efficiency and crop productivity.

Considerable savings in water (about 30 to 40 percent) can be achieved by adoption of sprinkler, drip/micro sprinkler irrigation in water scarcity areas having conditions conductive to their application (Crabtree et *al.*, 1985, Aujla *et al.*, 1991)

Scheduling of irrigation in relation to water availability is an important aspect of on–farm water management for optimizing production. Several criteria have been used to find out proper timing of irrigation to different crops (Parihar *et al.*, 1975). Efficient irrigation schedules for different crops have been developed (Parihar and Sandhu, 1987; Singh *et al.*, 1999).

1.9 CONSERVATION TILLAGE

More than 500 million hectares of land through out the world are seriously degraded by soil erosion. Intensive tillage accelerates soil erosion, the loss of soil water through increased evaporation and the loss of soil organic matter through increased biological activity and oxidative processes, Concerns over the loss of valuable top soil and potential contamination of surfaces water supplies from soil erosion along with the need to more efficiently utilize water resources, farmers are required to adopt conservation tillage which provides the best opportunity for halting these degradative processes and for restoring and improving soil productivity.

Conservation tillage may be defined as "any tillage and planting system that maintains at least 30 percent of the soil surface covered by residue after planting to reduce water erosion, or where wind erosion is the primary concern, maintains atleast 1000 kg per ha of flat, small grain residue equivalent on the surface during the critical wind erosion period" (CTIC, 1995)

The average reduction in soil erosion for conservation tillage system is about 50 percent of that for most conventional tillage practices (Browston, 1998). However, the range may vary from 30 to 90 percent depending on the particular type of conservation practice factors, such as soil type and previous tillage system.

Conservation tillage may mean different things to different people ranging from elimination of one or more tillage operations to complete no tillage systems. In general, it involves a reduction of the amounts of tillage that is used to control weeds and prepare the soil for the following crops. Maintaining residue cover through out the sensitive parts of the year is a major part of the conservation tillage system. Improved fertility management in conjunction with conservation tillage contributes higher yields.

1.9.1 Types of Tillage Systems

Some relevant type of tillage systems identified are no tillage, reduced tillage, mulch tillage, strip tillage, ridge tillage, conventional tillage and traditional tillage (Virmani and Eswaran, 1990, Lal and Kimble, 2000). The first three types, and to lesser extend strip tillage have application in dryland farming systems.

(a) **No–tillage**: It is a specialized type of conservation tillage in which the soil and surface residues are completely undisturbed from harvest to planting except minimally disturbed only during seed, plant and nutrient placement. Because there is no tillage, the surface residues are of critical importance for soil and water conservation. In most cases weed control is achieved with herbicides or in some cases with crop rotations.

Limitation of no-tillage system: Though the use of no–tillage system is increasing, farmer adoption of this practice has been slow. In USA, no tillage is practiced on less than ten percent (10%) of the farmland that is in some form of conservation tillage. Research results indicate that the success of no tillage is soil dependent and that the practice works better for some crops than others. Poorly structured and poorly drained soils appear to benefit from same form of tillage and, therefore, may not be well adopted to a no tillage system. The same is true, where hard pans or compacted sub-soil restrict root growth or water infiltration. Other limitations of no tillage include poorly understood biological factors that limit crop growth in surface residue system; increased plant diseases associated with surface residues; and problems associated with equipment and stand establishment in planting and fertilizing in trashy seedbeds and compact soils.

Karten *et al* (1990) defined soil tilth as " the physical condition of a soil described by its bulk density, porosity, structure, roughness, and aggregate characteristics as related to water, nutrient, heat, and air transport; stimulation of microbial and micro-fauna populations and processes, and impedance to seedling emergence and root penetration". The tilth formation is "the combined action of physical, chemical, and biological processes that bind primary soil particles into simple or complex aggregates and aggregate associations that create specific structural or tilth conditions." Supply and availability of air, water and nutrients which are taken care of by good soil tilth are as vital to sustenance of agriculture as air and water to the survival of mankind on this earth. Proper soil tilth is a condition in which the particles are completely aggregated; consistency is soft; air–water relations are equal and hard pan or mechanical impedance of continuing nature in the formation of columnar soil structure does not exist. Soil tilth is important because it affects all processes occurring in soil matrix including crop growth.

Maintenance of soil tilth is of paramount importance for higher yields which requires good physical conditions having proper air-water relations without any adverse effect, latent or express, which may either retard germination or inhibit the growth of crop plants. The development of root system should be conductive to meet moisture as well as proper nutrient requirement. Proper tilth is also conductive to transformation of and other nutrients form. Crop yield through fertilizers may not be realised if the applied plant nutrient carrier is not converted into forms most preferred by plants.

Temporary water-logging can cause asphyxiation and can instantaneously lead to the formation of toxic substances inside the plants like ethanol, alpha-amino-butyric acid, which can cause instantaneous death of the plants. For securing proper tilth is to perform primary tillage using soil turning implements, the direction for whose use should be changed every year. Any secondary tillage perhaps could be best done by small tools. The dust mulch thus created helps in conserving moisture, incorporating air in the soil, and controlling weeds.

(b) **Ridge tillage**: It is planting of row crops on a preformed ridge. This system has been gaining popularity as a conservation practice for maize (*Zea mays*. L) and soybean (*Glycine max*. L) production in the USA. Advantages are decreased erosion, reduced fuel and equipment costs. The row crops are sown in the spring on ridges formed by cultivation of the crop during previous growing season. The residues of the previous crop are usually shredded in the fall to provide an even distribution of surface cover during the non- growing season.

(c) **Reduced tillage:** The soil is disturbed as little as necessary prior to sowing by implements such as chisels, disks or any other appropriate equipment (e.g., cultivars, sweeps, blades) to break the crust, loosen the compacted layer and obtain good crop stand. Weed control is achieved through herbicides and secondary tillage operations.

(d) **Mulch tillage:** The objective of the conservation tillage system is to provide an adequate amount of crop residue mulch on the soil surface. Therefore, the soil is disturbed prior to sowing only by those equipment which do not bury the crop residue (e.g., chisels, field cultivars, disks, etc.). The desired weed control is accomplished through herbicides and/or cultivation.

(e) **Conventional tillage:** It involves the plough based tillage methods of soil inversion and those which leave less than 15 percent residue cover after planting or less than 500 kg per ha of small grain residue equivalent through out the critical wind erosion period, Conventional tillage usually involves the combination of primary, secondary and tertiary tillage operations.

(f) **Traditional tillage:** It refers to indigenous tillage methods based on specific tools developed and used in the region. There is wide range of traditional tillage tools depending on cultural, ethnic and social customs. Traditional tillage may be performed by manually operated tools, animal drawn equivalent and /or motorized implements.

Problems with traditional tillage system in India: In the Indian context, most traditional tillage implements perform a type of reduced tillage system. The basic problems with most traditional tillage systems in India (Lal and Kimble, 2000) are the following:

(i) All crops residues, including the roots and stumps, are removed and / or grazed.

(ii) Most soils are crusted, compacted and have massive structures that cannot be improved by shallow traditional tillage systems.

(iii) The compacted sub-soil, often with hard pan at the shallow depth of 15-20 cm, persists and is often reinforced and exacerbated with the traditional tillage system.

(iv) Supra–optimal soil temperatures (>30^0 C in the top 5 to 10 cm layer) and low available soil moisture accentuate the drought during the prolonged rainless period (>2 weeks) during the growing season.

(v) Low water infiltration rate leads to water accumulation on surface and high losses of water due to surface runoff and evaporation.

(vi) The low level of soil organic carbon (SOC) pool is further depleted by mineralization and accelerated soil erosion.

The solution of these problems lies in converting conventional and traditional tillage methods to an appropriate conservation tillage method which may have been an improvement in terms of alleviating soil compaction at shallow depth, increasing seed germination and crop stand establishment, facilitating application of fertilizers and other amendment and improving crop yields.

1.10 SOIL FERTILITY MANAGEMENT

1.10.1 Concept of Soil Fertility

The soil is very foundation of life. It supplies the people with food, fodder, clothing and shelter, which come from the fertility released by the soil. The fertility of the soil is, therefore, basic to all human existence and of civilization in the entire world.

Soil fertility may be defined as the inherent capacity of the soil to produce a certain yield of economic crops. Hence, it depends on those factors in soil that determine its crop production potential, such as, presence of essential plant nutrients in available forms and in a suitable balance; the proper micro-biological status of the soil to provide healthy environment for the release of plant nutrients; and freedom from any toxic or injurious agents, conditions or substances in soil. A soil may be fertile yet it may not be productive. Thus, a waterlogged soil may be highly fertile but may not produce good crops because of the unfavorable physical conditions. Similarly, a fertile soil may have saline, alkali or boron salts, which being toxic to plant growth restrict its potential power to produce crops. Conversely, a sandy soil poor in fertility may through the use of fertilizers and water, produce high crop yields. In brief, soil fertility is the potential status of a soil to produce crops; while soil productivity is the resultant of various factors influencing soil management.

The basic foundation of soil fertility, enunciated for first time by Dr. Lipman (USA) and Dr. Viswanth (India) at the same time by 1940, carry three essential components (Sankaram, 1996):

(i) All biological processes of soil fertility maintenance are the result of the activities of the living system of the soil.

(ii) The C/N ratio is an index of decomposability of organic matter added to the soil.

(iii) The living system for its normal functioning requires moisture, energy material–carbon source and other nutrients.

Hence, it is apt to say that organic matter is the elixir of plant nutrition.

1.10.2 Role of Organic Matter and Bio-fertilizers

The long-term studies demonstrated the ability of bulky organic manures to neutralize the rapid yield fall with the continuous use of chemical fertilizers. An early application of farm-yard manure at 10-15 tons per ha coupled with a recommended dose of NPK is able to sustain productivity (Lal and Mathur, 1989). Farm-yard manure (FYM) is a storehouse of several nutrients. Its regular application prevented the occurrence of Zn deficiency (Katyal and Abrol, 1989) and S deficiency (Nambiar and Abrol, 1989). More important, FYM is known to improve soil physical environment. It adds polysaccharides which are important in binding the soil particle together and for creating a stable soil structure. With the regular use of organic manures soil erosion is minimized and water retention characteristics improved.

The main functions of soil organic matter are: (i) physical soil improvement (increase in water and air capacity, protection against erosion), (ii) several chemical functions (increase of sorption capacity, mobilization of nutrients by mineralization of organic matter and from minerals), (iii) short-term immobilization into soil organisms (which, thus, compete with the crop for N), (iv) long-term fixation into stable humus substances (e.g., improvement of quality via C/N ratio), and supply of beneficial organic substances (growth promoters and antibiotics).

According to Beri and Meelu (1981), a 40 to 50 day old green manure crop can supply 80 to 100 kg N per ha. Even half of this N as crop usable, a green manure crop could substitute for 50 to 60 Kg fertilizer N per ha. Organic matter addition also favoured improvement in soil physical environment in several long term field studies conducted under diverse soil environment (Nambiar and Abrol, 1989).

Though bio-fertilizers can also increase yield significantly, its benefit in acid soil is limited. *Rhizobium* is beneficial for pulses like greengram, blackgram, peas, and oilseeds like soybean and groundnuts. But soil acidity is somewhat harsh for growth of *rhizobia*. Low base status of acid soils along with poor availability of P comes adversely in the way of efficient *rhizobial* culture. Blue-green algae (BGA) perform poorly when the soil is highly acidic. *Azospirillum* is appropriate for cereals, whereas *Azotobacter* for non-food grain crops, such as sugarcane, potato, cotton, etc.

1.10.3 Adverse Effect of Nutrient Accumulation

According to Abrol and Katyal (1990), fertilizer application on a continuous basis can bring about nutrient accumulation which in turn can prove toxic to plants or can interfere with the uptake of other essential nutrients. Nutrients, which are inefficiently used by the crops are relatively immobile to leave the feeding zone of crops, are of major concern.

Utilization of fertilizer P seldom exceeds 20 percent while Zn utilization efficiency is still poor, less than 5 percent accumulation of Zn and P occurred with their continuous application (Biswas and Benbi, 1989). A build up of P can prove antagonistic to Zn and Fe nutrition of crops. High levels of Zn in the soil can cause Fe chlorosis (Katyal and Randhawa, 1983).

1.10.4 Impact of Fertilizer Use on Environment

The threat posed by the nitrogenous fertilizers was observed by Commoner (1968) in United States. He found that intensive application N fertilizers had led to increased eutrofication and a potential hazard from nitrate poisoning (methemoglobinemia). This indicated that with an increase in fertilizer use, there is indeed a real danger of damaging the environment with untold detrimental consequences. The impact of fertilizer use on environment has focused primarily on:

(i) **Eutrofication of lake or river waters:** Eutrofication, the speedy growth and decay of aquatic biomass, is caused by high concentration of P in the waters of lakes and rivers. Fertilizer P can reach the water bodies through surface runoff, since leaching of P is generally limited though significant in some soils (Chhabra *et al.*, 1981)

Nitrogenous fertilizers have frequently been cited as responsible for ground water and environmental pollution. Crops are hardly able to use 50 percent of applied N. Nitrogen use efficiency with lowland rice is still poor, around 30-50 percent. This implies that 50 to 70 percent of the applied N does not contribute to crop growth and yields. Through surface run-off fertilizer N reaches waters of lakes and rivers, where the company of P accelerates eutrofication. In the form of nitrate, leaching of fertilizer N brings about ground water pollution. Transformation leading to evolution of gases into atmosphere leads to atmospheric pollution. Leaching and gaseous emissions are the principal pathway of N loss leading to low fertilizers use inefficiency and high environmental pollution (Abrol and Katyal, 1990).

(ii) **Methemoglobinemia due to nitrate pollution of the ground water:** A high nitrate ground water (> 10 mg $No_3 - N/L$) when used for drinking is involved in the incidence of methemoglobinemia-a condition incapacitating blood hemoglobin to carry oxygen to the body cells. Infants of less than one year age are more vulnerable to high nitrate levels than are the adults. High amounts of nitrate are also alleged to cause birth defects, gastric cancer and nervous system impairment (Aborl and Katyal, 1990).

(iii) **Destruction of Ozone layer due to gaseous emission of N in the form of nitrous oxide** (N_2O): Nitrous oxide is capable of promoting destruction of the stratospheric ozone layer which protects the biosphere from the harmful ultraviolet radiation. Denitrification is the source of nitrous oxide from agricultural lands. Maximum nitrous oxide is produced from anhydrous ammonia followed by urea and ammonium nitrogen containing fertilizers. With nitrate fertilizers, nitrous oxide production is minimum (Fung, 1988).

(iv) **Metal toxicities:** Among the heavy metal toxicities cadmium (Cd) deserves special mention because of its inadvertent additions through phosphatic fertilizers. In turn phosphatic fertilizers derive cadmium (cd) from phosphatic rock. Instances are on record when continuous phosphate application led to several folds higher Cd in soils (Mulls *et al*, 1980).

Fertilizers use related concern on environmental degradation should not be the plea to axe the rising trend in fertilizer consumption. Such as strategy will be catastrophic for a country like India where food grain production and fertilizer use run parallel (Dhar, 1989) Dangers of environmental pollution are genuine and need to be diminished by protecting the fertilizer N from various losses.

1.10.5 Nutrient Imbalance

Results from fertilizer experiments conducted at Ranchi (Jharkhand), India highlighted the weakness to the single nutrient application approach. Initially N applications enhanced productivity, but with time it could hardly sustain it (Table 4).

TABLE 4: Long term effect of selected fertilizer treatments on the yield of maize grain in maize-wheat rotation on an alfisol (acidic, red loam)

	Average yield (q/ha)			
Treatment	**1 – 12 year**	**13 – 18 year**	**19 - 24 year**	**25–28 year**
No Po Ko	8.3	8.3	5.5	5.3
N	17.0	11.5	0.4	0.0
NP	24.0	28.5	4.0	0.0
NPK	31.7	32.3	8.2	1.2
FYM (=N)	27.7	25.9	27.5	25.9
NPK + Lime	33.7	39.7	37.9	36.7

1. Following years 1 –12 of cropping local maize variety was replaced with a hybrid and the N:P:K does increased from 40:40:40 to 100:80:60
2. Following years 13 – 18, N:P: K does was increased to 110:90:70

Source: Lal and Mathur (1988)

Mitscherlich' lae became operative and the deficiency of the most limiting nutrients element started determining the productivity level Maximum deleterious effect of continuous N application was encountered in law CEC in acidic soils, Phosphorus application together with N salvaged the yield decline. Over a period of time K became critical. In its absence, decline in yield with NP application was increasingly observed in many soils (Nambiar and Abrol, 1989)

Intensive continuous cropping, as is being practiced in several regions of India, left sub- optimal levels of available Zn and S in soils and in case of severe deficiency its application made a difference between yield and no yield (Katyal and Randhawa, 1983). Tandon (1989) reported that exhaustion of soil–S reserves through intensive cropping coupled with increasing dependence on S free chemical fertilizers (Urea, DAP, TSP, etc.) accelerated the depletion to deficiency level. Currently, S deficiency is estimated to limit crop yield in 100 districts of India.

1.10.6 Integrated Nutrients Management – Key for Sustainable Production

Integrated nutrient supply refers to supplementary as well as complementary use of organic manures, green manures, crop residues, bio-fertilizers and rural and industrial waste along with chemical fertilizers for meeting nutrient demands of agriculture production systems. Over the past three decades, the agricultural growth in terms of food grain production has been very inspiring. This achievement has been possible due to development of improved varieties of several crops, increased use of fertilizers and adoption of recommended agronomic practices resulting in increase in the cropping intensity. With the expansion of irrigation infrastructure, the fertilizers resulted in increased yields by 50 percent or more (Brown and Roy, 1983). With the increasing food demand and the introduction of high yielding crop varieties with increased dependency on chemical fertilizers, the soil fertility has deteriorated overtime and particularly the soil organic matter is on decline due to imbalance fertilizers use (Abrol and Katyal, 1990; Khanna and Gupta 2000). Therefore, use of organic manures in various cropping systems is a must to maintain the productivity of the crops. Only organic manure application cannot provide food for ever increasing population, particularly in the developing countries meaning thereby that the integrated nutrient management is the only alternative to maintain the soil productivity without influencing the total food grain production.

In fact the agricultural production systems have open plant nutrient cycles and gradually deplete soils of soil organic matter, essential plant nutrients and water making them unsuitable in the long run. A sustainable agricultural production system must economically produce the needed food, fiber and feed without depleting the natural resources and without detriment to the environment. Integrated nutrient management involving application of nutrient needs through organic manures including green manures and crop residues and use of bio-fertilizers reduced the rate of fertilizer application and holds the key for sustainable agricultural production (Prasad, 2002).

1.10.6.1 Concept of integrated nutrient management system (INMS)

The basic principles of integrated nutrient management system is the maintenance of soil fertility, sustainable agricultural productivity and improving profitability through judicious and efficient use of chemical fertilizers, organic manure, green manure, bio-fertilizers and other recyclable materials.

The concept underlying integrated management system is the maintenance or adjustment of soil fertility and of plant nutrient supply to an optimum level for sustaining the desired crop productivity through optimization of the benefits from all possible sources of plant nutrient in an integrated manner (Roy and Ange, 1991).

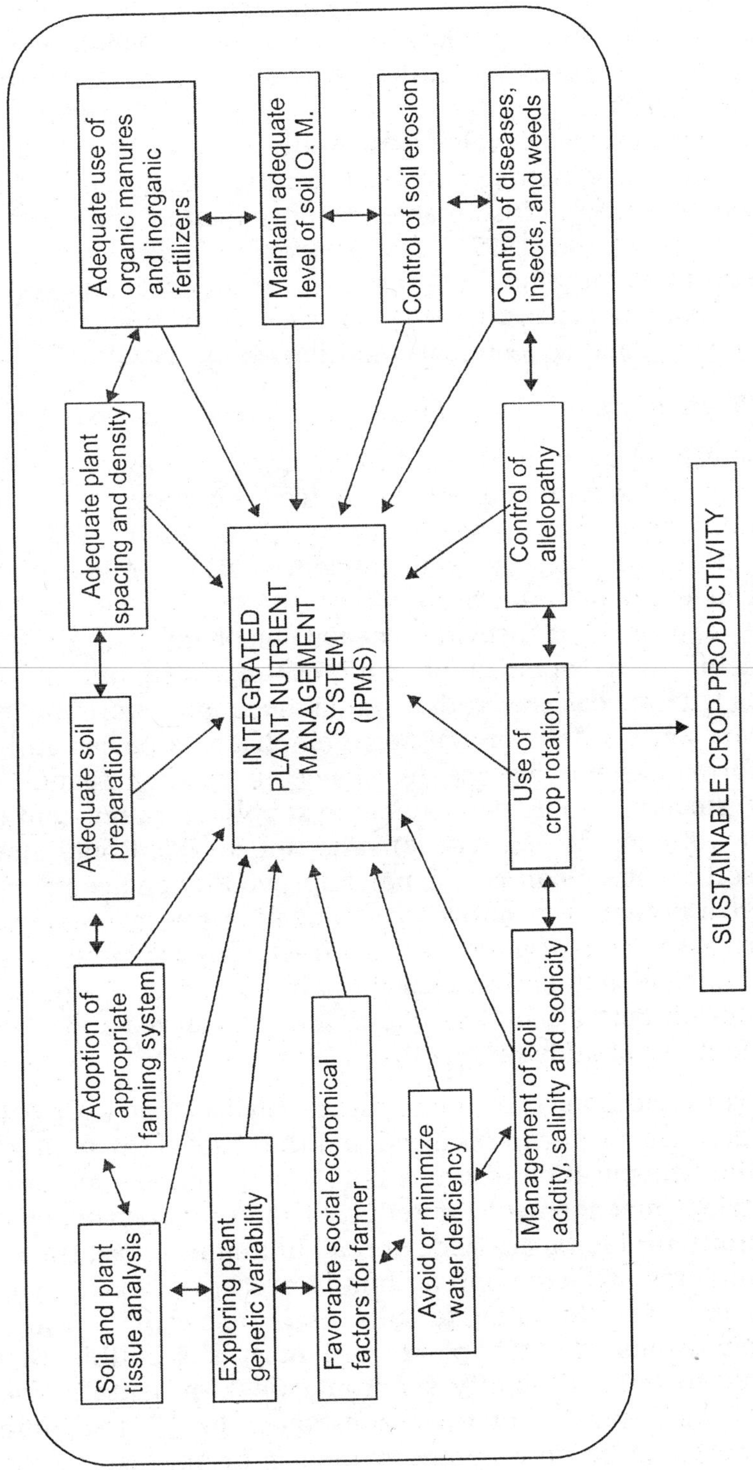

FIG. 3: Integrated plant nutrient management for sustainable crop production (from Fageria et. al., 1997).

The system (INMS) is ecologically, socially and economically viable and environmentally healthy technology, which can be adopted by the farming community to get higher productivity while maintaining fertility. Reliance on chemical fertilizers for sustainable agricultural development would have to be continued, inspite of the environmental threats. The integrated plant nutrient management system for sustainable crop production is depicted in Fig. 3 (Fageria *et al.*, 1997). The approximate combination of mineral fertilizers, organic manures, crop residues, compost or N-fixing crop varies according to the system of land use and ecological, social, and economical conditions (Roy, 1995).

1.10.6.2 Nutrient management for crops and cropping systems

Crops species and even varieties differ to a certain extent in their ability for mobilization and uptake of soil nutrients, especially under conditions of low nutrient supply. Since under low input condition the contribution of soil nutrient reserves (especially P and K) is of eminent importance (Syers, 1995). Hence, varieties should be chosen that provide special mechanisms, like root excretions for their mobilization and for improved uptake. The same holds true for cases of rather immobile single micronutrients (e.g. Fe, Mn, Zn, Cu deficiency). More emphasis is placed in high yield performance and disease resistance in relation to nutrient deficiency, since it is more economical to supply extra nutrient (for micronutrients even by leaf spray). The uptake from only slowly available soil reserves can be further improved by choice of crops with an extra potential of nutrient mobilization via mycorrhiza or rhizosphare associated organisms. The effect of this symbiosis is an important nutrient uptake in general, mainly a better P uptake of the roots due to better mobilization of soil P reserves and also to more efficient use of added P fertilizers (Barea, 1991).

For arid condition, it is important to realize that with better crop nutrition, less water is required and that especially with good K supply the transpiration of crops is reduced. Saline soils need a special nutrient management for reducing the uptake of unwanted Na. Soil acidity problems are wide spread in humid areas, the most striking problems are wide spread humid areas, the most striking example being the acid sulphate soils. Soils below pH 4.5 mostly produce low yields and the plants are most susceptible to the diseases due to the soil acidity syndrome (lack of some nutrient, especially P, and surplus of toxic substances like Al). In humid climates, there can be no cropping without nutrient losses (due to excessive leaching, runoff and soil erosion as well as evaporation).

The nutrients mainly affected by leaching are nitrogen (nitrate), sulphate, K and Mg. A reduction in N-losses means an increase of N-use efficiency. The main preventive measure against leaching of nitrate is (i) to keep a permanent vegetative cover on the soil, (ii) to restrict N-fertilization to the amount required for a realistic yield level (Velayutham and Bhattacharyya, 2000) and (iii) to apply N either in the form of slow release N-fertilisers or in split-doses at critical stages of plant growth when it is most needed.

1.10.6.3 Integrated nutrient management practices

According to Singh and Deka (1990) various integrated nutrient management (INM) practices clearly suggest for:

- more extensive use of animal and other organic manures along with mineral fertilizers to provide both organic matter and nutrients to the soil,
- more and more adoption of longer crop cycles in contrast to soil depleting mono-culture or short term rotations,
- more and more inclusion of legumes/pulses and green manure crops in cropping systems to add to N and organic matter to the soil,
- less use or in some cases no use of mineral fertilizers and increasing reliance on plant and animal residues,
- use of bio-fertilizer partially to substitute commercial fertilizers, and
- adoption of nutrient management practices to raise fertilizers use efficiency in cropping systems.

1.11 EFFICIENT CROPPING SYSTEMS

1.11.1 Concept of Cropping Systems

Cropping system is the kind and sequence of crops grown on a given area of soil over a period of time. It may be extensive *i.e.*, only one crop grown in year, or intensive i.e., regular rotation of more than one different crops in a year. The most general term for such cropping system is multiple cropping under which is the dichotomous classification of sequential cropping and intercropping. Sequential cropping is growing two or more crops in sequence on the same field per year. Crop intensification is only in time-dimension. There is no intercrop competion. Intercropping is space-dependent form of multiple cropping of growing more than one crop simultaneously on the same field. Crop interaction is both in time and space dimensions.

A proper cropping system is regarded as one of the important aspects of agronomic manipulation, but with the advent of high yielding photo-and thermo-insensitive varieties, the crops are grown in all seasons, which lead to perpetuation of insect pests and diseases, and other formidable enemies of crop plants. Continuous cultivation of groundnut, cotton and chickpea has given rise to serious pests like white grub, pink ball worm, wilt or premature drying and also the pod-borer which has become a pest in many crops. The crop plants possess their defense mechanism but this is broken because of faulty practices, and repeated sowing of the same crop year after year is one such issue which is the root cause of many problems like weeds, insect-pest and disease build up and ultimately of soil sickness. According to the phagocyclic theory of soil infertility, some of the crops secrete certain substances which can attract or repulse one or other bio-agents in the soil and can be useful or harmful. Proper crop rotation can also alleviate some of the nutrient disorders, apart from the incidence of pest and diseases.

1.11.2 Eco-friendly Management of Cropping System

Management practices that help in conserving the natural eco-system while ensuring sustainable system productively on a long-term basis may be called as eco-friendly practices. Cultivation of high yielding crop varieties with intensive use of chemical fertilizers, pesticides and irrigation water during past few decades, have helped to achieve self-sufficiency in food grain production but emergence of many new and unexpected problems have created an unsustainability in cropping systems. These problems are related mainly to emergence of new soil nutrient disorders, soil and environmental degradation, pollution, loss of biodiversity, development of new biotypes of pathogens and pests, and finally to human and animal health hazards (Yadav and Prasad, 2000). To feed the fast increasing population, more and more has to be harvested from shrinking land resource by intensive input use and sustain resources base for well being of future generations. To address these issues, development of alternative practices of cropping system can help in achieving a balance between the two. These practices are:

- Crop diversification
- Intercropping

- Precision agriculture
- Conservation tillage
- Organic farming

Conservation tillage has already been discussed in this chapter and organic farming is dealt in chapter IV.

(a) **Crop diversification:** A well designed diversified cropping system helps to achieve the goal of conserving natural eco-system through deriving the following advantages:

(i) Intermittent inclusion of deep rooted legumes between shallow rooted cereal based crop rotation improves the utilization of nutrients and water from different soil layers and recycles atmospheric nitrogen by symbiotic N-fixation. This results in saving of chemical fertilizers and irrigation water.

(ii) Inclusion of certain crops in sequence has been found to reduce incidence of some obnoxious diseases, insect pests and weeds to a considerable extent, thereby reducing the application of hazardous pesticides, insecticides and herbicides, otherwise necessary to achieve better crop yields. Such examples are reduced incidence of ***Phalaris minor*** in rice –wheat system by introduction of crops like sugarcane, sunflower and *berseem*, and of soil borne diseases by rotating cereal-legume crops.

(b) **Intercropping:** Intercropping is a practice of crop intensification in spatial and temporal dimensions. In intercropping, two or more crops generally dissimilar in growth habits are grown simultaneously in the same field. The main advantages of intercropping are:

(i) To enhance the efficiency of resource used, i.e., land water, nutrient and light.

(ii) To reduce ecological hazards.

Some examples, which indicate the possibilities of reducing herbicide and pesticide load in cropping system are reduced weed growth in sorghum based intercropping as compared with sole sorghum, prevention of top-borer attack in autumn planted sugarcane, intercropped with garlic, fennel or coriander, and reduced incidence of root-rot in cotton intercropped with red-gram.

(c) **Precision Agriculture:** The objective of precision agriculture is to improve the control of input variables, such as fertilizers, seeds, chemicals, or water with respect to the desired outcomes of increased profitability, reduced environmental risk or better product quality. The practice consists of four stages: (1) information (ii) acquisition (iii) interpretation, and (iv) evaluation and control. Precision agriculture offers a potential change in production efficacies. However, it is still in its infancy stage and requires much research and development.

1.11.3 Conservation Cropping Pattern for Sustained Production

1.11.3.1 Cropping pattern suited to land capabilities

It is always desirable to use the land as per its inherent capabilities to achieve the maximum potential with minimum cost. Very often, degraded lands including hill tops are put to crop production and some times goods lands are left fallow or diverted to tree planting or some other uses. Besides, there are vast cultivable waste lands which remain unutilized.

No doubt, rice and water will continue to be the main staple crops in India, vast rainfed areas including uplands are sown under these crops which are not quite suitable for them. There are other crops, e.g., pulses, and oilseeds which could exploit the rainfed situation in upland more effectively and bring higher remuneration to the farmer as well as diversify his food basket and also maintain the sustainability.

1.11.3.2 Crop management for rainfed condition

For sustained and enhanced productivity of drylands, management techniques related to conservation and utilization of rainwater and soil resources are most important. Crop and agronomic management practices for rainfed areas include primary tillage, early seeding, optimum plant population, timely weeding and use of appropriate crops and varieties suited to available growing duration and water regime, past management, and use of fertilizers.

Under rainfed conditions, the rain could be delayed, recede early or have small or long breaks during monsoon period. Strategies to meet such uncertain conditions should be worked out for different situations and agro- climatic regions.

A. **Contingent crop production strategy for rainfed farming in India:** Rain is the principal source of water for rainfed crops, which is received during the monsoon period. During this period, "bursts" of rain alternated with "breaks" are not uncommon. The important aberrations in the rainfall behaviour (Venkateshwaralu *et al*, 1991) are given below:

- The commencement of rain may be quite early or considerably delayed.
- There may be prolonged 'break' during the cropping season.
- There may be spatial and/ or temporal aberrations.
- The rains may terminate considerably early or continue for longer period.

(a) **Onset of monsoon:** The early or delayed onset of moonson is certainly aberrant at the result of which the rainfed farming has to bear pangs of vagaries of rain. These situations require such correction measures as change of crops or varieties form those being already sown in the region or adoption of other agronomic practices.

(b) **'Breaks' in the monsoon rains:** This can be of different durations. The 'breaks' of long duration of 2-3 weeks or even longer create the situation of plant-water stress leading to reduction in agricultural production. Several crops are quite sensitive to the drought situation and these 'breaks' cause heavy reduction in the yield of crops. In the soil of low moisture storage capacity with 5 days rainless situation, willing symptoms may be clearly visible. To meet such situations, the crops and varieties may be selected in such a way that the water sensitive stage(s) of the crop growth does not coincide with the 'breaks' most probable and most frequent.

(c) **Aberrant withdrawal of south–west monsoon:** Such situations call for attention of agricultural planners and the scientists mainly on two fronts:

(i) If the monsoon rains do terminate at a date(s) much earlier from the one considered to be normal, what measures are required to be taken to save the crop since by that time the crop has not reached the maturity stage and some soil moisture is still required till its maturity.

(ii) Whereas crop and varieties in any given region are selected and grown based on length of the growing season generally considered to be normal, the persistence of the rains much beyond the normal dates creates an extraordinary situation in the sense that more unusable water becomes available, but action plan may not exist to capitalise this amount of water.

(iii) Uneven distribution of monsoon rains in space and time: Such situations are encountered almost every year in one or another part of the country during south-west monsoon period leading to periodical draught and flood situations. As a matter of fact, the high variability of rainfall (or more precisely the soil-water) is the single factor which influences the high fluctuations in the crop yield in difference parts of the country.

B. The contingent crop production strategy required to be evolved for stabilizing rainfed agriculture under different weather conditions are discussed under the following heads:

(i) **Choice of suitable crops:** The first priority of a farmer continues to be production of food for his family and feed for his animals. Hence, selection of crops suiting the environment leads to increased and stabilized production. For example, the best environment for growing rice would be at least an average rainfall of 200 mm per month over a period of three months with 'breaks' of not more than seven days in the monsoon period.

(ii) **Choice of suitable varieties:** With the era of hybrids and high yielding varieties of crops particularly cereals, a replacement of the existing locals with these varieties or hybrids act as a 'primers' for ushering in better crops in drylands, They may not do any better than the traditional varieties in years of poor rainfall, but these are capable of capitalising on the better seasons so that in the long run a farmer is better off.

(iii) **Alternate crop strategies:** The alternate crop planning to meet weather aberrations in different regions have been evolved which is presented in Table 5.

Table 5. Alternate crop planning to meet weather aberrations (Venkateshwarlu *et.al.*, 1991)

Region	Alternate Crop strategy	
	Normal	**Delayed**
(a) Sub-humid red soils (Ranchi)	1. Rice 2. Finger millet+Pigeonpea 3. Groundnut	1. Transplanted millet 2. Maize+Pigeonpea 3. Rainy season potato
(b) Deep black soil semi – arid (Bellary)	1. Sorghum 2. Safflower	1. Sorghum (fodder) 2. Safflower 3. Chickpea
(c) Alluvial soils (Varanasi)	1. Rice 2. Maize 3. Pearl millet 4. Blackgram 5. Greengram 6. Sesame (to be followed by safflower or chickpea)	1. Rice (short duration) 2. Pearl milled 3. Blackgram 4. Greengram (to be followed by chickpea)
(d) Black soil (Indore) Monsoom and Post – monsoon crop	1. Sorghum or maize or Soybean followed by safflower of chickpea	Sorghum + Pigeonpea Sorghum + Soybean Soybean alone in post – monsoon if monsoon is further delayed
(e) Seirozemic soils (Hissar)	1. pearl millet 2. Green gram 3. Cluster bean	1. Transplanted millet 2. Greengram 3. Cluster bean 4. Cowpea
(f) Submontane soil (DehraDun)	Short duration rice or maize followed by wheat or chickpea.	Transplanted rice or maize followed by chickpea, rapeseed in September.
(g) Solapur	1. Pearl millet 2. Sunflower	1. Horse gram / moth bean

(iv) **Mid-season corrections:** Once the crop is growing, there could be moisture stress. If it is in a very early stage, i.e., within a week to 10 days of sowing it is better re-sow with subsequent rains than allowing inadequate plant stands to persist and yield a poor harvest. On the other hand, it the crop grows for 40-50 days, other means like ratooning or thinning need be considered. Thinning of plant population by removing every third row was found to be advantageous to mitigate moisture stress (Venkateshwarlu *et al., 1991*).

(v) **Crop life saving measures:** In order to increase and stabilize production in rainfed farming the runoff either can be harvested *in situ* or into dugouts at the lower reaches in a given watershed. Use of mulches helps in storing moisture in the soil profile over extended periods for better growth. Off season tillage and fertilizer use are also advantageous.

In situ water harvesting can be achieved by land configuration into ridges and furrows on grade formed prior to sowing or with eventual cultural operations. Also broad beds on grade could be formed. Graded border strips, dead furrows, inter-row water harvesting and interplot water harvesting are other examples.

Runoff varies from 10-40 percent of the total rainfall in the tropical and sub-tropical climates of our country. Of this, at least 30 percent could be harvested into dugouts and water so stored could be recycled to the donor area. The pay-off is very high and works out to 2 q per cm per ha of critical irrigation.

1.11.3.3 Cropping through a watershed management approach

Integrated watershed managent approach has been accepted as the most rational approach in preventing deterioration of ecosystem, restoration of degraded lands, efficient water harvesting and improving overall productivity.

Watershed is a hydrological unit. It covers the total catchment land area, starting from the highest point (ridge line) to the lowest point of the area through which water flows into the outlet of a natural stream. In physical terms, a watershed refers to the area lying above a given drainage point. The main thrust of watershed development is to rationalize conservation and utilisation of natural resources and prosperity of the area. The main components need to be integrated in watershed development programmes are (i) Conservation, management and development of soil and water resources, and (ii) efficient crop management improved cropping intensity and (iii) alternate land use system according to land capability.

The watershed approach aims at optimizing the use of land, water and vegetation in an area in the integrated way and thus, helps alleviate drought, moderate floods, prevent soil erosion, improve water availability, increase fuel, fodder and food on sustained basis.

1.11.3.4 Cropping in acid soil

Lime is usually recommended for ameliorating acid soils and dose of lime will depend on the nature of soil. In order to realize maximum benefit from liming, crops sequence in the acid soils should be rationalised. Mandal and Jha (1969) reported that effect of lime applied in full dose lasts for 4 to 5 years which gradually declines in the last 2 years. This information can be utilized for selecting suitable crops in acid soils. On the basis of degree of response to lime, crops are grouped into following three classes (Mandal *et al.*, 1969).

(i) High response class: Pigeonpea, soybean, cotton.

(ii) Medium response class: Chickpea, lentil, peas, groundnut, maize, sorghum, and wheat.

(iii) Low or no response class: Barley, finger, millet, little millet, rice, mustard and potato.

Studies by Mandal *et al.* (1969) at Ranchi (India) revealed that cotton, gtroundnut, soybean and pigeonpea grew best only at pH 6.5 to 7.0 and as such may be grown in the first two years of lime application. This should be followed by medium responsive crops at soil pH 6.0 to 6.4, such as maize, sorghum, blackgram and greengram, etc. In the 5th or 6th year when the effect of lime is gradually reduced low responsive crops like rice, small millet and potatoes may be grown at pH below 6.0.

Kanwar and Joshi (1964) observed that wheat, maize and potatoes responded to liming in unfertilized soils, but in presence of fertilizer, the margin of difference narrowed. However, Gupta *et al.* (1989) in acid soils of Sikkim found that liming significantly increased the yield of wheat and to a lesser extent the yields of soybean, barley and maize.

On the basis of trials on intercropping and successful cultivation of linseed in winter season, the following crop sequences are suggested for rainfed upland situation of Ranchi (India):

1st year : Groundnut-Linseed
2nd Year : Maize + Pigeonpea
3rd Year : Maize + Pigeonpea
4th year : Direct sown rice – Linseed

5th year : Rainy season potatoes – Linseed

6th year : Rainy season potatoes – Linseed

As per choice of the farmer, linseed may be replaced by safflower.

It has been reported that liming applied to the pulse crop like blackgram either directly or indirectly has adequately benefited the following wheat or groundnut crop in lateritic soil (Panda, 1991).

1.11.3.5 Cropping in alkali and saline soils

Plants differ in their capacity to tolerate salts and alkali. Even the different varieties and strains of a particular species may exhibit differential tolerance in this regard Successful production in such soil would require selection of suitable crops that can produce satisfactory yield under various levels of salt and alkali accumulations.

Uppal el al (1961) prepared a list of crops based on tolerance to salt as given below:

(a) High salt tolerant crops

Kharif (rainy season): Dhaincha (*Sesbania aculeata)* rice and sugarcane

Rabi (winter seaaon): Oats, berseem, Lucerne, *senji (Trigonella spp.), methi (Fenugreek)* and barley

(b) Medium salt tolerant crops

Kharif (rainy season): Castor, cotton, sorghum, peral millet and maize

Rabi (winter season): Mustard and wheat

(c) Low salt toterant crops

Kharif (rainy season): Greengram, blackgram, pigeonpea and sunhemp

Rabi (winter season): Chickpea, peas, Linseed and sesamum

As far as possible, crops are to be taken in the order given above.

Crop yields are seriously affected if the electrical conductivity of saturation extract is more than about 12, 6 and 4 millimhos per cm in respect of tolerant, semi-tolerant and sensitive crops, respectively (Abrol and Fireman, (1977) as given in Table 6 below:

TABLE 6: Relative tolerance of crops to salinity (Saline soils)

Tolerant	Semi-tolerant	Sensitive
Date Plam	Pomogrante	Citrus
Barley	Wheat	Cowpeas
Sugarbeet	Oats	Gram
Spinach	Rice	Peas
Rape	Sorghum	Groundnut
Cotton	Maize	*Guar*
	Sunflower	Lentil
	Potatoes	*Mung*

Note: Tolerance in each column decreases from top to bottom.

Crop yield are seriously affected if the Exchangeable Sodium Percentage (ESP) is more than about 55, 35 and 10 in respect of tolerant, semi-tolerant and sensitive crops, respectively, as given in Table 7 (Abrol and Fireman, 1977).

TABLE 7: Relative tolerance of crops to exchangeable sodium (Alkali soils)

Tolerant	Semi-tolerant	Sensitive
Bermuda grass	Wheat	Cowpeas
Para Grass	Barley	Gram
Rice	Oats	Groundnut
Sugarbeet	**Raya**	Lentil
	Senji	*Mash*
	Berseem	*Mung*
	Sugarcane	Peas
	Millets	Maize
	Cotton	Cotton (at germination)

Note: Tolerance in each column decreases from top to bottom

1.12 INTEGRATED PEST MANAGEMENT AND STRATEGIES

A pest is an organism that inflicts economic damage to agricultural crops or produce or to human resources in the process of their multiplication. According to one estimate annual crop losses are inflicted by pests in India of which weeds account for 33 percent, diseases 28 percent insects 20 percent, rats 6 percent, and birds and others account for the rest. Pesticides or agrochemicals usage in India is increasing from year to year.

Indiscriminate use of pesticides adversely affects the ecological balance resulting in pest resurgence, pesticide resistance, pollution of environment and accumulation of pesticides residue at toxic levels in the food chain. It is increasingly evident that chemical pesticides alone cannot provide desired protection against all pests. Hence, Integrated Pest Management (IPM) involving all available pest control measures, namely host plant resistance, pest specific-bio-control agents, botanical pesticides and cultural practices need to be propagated in the interest of cost effectiveness and ecological security. While production of good quality pesticides is to be encouraged in the country including efforts for their export, it is necessary that efforts are made to minimize dependence on pesticides for the control of diseases and pests, since their continuous use may either increase and / or pollute the environment around us. Many a time weed may develop tolerance to herbicides and so is the case with other biotic agents.

From a state of over kill by pesticide, the approach has moved towards the ecologically sound Integrated Pest Management (IPM). In IPM a series of steps are followed to keep the population level of the pest below the economic threshold and demands a judicious use of pesticides, disease free seeds of resistance varieties, biological suppression of the pest population, usage of pheromones, timely spray of pesticides etc., to form part of plant protection umbrella.

Nevertheless, the use of agrochemicals is necessary in many a situation. The use should always be made when it becomes inevitable, and of the best strategies to ward off the bio-agents will be to go for trap cropping, ploughing immediately after the harvest of the crop, crop rotation and deferred growing in a number of situation.

To make pest control effective, timely pest surveillance are required covering the major crops, such as wheat, rice, pulses, oilseeds, sugarcane, cotton, etc. The periodical visits of the farms are essential to study the symptoms and evaluate the damage and whenever necessary, after advice concerning suitable biological control measures as well as judicious usage of pesticides. This exercise has proved highly successful in case of rice and cotton in Tamilnadu (South India) and wheat and sugarcane in the North-west India (Paroda, 1988; 1990).

Biological control of weeds, insects and even plant pathogens has proved to be highly successful. The control of water hyacinth (*Eichhornia crassipes*) through a weevil *(Neochitina bruchi)*, and

Coconut *Rhytocerus* beetle by the baculovirus are some success stories. Table 8 on important parasitoids and predators of crop pests and weeds and Table 9 on important pathogenic micro-organisms which can be used for pest/ disease suppression amply substantiate the fact that biological control of diseases/ pests and weeds in many crops can be done (Paroda, 1994). Submerged aquatic weeds could successfully be controlled by grass carp fish (Mukhopadhyay, 1986).

TABLE 8: Important parasitoids and predators of crop pest and weed

	Crop	Pest/ weed	Parasitoid/Predator
1.	Cotton	*Earias vitella*	*Chelonus blackburni*
		Pectinophora gossypiella	*Bracon kirkpatricki*
		Heliothis armigera	*Bracon hebetor*
			B. kirkpatricki
			Chelonus blackburni
			Trichogramma chilonis
2	Paddy	*Nilaparvata lugens*	*Cyrtorhinus lividipennis*
		Seripophaga incertulas	*Trichogramma japonicum*
		Cnaphalocrocis medinalis	*Trichogramma chilonis*
		Orselia oryzae	*Platygaster oryzae*
3	Pulses	*Heliothis armigera*	*Campoletis chlorideae*
			Hyposoter didymator
			Trichogramma chilonis
4	Sugarcane	*Chilo infuscaatellus*	*Sturmiopsis inferens*
		Melanaspis glomerate	*Adelencyrtus mayurai*
			Aphytis costalimae
		Phyrilla perpusilla	*Epiricania melanoleuca*
			Tetrastichus pyillae
5	Tabacco	*Spodoptera litura*	*Telenomus remus*
		Myzus persicae	*Menochilus sexmaculatus*
6	Castor	*Achaea janata*	*Telenomus proditor*
7	Weed	*Salvinia molesta*	*Cyrtobagous salviniae*
		Eichhornia crassipes	*Neochetina bruchi*
		Parthenium hysterophorus	*Zygogramma bicolorata*

Source: Paroda, 1994

In the bioherbicide approach, microbial plant pathogens are applied to target weeds in a manner similar to chemical herbicides. The term 'bioherbicides' refers to microbial plant pathogens (bacteria, virus, fungi) which are applied as sprays that uniformly kill or suppress the growth of weeds (Templeton and Smith, 1977). A formulation of soil-borne fungus *Phytophthora palmivora* Buller was registered in 1981 as the first selective mycoherbicide for the control of strngler (milkweed) vine (*Morrenia adorata* Lindl. in Florida and citrus groves under the trade name *De vine* (Abbott Laboratories), Chicago in USA (Kenney,1986).

TABLE 9: Important pathogenic micro-organisms which can be used for pest/disease suppression

	Pathogen	Pest/ Diseases	Crop / Commodity
1.	Nuclear poluhedrosis virus	*Helicoverpa armigera* *Spodoptera litura* *Amsacta albistriga* *Agrotis ipssilon*	Cotton & Pulses
2	Granulosus virus	*Chilo infuscatellus* *Phthorimaea operculella*	Sorghum & Sugarcane
3	*Metarhizium anisopliae*	White grubs	Sorghum & Coconut
4	*Beauveria bassiana*	*Epilachna beetles*	Sugarcane
5	*Trichoderma harzianum*	Soil borne and aerial plant diseases	Tobacco & Pulses
6	*Bacillus thuringiensis*	Various lepidopterans	Cotton & Tobaco
7	*Bacillus popilliae*	White grubs	Groundnut & Sorghum

Source: Paroda, 1994

Usage of resistant verity is one of the important components in IPM. In an approach to breeding for resistance, creation of artificial epiphytotics has permitted proper selection and advancement of early generation material. Resistance breeding has also led to the identification of multiple disease/ pest resistant varieties and hybrids in many crops.

Number of plant products, such as *Pyrethrum* nicltine sulphate, *neem* kernel extract, leaf extracts of *Pongamia* and *Ocimum* species exhibit insect and nematode repellant properties. Many of them also suppress fungal and bacterial plant pathogens. Heavy Molecular glycoprotiens from *Thuja* and *Baugainvillea* exhibit anti-viral properties (Paroda, 1988).

Pest management strategies: Pest management in intensive cropping systems involves diversity of methods. Van Emden (1974) has outlined strategies suitable for tropical small farmers as given below:

(i) **Suppression** preventing the population form increasing.): This is achieved through judicious use of pesticides, crop rotation, destruction of crop residues, change in sowing dates and other such cultural melhods Pesticides applied to one crop may be carried over to another crop to have residual toxicity. Pest escalation is possible due to presence of alternate hosts in intensive cropping. Manipulation in planting date, pattern and plant population can reduce this problem. There is evidence that bare ground in cropped area is attraction for the pests. It might be useful to include crop with good ground coverage, such as, legumes. Growing crop of various heights and inclusion of non- attractive crop in intercropping might reduce the spread of pest.

(ii) **Regulation** (preventing the population from exceeding the point of control of natural enemies): Increasing the crop diversity in intensive cropping helps to maximise the action of natural enemies and minimise the pest build up. Integrated pest and disease management is a logical approach to be applied in multiple cropping systems. The need for basic information on the biology of insent and their hosts and natural enemies is obvious. Careful study of these interactions can lead to new management approaches to control.

(iii) **Resource limitation** (making the population crashing from its peak density): Planting genotypes having resistance or at least field tolerance to the problem pests is a better way of reducing the pest damage. The genotypes which yield well after the pest damage or which compensate the damage by re-growth are preferable. In developing strategy of pest management for intensive cropping, the approach should be on crop management rather than pest population is given importance.

1.12 STRATEGIES FOR SUSTAINABLE AGRICULTURAL DEVELOPMENT AND ENVIRONMENT CONSERVATION

The following strategy elements are suggested (Singh, 1998) for promoting sustainable agricultural development and environmental conservation at the farm and community lrvel in low potential areas:

(i) Perception of the environment as fundamental to present and future livelihoods.

(ii) Development of labour-saving and time-saving technology for fuelwood and water collection, food preparation and post-harvest storage.

(iii) Substitution, where possible, of farm-grown inputs which make little demand on household finances, such as integrated post management, biological nitrogen fixation, organic waste recycling and composting, and biogas production where feasible.

(iv) Creation of non-farm income opportunities to support, not undermine, sustainable farming systems.

(v) Search for other means of supporting household livelihood when common access to resource, such as, grazing land leads to increased degradation of the resource.

(vi) Improvement of linkages between researchers and their target group of farmers and rural people to develop sustainable agricultural technologies for marginal lands and smallholders.

(vii) Adoption of policies that seek to fill those gaps in the food system of critical importance to poor people in low resources areas, such as post-harvest technology to avoid food loss, agroforestry, decentralized marketing, better biomass utilization, alternative sources of income generation, which require the development and dissemination of improved agricultural technology.

(viii) Encouragement of the integration of tree growing in farming system, as well as the integration of food fodder and fuel wood systems.

(ix) Better environment monitoring (satellite predications, remote sensing) to improve planning and assessment of land and water resources.

The high potential areas require many of the management and conservation measures described for low-potential areas, but they can generally sustain intensive cropping using existing technologies, if care is taken to maintain the soil's regenerative capacity and external inputs, such as agrochemicals are carefully managed. In general, these areas are either irrigated or have reliable, adequate rainfall and good soil fertility.

For future, policies need to be addressed, the following reorientations (based on papers presented for vision 2000 World Conference on Food, Agriculture and Environment held at International Food Policy Resource Institute, Washington DC. during June, 1995) are suggested:

- Invest more in research and extension and increase farmer input and feedback into technology generation and the dissemination process.
- Shift fertilizer policy from an emphasis on increasing the level of use to improving efficiency of nutrient balance and timing and placement of fertilizer.
- Shift crop protection policy from dissemination of chemical pesticides to use of integrated pest management.
- Shift emphasis in irrigation policy from investment in new system to improvement of water-use efficiency and productivity in existing systems.
- Adopt appropriate economic incentives through price policies that keep domestic prices in line with long-term price trends.
- Reform trade and microeconomic policy regimes that penalize agriculture to stimulate production by improving short-term input-use- efficiency and encourage long-term investment and technology change in the agriculture sector.

REFERENCES

Abrol, I.P. and Fireman. 1977. Alkali and Saline Soils. ICAR Bulletin No.4, Central Soil Salanity Research Institute, Karnal, pp: 31-32.

Abrol, I.P. and Katyal, J.C. 1990. Managing soil fertility for sustained Agriculture. Sustainable Agriculture: Issues, Perspective and Prospects in Semi-Arid Tropics. Indian Society of Agronomy, New Delhi, pp: 235-263.

Aujla, M.S., Singh, C.J., Vashist, K.K., and Sandhu, B.S. 1991. Evaluation of methods for irrigation of cotton in canal irrigated area of south –west Punjab (India). Arid Soil Research and Rehabilitation, 5:225-234.

Barea, J.M. 1991. Vescular-arbuscular mycorrhizae as modifiers of soil fertility. Advances in Soil Science, 15:2-40.

Beri, V. and Meelu, O.P.1981. Substitution of nitrogen through green manure in rice. Indian Farming (May issue), pp: 34.

Bhardwaj, S.P. 1994. Bunding and strip cropping for erosion control in agricultural lands of Doon Valley. Indian Journal of soil conservation, 22(3): 15-19.

Biswas, C.R. and Benbi, D.K. 1989. Long term effects of manure and fertilizer on wheat based cropping systems in semi-arid alluvial soils. Fertilizer News, 34(4): 33-38.

Braun, H. and Roy, R.N. 1983. Maximizing the efficiency of mineral fertilizers. Development of Plant and Soil Science, 10:251-273.

Brosten, D.1988. How much can we lose? Agrichemical Age. Paul. HBJ Farm, California, USA (Nov. issue), pp:7-8.

Cabtree, R.J. Yassin, A.A., Karagousou, I. and Mc New, R.W. 1985. Effects of alternate furrow irrigation: water conservation on the yield of two soybean cultivars. Agriculture Water Management, 10: 253-264.

Chhabra, R., Abrol, I.P. and Singh, M.V.1981. Dynimics of phosphorus during reclamation of sodic soils. Soil Science, 132:319-324.

Chatterjee, B.N. and Srivastava, V.C. 1990. Diversification of crops and cropping systems. Agronomic Research Towards Sustainable Agriculture. Publ. Indian Society of Agronomy AIRI, New Delhi: 81-95.

Commoner, B. 1968. The killing of a great lake. World Bank Year book, Field Enterprises. Education corporation Chicago, USA.

CTIC. 1995. National crop residue management survey. Executive summary Conservation Tillage Information Centre, West Lafayette. In., 21.

Engelman, P. and Le Roy, P. 1993. Sustaining water population and future of renewable water supplies. Population Action International. Washington DC.

Dhar, B.K. 1989. Impact of weather factor on fertlizer consumption in India. Fertilizer News, 34(8): 37-41.

El-Swaify, S.A. Walker, T.S. and Virmani, S.M. 1984. Dryland management alternatives and research needs for alfisols in the semi- arid tropics. Workshop on the State of the Art and Management Alternatives for optimizing the productivity of SAT alfisols and related soils. ICRISAT, Patancheru, Hyderabad, India.

Fageria, N.K. Baligar, V.C. and Jones, C.A. 1997. Growth and mineral nutrition of field crops (2nd edition). Marcel Dekker, New York, USA.

FAO. 1983. Guidelines for the control of soil degradation. United Nations Environment Programme, Food and Agriculture Organisation, of the United Nations, pp: 38.

Fung, L. 1988. Agriculture emission coefficients estimate. In Proc. Workshop on Agriculture and Climatic Change. USEPA, Washington D.C., USA.

Greacey, C.A. 1990. Sustainable Agriculture: Vision or division. Better crops with Plant Food, 72 (2): 3-5.

Gtegory, P.J.1988. In. Proc. International Conference on Dryland Farming held on 15-19 August, 1988. Amar. llo/ Bushland, Texas, USA, pp: 171-175.

Herdt, R.W. 1993. Measuring sustainability using long-term experiment. Proc Conference held at Rothamstead Experiment Station on 29-30 April, 1993, Rothamstead, UK.

Holt, D. 1989, A long affair with LISA, Better Crops with Plant Food. 73(3):4-7.

Kanwar, J.S. and Joshi, M.D. 1964. Liming in acid soils. Research Bulletin No., 1 Ludhiana, Punjab (India).

Karten, D.L., Erbach, D.C., Kaspar, T.C., Kolvin, T.S., Berry, E.C., and Timmons, D.R. 1990. Soil tilth a review at past perceptions and future needs. Soil Science Society of America Journal, 54. In Sustainable Agriculture Vol.2, Indian Society of Agronomy, New Delhi, India, 2:88.

Katyal, J.C. and Randhawa, N.S. 1983. Micronutrients. Fertilizer and Plant Nutrition Bulletin 7. Food and Agriculture Organisation of United Nation, Rome, Italy.

Kenney, D.S.1986. Weed Science, 34(Suppl.1): 15-16.

Khanna, S.S. and Gupta, A.P. 2000. Integrated nutrient management for increasing productivity and maintenance of soil fertility. International conference on 'Managing Natural Resources for sustainable Agricultural Production in the 21st Century' held on Feb. 14-18,2000, New Delhi, India: 141-154.

Khepar, S.D., Taneja D.S. and Sondhi, S.K. 1996. Agriculture ground water recharge in India- need, scope, present status and research priorities. Key Note Address at Silver Jubilee Workshop of AICRP on Optimization of Ground Water Utilization through Wells and Pumps, Dec. 4-5, 1996 held at JNKVV, Jabalpur, India, pp:13.

Khola, O.P.S., Dubey, R.K. and Sharma, N.K. 1999. Conservation ability of maize (*Zee mays)* – legume inter cropping systems. Indian Journal of Agronomy, 44 (1): 53-57.

Kumar Swamy, A.S. 1990. International symposium –a delegates impression. Agromy News. Indian Society of Agronomy, New Delhi, India, 6:2

Lal, R. and Miller F.P.1990, Sustainable farming for tropics. In Sustainable Agriculture: Issues, Perspectives and prospects in Semi-Arid Tropics, (ed.) R.P. Singh, Publ. Indian Society of Agronomy, New Delhi, 1:69-89.

Lal, R. 1991. Soil structure and sustainability. Journal of Sustainable Agriculture, 1:67-92.

Lal, R. 1993. Tillage effects on soil degradation, soil resilience, soil quality and sustainability. Soil and Tillage Research, 27:1-8.

Lal, R. and Miller, F.P.1993. Soil quality and its management in humid sub-tropical environments. Proc. of 26th International Grassland Conference, Massey, New Zealand.

Lal, R. and Kimble, J.M. 2000 Conserving tillage: prospects for the future. International Conference on Managing Natural Resources for Sustainable Agricultural Production in the 21st. Century held on Feb. 14-18, 2000, New Delhi, India, pp: 116-125.

Lal, S. and Mathur, B.S. 1988. Effect of long-term manuring, fertilization, and liming on crop yield and some physico-chemical properties of acid soil. Journal of Indian Society of Soil Science, 36:113-119.

Lockeretz, W. 1988. Open questions in sustainable agriculture. American Journal of Alternative Agriculture, 3:174-181.

Mac Rae, R.H., Hill, S.B., Mehuys, G.R. and Henning, J.1990. Farmscale agronomic and economic conversion from conventional to sustainable agriculture. Advances in Agronomy, 43:155-192.

Mandal, S.C. and Jha, K.K. 1969. Improvement of acid soils in Bihar by liming. Technical Bulletin No.1. Agricultural Research Institute, Kanke, Ranchi (India).

Mandal, S.C., Jha, P.P. and Rao, M.S. 1969. Crop adjustment in plateau region of Bihar, Technical Bulletin No.3 Agriculture Institute, Kanke, Ranchi (India)

Mittal, S.P., Singh. P. Agnihotri, R.C. and Sud, A.D. 1988. Effect of strip cropping of maize and legunmes on runff, soil loss and productivity. Indian Journal of soil Consservation, 16(2): 12-16.

Mukhopadhyay, S.K. 1986. Aquatic weed control in India. Indo- Pak-US Weed Control Workshop, Jaipur, India, Pesticides, November, 1986, pp: 15-22.

Mulla, D.J., Page, E.L., and Ganje, T.J. 1980. Cadmium accumutation and bio-availability in soils from long-term phosphorus fertilization. Journal of Environ. Qual., 9: 408-412.

Nambiar, K.K.M. and Abrol, I.P. 1989. Long-term fertilizer experiments in India an overview. Fertilizer News, 34(4):11-22.

Oldeman, L.R. 2000. Impact of soil degradation: a global scenario, International Conference on "Managing Natural Resources for Sustainable Agriculture Production in the 21st century" held on Feb.14-28, 2000 in New Delhi, India, pp: 79-86.

Panda, N.1981. Crop planning and crop management in acid soils. Second FAI Training Programme on management of acid soils, OUAT, Bhubaneswar.

Paroda, R.S. 1988. Research advances in crop sciences. 40 years of Agricultural Research and Education in India. ICAR, New Delhi, India, pp: 35-80.

Paroda, R.S. 1990. Present scenario and future strategy for crop improvement Research in India. 2nd Assam Company Lecture, Nov 12,1990. Assam Agricultural University, Jorhat, Assam, India.

Paroda, R.S. 1994. Crop improvement for sustainable agriculture. Invited Lecture at CCS Haryana Agricultural University, Hisar, Haryana, India, pp: 1-8.

Paroda, R.S. 1998. New resource management for sustainable agriculture: a new paradigm. Plenary Lecture in "First International Agronomy Congress held on 24 Nov 1998", in New Delhi, India

Prasad, R.N. and Biswas, P.P. 1998. Soil resources of India. In 50 years of Natural Recourse Management Research (eds. G.B. Singh and B.R. Sharma), ICAR, New Delhi, India.

Prasad, R. 2002, Sustainable agricultural production and intergraded nutrient management. Food and Environment. Second International Agronomy Congress on Balancing food and Environmental Security – A Continuing Callange, November 26-30, 2002, New Delhi. India, pp: 57-66.

Prasad, S.N., Ram, S. and Khybri, M.L. 1980. Effects of crops and dates of sowing on canopy development and splash erosion. Annual Report of Central Soil and Water Conservation Research and Training Institute (CSWCRTI), Dehradun, pp: 115.

Parihar, S.S., Sandhu, B.S., and Singh, N.T. 1975. A critical appraisal of research on irrigation scheduling to crops. Proceedings of second world congress. International water Resource Association, December, 1975, New Delhi, India, pp.142.

Parihar, S.S. and Sandhu, B.S. 1987. Irrigation of Field Crops-Principles and Practices, ICAR, New Delhi, India, pp: 142.

Papendick, R.I. Mornick, S.B., Youngberg, I. Gand Meyer, R.E. 1989.

Parr, J.F., Proceedings of International Symposium on Sustainable Agricultural System. Soil and Water Conservation Society, Ankeny, Iowa, USA.

Parr, J.F. and Hornick, S.B. 1990. Recent development in alternative agriculture in the United States. In proceedings of international Conference on "Kyusel Nature Farming", University of Khon Kaen, Thailand. October 17-21, 1990.

Parr, J.F. Pependick, R.I., Hornick, S.B., and Meyer, R.E. 1990. Conservation tillage system for sustainable agriculture. Sustainable Agriculture Issues Perspective and Prospects in Semi-Arid-Tropics, Indian Society of Agronomy, New Delhi, pp: 19-48.

Rosengrant, M.W. 1997. Water Resources in Twenty First century: Challenges and Implications for Action. 2020 vision for Food, Agriculture and the Environment. Discussion paper 20. International Food Policy Research Institute, Washington DC.

Roy, N.R. 1995. Integrated plant nutrient systems. FAO Fertilizer and plant Nutrition. Bulletin 12, pp: 49-66.

Roy, R.N. and Ange, A.L. 1991. Integrated Plant nutrition system (IPNS) and sustainable agriculture. Proceedings of FAI Annual Seminar, Fertilizer Association of India, New Delhi, pp: 1-12.

Samra, J.S., Sharma, A.R. and Khola, Q.P.S. 1998. Sustainable crop production through resource conservation: Indian experience. First international Agronomy Congress held on November 23-27, 1998 in New Delhi, India. Indian society of Agronomy, ICAR, New Delhi, pp:117-134.

Sarkar, A.K. 2001. Nutrient mining in agro-climatic zones of Jharkhand. Fertilizer News, 46(4): 103-107.

Sankaram, A. 1996. Soil fertility management for reconciling sustainability with productivity. Journal of Indian Society of Soil Science, 44(4): 593-600.

Silva, J.A. and Uehara, G.1985 Transfer of agro technology. Benchmark of Soil Project. Hawaii Institute of Tropical Agriculture and Human Resources, USA, pp: 2-11.

Singh A. 1998. Hydrological behaviour of experimental watershed, pp 169-171. Annual Report, ICAR Research Comples for NEH Region, Barapani, Shillong.

Singh, A. 2000. Some thoughts and observations on the management of cropping systems. International Conference on 'Managing Natural Resources for Sustainable Production in 21st. Century', held on February 14-18, 2000, New Delhi. India, pp.504-509.

Singh, K.N. and Deka, J.1990. Integrated nutrient supply system for sustainable crop production. Agronomic Research Towards sustainable Agriculture (eds. K.N. Singh and R.P. Singh) Indian society of Agronomy, IARI, New Delhi, India, pp.35-56

Singh, H.G. 1992. "Workshop on Agronomic strategy for sustainable crop production" held in Rajasthan University, Bikaner, Rajasthan, India on February 25-28, 1992. Indian Society of Agronomy, New Delhi, India, pp: 7-14.

Singh, M. 1990. Education for sustainable agriculture. Indian Journal of Agronomy, 35 (1&2): 1-22.

Singh, P. 1998. Towards sustainable agricultural development: challenges in 21st century. Proceedings of First international Agronomy Congress held on 23 Nov, 1998. Indian society of Agronomy, ICAR, New Delhi India, pp I-XXIII.

Singh, S.R., Choudahary, T.N., Batt, R.K., Bhatnagar, P.R., Saha, B., and Patil, N.G. 1999. Irrigation water management. In: 50 years of Natural Resurce Management Research, (eds. G.B. Singh and B.R. Sharma), ICAR, New Delhi, pp: 265-310.

Srivastava, V.C. and Mohsin, M.A. 1989 Effect of conventional and zero-tillage with and without mulch and weedicides on wheat and barley under rainfed upland situation. Journal of Research (BAU), 1(1): 9-14.

Srivastava, V.C. and Tripathi, S.N. 1989. Relative efficiency of slow release nitrogen sources in lowland rice and their residual effect on succeeding wheat. Journal of Research (BAU), 1 (2): 21-26

Stewart, B.A., Lal, R and El-Swaify, S.A. 1990. Sustainable resource base of an expanding world agriculture. In Machinery for a productive and sustainable resource base. Soil and Water Conservation Society, Ankeny, IA.

Syres, J.K. 1995. Contribution of soil reserves to plant nutrient supply. In Integrated plant nutrition systems. FAO Fertilizer and plant Nutrition Bulletin, 12:243-252.

TAC. 1988. Sustainable Agriculture Production: Implications for International Agriculture Research. Technical Advisory Committee (TAC), FAO, Rome, Italy.

Templeton, G.E. and Smith, R.J. Jr. 1977. *In* Plant Disease - An Advance Treatise (eds. J.G. Horsefall and E.B. Cowling). Academic Press, New York, USA 1:167-176.

TOI. 2001. Sunday Magazine, Times of India, 27 May, pp: 2

Tripathi, K.P. and Samraj, P. 1994. Problems of soil erosion and conservation strategies in the southern hill region with particular reference to Nilgiris. Indian Journal of Soil Conservation, 22(122).94-101.

Uppal, H.L., R.R. Agrawal and Kibe, M.1961. Reclamation of saline and Alkaline lands. Farm Bulletin, 66:1-30. Directorate of Extension, Ministry of Food and Agriculture, New Delhi, India.

Van Emden, H.F. and Williams G.F. (1974) Insect stability and diversity in agro-ecosystems. Ann. Rev. Entomology. 19:455-475.

Venkateshwarlu, J., Vishnumurty. T and Padmanabhan M.V. 1991. Contingent crop production strategy in rainfed areas under different weather conditions. Central Research Institute for Dryland Agriculture (CRIDA). Hyderabad, India, pp: 1-15.

Velayuthum, M. 1998. Mapping soil resources for optimal land use planning. In 50 Years of Natural Resource Management Research (*eds.* G.B. Singh and B.R. Sharma), ICAR, New Delhi, pp:83.

Velayuthum, M and Bhattacharya, T.2000. Soil resource management. International conference on 'Managing Natural Resources for Sustainable Agricultural production in 21st century', held in New Delhi on February 14-18, ICAR, New Delhi, India (*eds.* J.S.P. Yadav and G.B. Singh), pp 1-135.

Verma, U.N. and Srivastava, V.C. and Prasad, R.B. 1989. Effect of Zero –tillage on rice (*Oryza sativa)* atubble mulching on wheat (*Triticum aestivum).* Indian Journal of Agricultural Science, 59(10): 669-70.

Verma, U.N. and Srivastava, V.C. 1989. Weed management in wheat under zero and optimum tillage conditions. Indian Journal of Agronomy, 34(2): 179-89.

Verma, U.N. and Srivastava, V.C. 1994. Research Review: production technology of non- tilled wheat after puddle transplanted rice Indian Journal of Agricultural Science. 64(5): 277-284.

WCED.1987. Our common future. Word commission on Environment and Development (WCED), Oxford University Press, Oxford.

WWIR.1993. World resources (1992-93) Word Watch Institute Report (WWIR). Washington DC., USA.

Yadav, R.L. and Prasad, K.2000. Modern concepts for eco-friendly management of cropping systems. International conference on 'Managing Natural Resources for Sustainable Production in 21st Century' held on February 14-28, 2000, New Delhi, pp:481-482.

York Jr., E.T. 1989. Research for a sustainable agriculture. Better crops with plant Food.73(3): 8-11.

CHAPTER II

FARMING SYSTEMS

2.1 INTRODUCTION

2.1.1. Concept and Definition

Farming system is not a new concept. The marginal, small and medium farmers have been adopting different enterprises since long without any scientific base. The growing dependence of farmers mainly on crop production invariably subjected to a high degree of uncertainty in income and employment. Thus, in order to increase food production immediately and substantially to meet the demand of increasing population, and to ameliorate soil degradation, environmental problems of water quality, and the global threat of "green house" effect, there is growing awareness and enthusiastic interest in modern farming systems concept which gives scientific touch to the existing practice by integrating ancilliary propositions with crop production tailored to individual farming situations providing the farmers some option to suit any contingency.

Farm: Literally the word 'farm' means a piece of land cultivated by its owner or tenant. However, scientifically its understanding is a little more which includes cultivated land, homestead used for housing the farmer, his family, farm animals and implements used for farming.

Farming: Farming is the process of harnessing solar energy in the form of economic plant and animal product.

According to Ruthenbery (1990), farming is an activity carried out by household on holdings that represent managerial units organized for the economic production of crops and livestocks.

The farming activities are carried in a systematic and well arranged pattern in a community centering around a household-the unit of activity for producing a dependable and continuous food supply and many of their needs, such as, clothing and shelter and surplus for sale. The owners of the households, i.e., the farmer have at their disposal certain physical, biological and socio-economic resources on the basis of which they can arrange their farm production with the technology they have.

System: The whole activities of the farm for the purpose of production and consumption result in a system.

A system is defined conceptually any set of elements or components that are inter- related and interact among themselves.

A system implies inter–related practices/ processes organized into a functional entity, i.e., an arrangement of components/parts that interacts according to some persons and transforms inputs into outputs (Fresco and Wastphal, 1988).

A system is a group of interacting components operating together for a common purpose capable of interacting as a whole to external stimuli. It is unaffected directly by its own outputs and has a specific boundary based on the inclusion of all significant feedbacks (Spedding, 1988).

Farming System: A farming system is the result of a complex interaction of a number of interdependent components – soils, water, crops, investock, labour and other resources within an environmental setting.

Farming system is defined by its physical, biological and socio- economic settings and by the farm families goals and other attributes, access to resources, choice of productive activities (enterprises), and management practices. It has been defined in different ways as follows:

1. According to Consultative Group on International Agricultural Research (1976), a faming system is not simply a collection of crops and animals to which one can apply this input or that and expect immediate results. Rather, it is a complicated interwoven mesh of soils, plants, animals, implements, workers, other inputs environment influences with the strands held and manipulated by a person called the farmer, who given his preferences and aspirations, attempts to produce outputs from the inputs and the technology available to him. It is the farmers unique understanding of his immediate environment, both natural and socio-economic, that results in his farming systems.

2. A farming system is a collection of distinct functional units, such as, crop, livestock, processing investment and marketing, which interact because of the joint use of inputs they receive from the environment, which have the common objective of satisfying the farmers (decision makers) aims. The definition of the borders of the system depends on the circumstances, often it includes not only the farm (economic enterprise), but also the household (farm-household system) (Ruthenberg, 1980).

3. Farming system (farm system or whole farm system) is the production and consumption activities used by a person called a farmer to derive benefits from land and other inputs through crop growth and the use of technologies available to him under specific environmental conditions (Zandstra *et al.*, 1981).

4. A farming system is a unique and reasonably stable arrangement of farming enterprises that the household manages according to well defined practices in response to physical, biological and socio-economic environments and in accordance with the households goals, preferences, and resources (Shaner *et al.*,1982)

5. Farming system integrates entire complex of development, management and allocation of resources as well as decisions and activities which within an operational farm unit or combination of units results in agricultural production and the processing and marketing of the products (Krantz *et al.*, 1975).

6. Farming system represents an appropriate combination of farm enterprises viz., corpping systems, livestock, fisheries, poultry and the means available to the farmer to raise them for profitability. It interacts adequately with environment without dislocating the ecological and socio-economic balance on one hand and attempts to meet national goals on the other (Pal *et el.*, 1985).

7. A farming system, as defined by Singlachar (1987), is a complex inter-related matrix of soils, plants, animals, implements, power, capital and other inputs controlled in part by farming families and influences to varying degree by political, economical, institutional and social forces that operate at many levels. The farm is viewed in a holistic manner.

8. Farming system is a resource management strategy to achieve economic and sustain agricultural production to meet diverse

requirements of the farm household while preserving the resource base and maintaining a high environmental quality (Lal and Miller, 1991).

Thus, the farming system designates a set of agricultural activies organized into functional unit (s) to profitably harness solar energy while preserving land productivity and environmental quality and maintaining desirable level of biological and ecological stability.

2.1.2 Scope and Objectives

2.1.2.1 Scope

The income from cropping alone from small and marginal farm is hardly sufficient now to sustain the farmer's family in underdeveloped and developing countries. With the decline in farm size due to population explosion, it would be increasingly difficult to produce enough food for the family. Therefore, the farmer, to be assured of a regular income for a decent living (above the poverty line), has to think of other enterprises which will compliment his cropping activity. A judicious mix of any one or more of these enterprises with agronomic crops should complement the farm income and help in recycling the farm residue/waste. The selection of enterprises must be based on the cardinal principle of minimizing the competition and maximizing the complimentary effect between the enterprises.

The various farm enterprises that could be integrated in farming system are crop, livestock, poultry, fish, plantation crops' mushroom cultivation, apiculture, sericulture, silviculture, agro - forestry, agro based industries, food processing, etc. A combination of one or more enterprises with cropping, when carefully chosen, planned and executed, gives greater dividends than single enterprises specially for small and marginal farmers. It also ensures optimization of resources, minimization of risk, generation of employment and overall growth and stability in productivity and profitability. The intergration of farm enterprises depends on many factors, such as:

- Soil and climatic features of selected area
- Availability of the resources, land, labour and capital
- Present level of utilization of resources
- Returns from existing farming system

- Economics of proposed integrated farming system
- Managerial skill or the farmer

2.1.2.2 *Objectives*

The sustainability is the main objective of the farming system where production process is optimized though efficient utilization of inputs without impairing the quality of environment with which it interacts.

A farming system must have the following objectives:

- Technically feasible and economically viable system.
- Integration of cropping with allied complementary enterprises for different farming situations by adopting improved management skills.
- Ensuring optimum utilization and conservation of available resources, effective recycling of farm residues within the system and application of judicious levels of purchased inputs.
- Maintaining sustainable production system without damaging resource base and environment, i.e., decreasing risk of soil and environmental degradation, and restoring productivity of degraded land.
- Capable of providing stability in production, generating employment throughout the year and income at periodic intervals by raising overall profitability of farm household for decent living by complementing main and allied enterprises with each other.

2.2 *Plant Interactions*

The interaction is mainly due to response of one species to the environment as modified by the presence of other species. In cropping system, the crops are grown in association (intercropping) or sequence (sequential cropping). In both the situations, interaction between the component crops may occur. However, the chance of competition for light, moisture, nutrient and / or space in sequential cropping is minimum. In relay cropping, crop under short span of over lapping at the seedling stage may have competition for light. The interaction between the component crops may be of the following types:

(i) **Non-Competitive:** In case of crops grown in association but the growth of either of the concerned species is not affected, such type of interaction is called as non-competitive interaction.

(ii) **Competitive:** Competition is a kind of interference. It occurs (a) when two plant draw a requirement (e.g., nutrient, moisture, light, CO_2, space) from the same limited pool (Donald,1963), (b) the supply of this requirement to atleast one of the plant is reduced by the presence of the other, and (c) the reduced supply of requirement influences, the growth, reproduction or survival of the plant.

In competitive relationship, one species may have a greater ability to use the limiting factor and will gain at the expense of the other. In other words, output of one crop would be increased through a decline in the production of the other. This is also termed as 'Compensation'. Willey (1979), has referred the two species as 'dominant' and 'dominated'.

(iii) **Complementary**: In this type, the growth of the either of concerned species grown in association, is not affected. In complementary interaction, an increase in output of one crop helps to bring about an increase in output of the other. This is referred as "mutual cooperation" (Willey, 1979). According to the Ludwig (1959), 'annidation' is the complementary use of resources by exploiting the environmental supplies in a different ways by the components of a community.

(a) **Annidation in space:** The leaf conopy of component crops may occupy different vartical layers with the taller component tolerant of strong light and high evaporative demand and the shorter component favouring shade and high relative humidity. For example, multi-storeyed cropping in coconut and planting shade trees in cocoa and tea plantations (Aiyer, 1949; Nair, 1979).

(b) **Annidation in time:** The peak demand for nutrients, moisture and light are likely to occur at different periods of component crops planted at widely varying duration, thus reducing competition. For example, component crops of early maturing greengram or upland direct sown rice and late maturing pigeonpea in a intercroppy system (Ahmad anc Srivastava, 1993).

(c) **Other complementary effects:** Beneficial effects of legumes on non-legumes have been reported (Palaniappan *et al.*, 1976; Ofori and Stern, 1987). Part of

N fixed in the root nodule of the legume may become available to the non-legume components in a intercropping system. Radke and Hagstorm (1976) reported that taller components may act as wind barriers protecting the shorter components from lodging. Physical support by one species to the intercnpped climbing species may improve to the yield of the climber, for example maize + beans, coconut + pepper.

(iv) **Supplementary:** In the case output of one crop may be increased without having any influence on the output of the other. This situation usually arises when the maturity of two component crops or their duration widely differ.

(v) **Mutual inhibition:** This happens when the actual output of each species is less than expected.

(vi) **Allelopathy:** Swiss botanism M.A.P. de Candole was first to report in 1932 involvement of plant produced chemicals in plant-plant interaction. He suggested the possibility that some plants excreted something from their roots which was injurious to other plants and proposed the theory of antagonism. The term 'allelopathy' derived from Greeek word was coined by Hans Molisch in 1937 to include both harmful and beneficial biochemical interactions between all types of plant and interactions involving microorganisms (Rice,1974 1984). Evidence indicates that alletopathic compounds get out of plants by volatilization, exudation from roots, leaching from plants or residues by rains or decompositions of residues.

Allelopathy, thus may be defined as any direct or indirect harmful effect that one plant has on another through the production of chemical substances that escape into the environment. In intercropping systems, some crops may be unsuitable to be grown as intercrops because they may produce and excrete toxins into the soil which are harmful to the associated components. Trenbath (1976) reported that allelochemicals produced from the leaves of *Eucalyptus globulus* drastically reduced the germination of mustard (*Brassica spp.*) seeds sown underneath. The soil sickness in monocropping is also due to allelopathy since many toxins are added to the soil by the crop. The citric acid produced in cool, wet, anaerobic conditions inhibits germination of wheat, barley, oats and maize. Verma (1995) observed that the number of ratoons in sugarcane is restricted due to toxic substances in the rhizosphere form fungi like *Fusarium oxysporum* and *Trichoderme harzianum*.

Some distinctions are made to separate inhibiting chemicals produced by higher plants and microorganisms. Phytoncides are inhibitors produced by higher plants against microorganisms. Marasmins are the compounds produced by microorganisms and are harmful to the plants, while kolines are chemical inhibitors produced by plants against plants. Antibiotics like patulin exhibit marked toxicity to higher plants. It may be noted that phytoncides, marsmins and kolines harm cultivated plants, weeds and microorganisms (Hunsigi and Krishna, 1989). Rice (1984) has classified these allelochemicals in five categories, i.e. phenyl propanes, acetogenins, tarpenoids, steroids and alkaloids.

2.3. COMPONENTS OF FARMING SYSTEMS

The farming system is a part of largar system, e.g. the local community. This can be divided into subsystems like cropping systers, fish production systems, agro-forestry production system, etc. The flow diagram showing components and sub systems of a farming system is given in Fig.4. Example of a classification of world farming system and a village system are shown in Fig. 5 and Fig.6, respectively, Resource flows in a farming system is depicted in Fig.7.

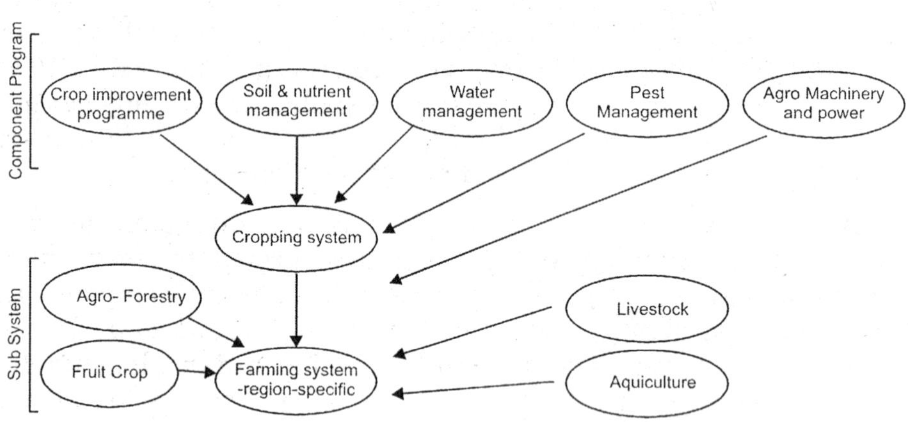

FIG. 4: Flow diagram showing components and sub systems of a farming systems.

2.3.1 Management of Animal Components

2.3.1.1 Cattle / Dairy management

Based on use, the cattle are classified into three groups:

(i) Draft purpose, e.g., the bullocks of Hariana, and Nagore breeds

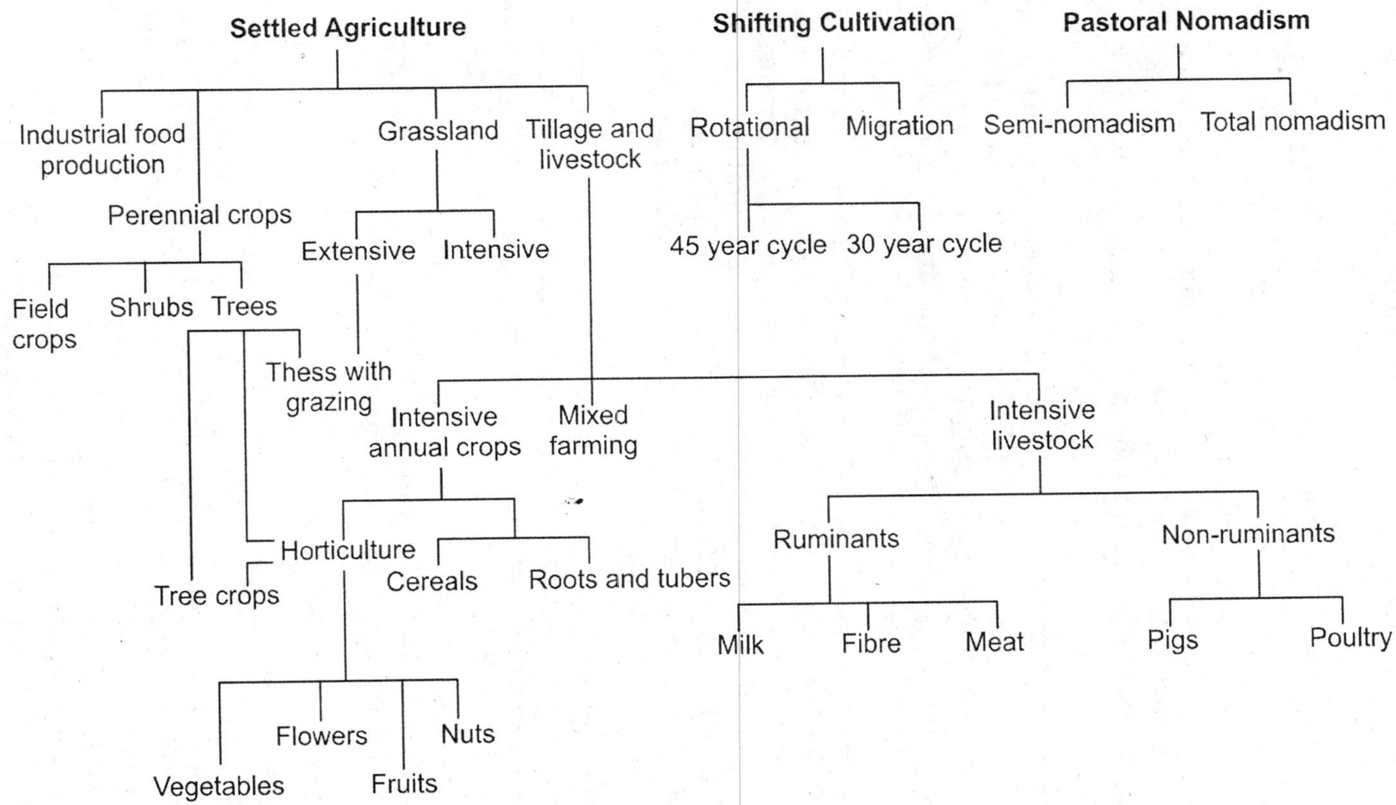

FIG. 5: Example of a classification of world farming systems. (Spodding, 1998).

(ii) Dairy purpose, e.g., the cows of Hariana, Sahiwal, Sindhi and Gir breeds

(iii) Dual purpose, e.g., Hariana, Ongole and Kankrej.

The exotic breeds of high milk yielders are Jersey crossbred cows (F_1 Crosses) and Holstein-Friesian. Jersey breed of cows consume less feed and fodder, give more milk with high fat content and are comparatively better disease resistance, whereas Holstein-Friesian are high milk yielder in cooler climate as they lack heat tolerance.

Important dairy breeds of buffalo like Murrah crosses and even on-descript types, Nili Ravi, Godawari could be reared for milk production and higher profit. Buffaloes can digest more roughages than cows and thrive well on dry fodder. Buffaloes contribute nearly 60 per sent of milk to India's national milk grid.

The space required by each cow and baffaloes is 12 to 18 sq. m. with good drainage facility. The cows remains in milk for 9 to 10 months, the average calving interval being 16 to 18 months. The gestation period is about 280 days. In case of baffaloes, the milk period last for 7 to 9 months. The gestation period is about 310 days. Baffaloes require abundant water and wallowing. They also need regular de-worming.

Cattle feeding consists of roughages and concentrates. Roughages include legumes, non-legumes, hays, straw and silage of legumes and grasses. Grains of maize, sorghum, oats, barley and cakes of groundnut, sesame, rapeseed, soybean and linseed constitute the concentrates. The ration per animal per day normally includes concentrate at the rate of 1 kg for 2 litres of milk, green fodder (20-30 kg), straw (5-7 kg) and water (32 litres).

2.3.1.2 Sheep and goat rearing

Based on the geographical division, three types of sheep are found in India. The breeds of Gurez, Karanah, Bhakarwal, Gaddi, Rampur-Bushiar are found in the temprate Himalyan region. In dry western region, breeds of Lohi, Bikaneri, Marwari, Katchi, Kathiawari are found. Southern regions are dominated by Deccani, Nellore, Bellary, Mandya and Bandur types. One ram is sufficient for maintaining 40 to 50 ewes, The duration of heat period for ewe will range from 1 to 3 days. The gestation period will vary from 142 to 152 days. Sheep grow fully at two years of age. On an average aves produce about five crop of lambs.

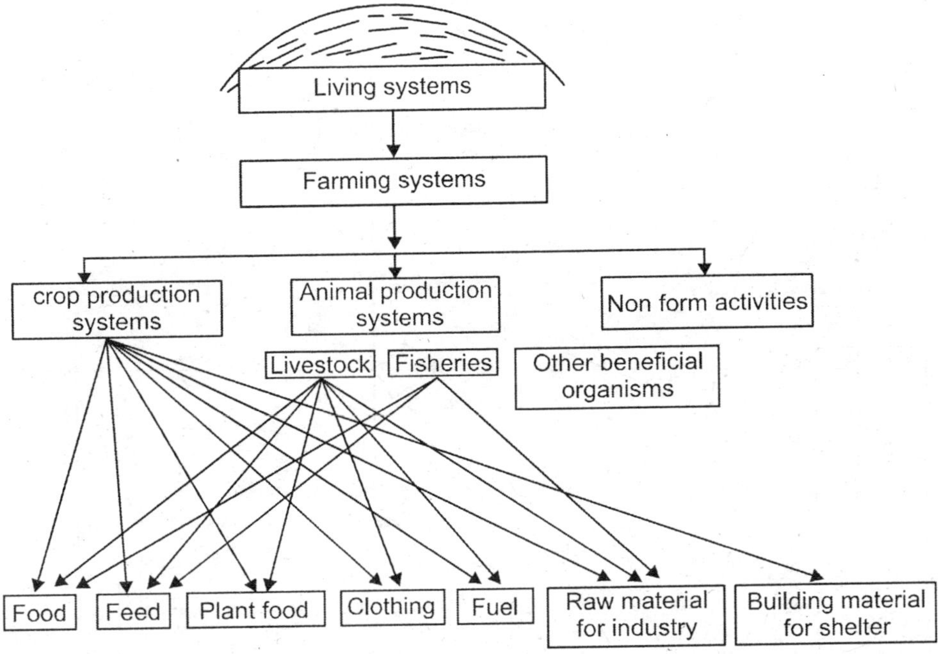

FIG. 6: A village system.

About 1 to 2 kg of leguminous hay per day is required by a sheep. Concentrate in the form of 110 to 225 g cake of groundnut, seasame or safflower may supply proteins to a sheep. To meet the mineral deficiency common salt, ground limestone and sterilized bonemeal may be fed to sheep.

Goat is mainly reared for meat, milk, hide and skin in different kind of environments. The space requirement for a goat is 4.5 to 5.4 sq. m. dry, comfortable, safe and secure place. Marshy or swampy ground is unsuitable for a goat. Kitts are kept under inverted baskets. Goat matures in about 6 to 7 months. It gives birth to 1 to 3 kitts per time. Gestation period is 145-155 days. They eat tree leafs and other grass species. Goats should be fed with maize, wheat, horseqram, qroundnut cake, fish meal and wheat bran. Common salt should also be added. Clean fresh water should be made available to the goats.

2.3.1.3 Poultry and duck rearing

Poultry meat accounts for 27 per cent of the total meat consumed in world. The stock used for egg production are crosses involving the strains of White Leghorn, Rhode Island Red, California Grey and Astralop.

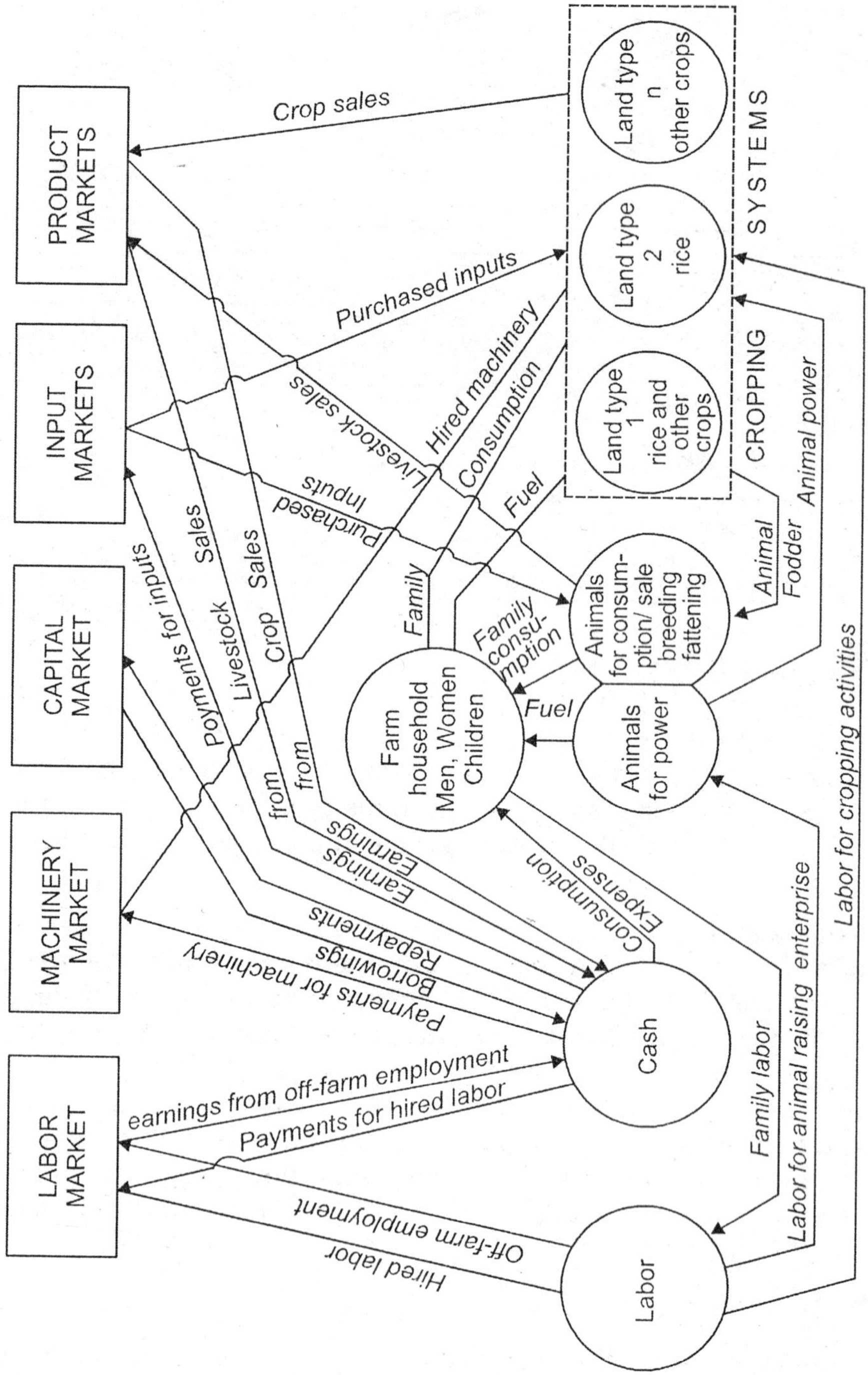

FIG. 7: Resource flows in the farming system.

Heavy breeds, such as, White Playmouth Rock, White Cornish and New Hamshire are used for cross-bred broiler chicken. Floor area of 0.2 to 0.4 sq. m. per bird is required. The house should be located on well-drained ground. It should be ventilated and reasonably cool in summer and warm during winter. The feed stuff used for making poultry ration are cereals (e.g., maize, barley, oats, wheat, pearl millet, sorghum and broken rice), cakes and meals (e.g., Oil cakes, fish meal, and maize-gluten-meal) and minerals (e.g., common salt, limestone, oyster shell). Feed may be given 2 to 3 times. The vaccination should be done against Ranikhet disease within first 6 to 7 days of age. At the optimum body weight of 1 to 1.5 kg around six weeks, they can be marketed for broiler. It may not be economical to keep the hens beyond one and half years. They should be disposed off for table purpose. One hen is capable of laying 180 to 230 eggs in a year starting from 6 months.

Ducks are predominantly of indigenous type and reared for egg production on natural foraging. It accounts for about 7 per cent of the poultry population in India. Ducks are quite hardy and resistant to common aviation diseases. Duck farming can be done in places like marshy river side, wetland and barren moors where chicken do not flourish. Khaki Compbell and Indian Runners are the most popular breed for egg laying with production potention of 250 to 300 eggs per bird per year. White Pekin, Muscovy and Aylesbury breeds are known for meat production. Indian breeds are Sylhet Mete and Nageswari with production potential of 130-140 eggs per bird per years. Ducks can be raised without artificial shelter since they prefer to stay outside.

One nest of size 0.3 x 0.3 x 0.45 m to every 3 ducks is sufficient. The duck house should be well ventilated, dry, leak and rat proof. Duck supplement their feed by foraging, eating fallen grains, small fishes and other aquatic materials in lakes and ponds. Ducks prefer wet mash due to difficulties in swallowing the dry mash. A duck or broody hen can take care of 10 to 15 ducklings. The ducks come into production at 16 to 18 weeks of age.

2.3.1.4 Piggery

Pigs are maintained for the productions of pork. Breeds of Large White Yorkshire and Land Race are very popular in India. A good boar weighs 90 kg in about 5 to 6 months. One pig requires about 2.7 sq m with a wall of 1.2 m height. Eight boars can be kept in 2.7-4.5 sq m area with 2.4-6.0 sq m open space. The house should not be damp and over heated. The main ingradients of swine ration are cereals and millets and their byproducts. The fibre content in feed should be low (5 to 6%) for better feed utilization efficiency. Mixed ration should also contain salt. The consumption of feed is 3.5 per cent of total weight. Feed allowance is calculated as 2.5-3.0 kg per 100 kg body weight plus at the rate of 0.25 kg feed per piglet with the lactating mothers. The gestation period is on an average 114 days. The boar-sow ratio should be 1:15. Mortality in new born piglet is maximum when farrowing takes place in acute cold or hot climate.

2.3.1.5 *Rabbit rearing*

Rabbit is reared for meat and wool all over the world. Though its farming has been introduced recently in India, rabbit production on commercial lines has gained momentum because it involves low production cost and easy management. Rabbit meat, a White meat, is rich in protein and low in fat and cholesterol content. New Zealand, White New Zealand Red, California White, Great Giant, White Giant, Black Brown. are some of the important breeds of broiler rabbits. Breeds for wool are Angora, Dan Florida and Polish Bolamina. Rabbits of 5-6 months age with body weight of 2.0 kg are ready for breeding. Female rabbit is to be introduced to male rabbit cage or section and is to be separated immediately after crossing. The pregnant female is to be raised separately. The gestation period is 28-32 days. After one month, the young ones are separated from the mother. In one year, it gives birth five times, each time producing about six numbers. 1.8 sq. m of area is needed for one kg body weight of rabbit. Outside it can be kept in hutches and inside in cages with welded wire mesh at on floor and on sides with a door of 30 × 30 cm. Rabbit are non-ruminant herbivorous and can consume different types of green roughages like alfalfa, berseem, lucerne, cowpea etc. The feed ingredient may constitute alfalfa (25%), maize (20%), wheat bran (40%), groundnut cake (12%), mineral mixture (2%) and salt (1%). Root crops like carrot, turnip, etc., are also much relished by rabbits. The requirement of concentrate feed ranges from 50 to 150 g, consisting 50 to 70 g of vegetable and 150 to 200 g of green fodder.

2.3.2 Management of other allied components

2.3.2.1 *Aquaculture (Fish culture)*

In order to improve traditional management of fish culture resulting low yield per ha, poly fish culture with stocking density of 7500 fingerlings and supplementary feeding with cereal bran and oilseed cakes will boost the total biomass production. The phytophagous fish (Catla, Rohu and Mirgal) can be combined with omnivourous (common carp), plankton feed (Silver carp) and mud eaters (Mirgal and Calbasu), in a composite fish culture system.

Among the major species of Indian fish, Catla (*Catla catla*) is a fast growing, surface and column feeder of vegetation and decomposed plants. Rohu (*Labeo rohita*) is column feeder of growing plants, algae, decomposing vegetation etc. Calbasu (*Labeo calbasu*), Mirgal (*Cirrhimus mirgale*) and common carp (*Ciprinus carpio*) are bottom feeders of algae and higher plants. Common carp is omnivorous. Silver carp (*Hypophthalmichlythus motitrix*) is a surface feeder on microplants. Grass carp (*Cyernus carpia*), a exotic fish is a specialized feeder on aquatic plants, cut grass and vegetable matters.

Clay soils with 1.5 to 2.0 m depth of water in pond is idle to help effective photosynthesis, temperature maintenance and growth of fish. For N, nitrogenous fertilizers like ammonium sulphate and urea and for P, superphosphate can be added. Organic manures, such as, farm yard manures and poultry droppings may also be applied to promote growth of fish culture. Pond should have appropriate proportion of phosphate (0.2-0.4 ppm), nitrate (0.06-0.1 ppm) and dissolved oxygen (5.0-7.0 ppm) with pH of 7.5 to 8.5 pH may be adjusted with addition of lime at the interval of 2 to 3 days, if it falls below 6.5.

2.3.2.2 Apiculture

Apart from producing honey, honeybees play a vital rote in nature by pollinating many plants and therby ensuring their survival. Bee keeping is one of the most popular agro-based industries where nectar and pollen from flowers are the row materials. There are two species of bee-*Apis mellifera* suitable for plains and *Apis cerana indica* acclimatized to higher altitudes. The most important sources of nectar and pollen are maize, great millet, sunflower and palm. A minimum of two colonies for bee keeping is recommended for the beginner. The hive may be single or double walled available in the market. The late spring or early summer is the most suitable time for starting bee keeping when bees divide their colonies, known as swarming. The nectar collected by bees is processed and placed in comb cell for ripening. During the ripening sucrose is converted into glucose and fructose by an enzyme called invertase which is added to it by the bees. Honey is an excellent energy food with an average of 3500 calories per kg (Balasubrananiyan and Palaniappan, 2002).

2.3.2.3 Sericulture

Combined practice of mulberry cultivation, silkworm rearing and silk reeling is known as sericulture. Cultivation of mulberry plant is called as moriculture. The commonly cultivated species of mulberry are *Morus alba*, *Morus indica*, *Morus serrata* and *Morus latifolia*. The yield of mulberry leaves is 30 to 40 tons per hectare per year for approximately 12 years. After this period fresh planting is done. Organic manure is applied at the rate of 20 t pet ha for irrigated crops and 10 t per ha for rainfed crop. Planting is done by using stem cutting perpared from branches of 8-12 months old trees. The cuttings should be 15-20 cm long with 3 to 4 buds and 1.5 cm thick and should have 45° slanting cut in both sides. The recommended fertilizer dose of N, P and K is 300: 120: 120 kg per ha, respectively. Irrigation may be given at 15 days intervals. Mulberry being a perennial crop usually cultivated in wider spacing of 60 × 60 cm to 90 × 90 cm. After each pruning of mulberry crop, the inter row space can be utilized for growing short duration crop like legumes and vegetables. In temperate region (e.g., Jammu and Kashmir) because of growing inter crop like mustard and beans, an substantial additional income could be earned with the income delivered by utilizing the mulberry leaves for rearing silkworms. The silkworm larval litter incorporated along with cow-dung and old slurry in the proportion of 1:2:2 gave the maximum quantity of 1524 cc of gas per day on an average over a period of 12 weeks. The silkworm cocoon wastes and composite mixture of vegetable wastes could generate 910 and 1000 cc per day, respectively, when compared to control (972 cc per day).

Silkworm reared are of four types, e.g., Mulberry silkworm (*Bombyx mori*) eri silkworm (*Philosamia ricini*), tassar silkworm (*Antheraea mylitta*) and muga silkworm (*Antheraea assami)* The fertilized moth is covered with an-inverted funnel or cellule and eggs are allowed to be laid over cardboard. Parasite may be removed by brushing the egg masses with a fine brush. In a bamboo tray rice husk is spread. Tender chopped mulberry leaves are added to the tray. The hatched out larvae are transferred to the leaves. It is important to change the leaves every 2-3 hours during the first 2-3 days. the cocoon is constructed with a single reelable thread of silk. If the moth are allowed to emerge from the cocoons, the silk thread is cut into pieces. Hence, the pupa are killed 2-3 days before the emergence of moth and processed. The cocoons required for further rearing are kept separately and moths are allowed to emerge from them. The life cycle of the silkworm extends from 42 to 67 days depending on the race and the season. The optimum temperature for silkworm rearing is 20 – 28^0 C and optimum humidity is 75 per cent.

2.3.2.4 *Mushroom cultivation*

Mushroom is am edible fungi. It is fairly good source of vitamin C and good for blood pressure and diabetic patients since mushroom contains less of fat and carbohydrate and high in quality protein having 60-70 per cent digestibility and all amino acids. The cultivated mushroom varieties are oyster mushroom (*Plearotus sp.*), paddy straw mushroom (*Volvariella volvacea*) and while button mushroom (*Agaricus gisporus*).

For cultivation of oyster mushroom, paddy straus is cut into pieces of 3 to 5 cm length. The cut straw pieces are soaked in the water for 4 to 6 hours before boiling for half an hour. The straw is then dried in the shade. A polythene bag of 60 x 3 0 cm size with two holes of one cm diameter in the centre of the bag facing opposite side is taken. The bottom of the bag is made flat by tieing it with a thread. The paddy straws are filled in the bag upto the height of 10 cm and spawn is inoculated alternatively with 4 to 5 layers of straw. A temperature of about 22-28°C with relative humidity of 85 to 90 per cent is maintained by putting the bag in spawn running room. After 15 to 20 days when the spawn running is completed, the bag is opened and taken to cropping room. The mushroom is harvested after 7 days of growing. The approximate yield is 0.5 to 1.0 kg per bed.

For cultivation of paddy straw mushroom, the straw is cut into the long pieces of 60 to 90 cm before soaking it in water for 12 hours and sterilizing for 15 minutes. The straw is arranged in bundles. The moistured straw bundles are kept on slightly raised concrete floor or on wooden platform in layers of four bundles width. The beds are seeded simultaneously in each layer either by broadcasting or placing the grain spawn at different spots. Don't spawn below the top most layer. Maintain a temperature of 30-35°C. Harvesting should be done after 25-30 days. An approximate yield of 1 to 5 kg per bed is expected.

For production of white button mushroom, compost prepared by a complex method is used as substract. The tray is filled up with compost and spawning is done. After spawning, compost is pressed hard to make it compact. The trays are arranged in tiers which are covered with newspaper sheets sprayed with 2 per cent formalin. The temperature of 20 to 25°C and relative humidity of 90 to 95 per cent are maintained. After 15 to 20 days spawn running is completed and casing is done. Pin heads appear within 10 to 15 days after casing. Mushroom can be harvested at button stage after 60 to 75 days of cropping. Yield of 6 to 7 kg per m^2 is expected (Balasubramaniyam and Palaniappan, 2002).

2.3.2.5 *Biogas*

Biogas is an unpolluted alternate energy source in which cattle dung and other organic wastes available in the farm are efficiently utilized during anaerobic decomposition for generating gas by a complex biochemical process. The celluloitic material are broken down to methane and CO_2 by different groups of microorganisms. The biogas are used for cooking purpose, burning lamps, running pumps, etc. In the wake of mounting energy crisis due to shortage of fossil fuels and inadequate electricity its importance has further increased as a cheap energy source. Biogas plants are of two types-float dome type and fixed dome type. Economic, technical and climtological factors govern the selection of a particular type. The site should be close to the kitchen (or place of use) and nearer to the cattle shed. Plant should get clear sunshine during most part of the day. The amount of gas production is considerably higher in summer followed by rainy and winter seasons. One cubic meter capacity plant will need two to to three animals and 25 kg of dung to meet the gas requirement of a farmily of 4 to 5 members,. Maximum gas production is achieved at a temperature between 30 to 35°C. Slurry obtained after the production of biogas is an enriched manure.

2.3.3 Terms and Definitions

2.3.3.1 *Crop components*

- **Multiple cropping**: It refers to intensification of cropping both in time and space dimensions. It includes sequential cropping, intercropping and mixed cropping.

- **Monocropping / monoculture**: It refers to growing of only one crop on a piece of land year after year. It may either be due to climatological limitations or due to specialization by a farmer to grow a particular crop. In other words monocropping is the repetative growing of the same crop on the same land.

- **Intercropping**: it refers to growing of two or more dissimilar crops simultaneously on the same piece of land, base crop necessarily in distinct row arrangement. In **additive series** of intercropping the main or basecrop is sown with 100 per cent recommended population in pure stand and intercrop is introduced by adjusting or changing crop geometry. However, in **replacement series** plant population (planting density) of main crop is proportionately decreased as the plant population of intercrop is increased. The different types to intercropping are:

    i **Mixed intercropping**: Practice of growing two or more crops simultaneously intermingled in the same plot with no distinct row arrangement.

    ii **Row intercropping**: Growing two or more crops simultaneously in the same plot intermingled within a distinct row arrangement.

    iii **Strip intercropping**: Growing two or more crop simultaneously is strips wide enough to permit independent cultivation but narrow enough for the crops to interact agronomically.

    iv **Relay intercropping**: It refers to planting of succeeding crop before the harvest of preceding crop, Planting of succeeding crop may be done before or after flowering, before or after attainment of reproductive stage, completion of active life cycle, senescence of leaves or attainment of physiological maturity

- **Mixed cropping**: It is growing of two or more crop simultaneously on the same piece of land seeded either after mixing the seeds of the crops intended to be grown mixed or sowing alternate rows in various replacement ratios.

- **Sequential cropping**: Growing of the two or more crop in sequence in the same piece of land either in one calendar year or more than one year depending upon irrigation facilities and soil moisture storage and availability (a dependent function of precipitation).

    Same variations of sequential cropping are:

    i) **Double cropping**: Growing of two crops in a sequence on the same piece of land in a calendar year

Identical concept is applicable in triple and quadruple cropping involving three or four crops, respectively.

ii) **Triple cropping:** Growing three crops in sequence on the same piece of land in a calendar year.

iii) **Quadruple cropping:** Growing four crops in sequence on the same piece of land in a calendar year.

iv) **Relay cropping**: A significant part of life-cycle of the second crop overlaps with the cropping cycle of the first.

- **Multi–tier or Multi–storied cropping**: The practice of growing different crops of varying height, rooting pattern and duration.
- **Ratooning**: It refers to raising of regrowth coming out of roots or stalks of the preceding crop.
- **Cropping system**: It refers to principles and practices of cropping and their interaction with farm resources, technology, aerial and edaphic environment to suit the regional or national or global needs and production strategy.
- **Crop rotation**: It refers to recurrent succession of crops on the same piece of land either for a year or for a longer period of time. Component crops are so chosen that soil health is not impaired.
- **Cropping pattern:** It refers to relative arrangement of crops on a farm, region, province or country opportioning due consideration to natural features (soil and climate), crop efficiency, land capability, socio-economic structure, technological and extension infra-structure (chargeable) and the national agricultural policy.
- **Opportunity cropping:** The practice of placing primary emphasis on the unit of stored moisture when determining weather or not to establish a crop (Bateman and Rowling, 1980).
- **Recession farming**: It is a system in which crop are planted in flooded areas as the rainy season and water recedes. This system take advantage of thoroughly saturated soil profile. It also has advantage of silt and nutrient left behind by flood water. (Hall *et al.*, 1979).
- **Conservation Cropping**: It is a way of farming that aims to maximize the protection against erosion that can be achieved through soil and crop management practices.

- **Ley farming**: The definitions reported by researchers are given as below:
  - i) It is a rotation of arable crop requiring annual cultivation and artificial pastures occupying field for two years or longer (Rulhenberg, 1976).
  - ii) Ley farming is a system in which crops and pastures are alternated on the same field (Hall *et al.*, 1979).
  - iii) Ley farming is the growing of grass or legumes in rotation with grain crops as a soil conservation measure.

The concept of ley farming includes grass-legume mixture. The grass has the quality of improving soil structure while the legumes enrich the soil nitrogen stutus. Singh *et.al.* (1990) observed that *stylosanthus hamata* raised far a period of 2-3 years in a four years rotation helped in building up of soil N to the extent of 35 kg per ha. With this, there was no need to supply N to the following sorghum crop. This system fits well where moderate rainfall (500-700 mm) is received but the soil is deep (770 mm).

2.3.3.2 *Tree based components*

- **Shifting cultivation**: The shifting cultivation is any agricultural system in which fields are cleared by firing and are cropped discontinuously, Shifting cultivation (slash and burn) is widely practiced food production system on steep slopes. It consists of three phases: Land clearing, cultivation phases and restoration phase. The system involves use of fire for clearing the land, keeping the land fallow for regeneration for a number of years, use of human labour as main input, non-employment of draft animals, use of crude and simple implements, such as, dibbling sticks and scrapers, and all the crops grown are mixed together. The system is practiced extensively in north-eastern hill regions of India and to a limited extent in other tribal populated States. Shifting cultivation is, thus, defined as "the practice of cultivating clearings scattered is the reservoir of natural vegetation (forest, grass, woodland) and of abandoning them as soon as the soil is exhausted, and this includes the practice of shifting homesteads in order to follow the cultivator's search for new fertile area."
- **Taungya system**: This system is method of establishing forest species on temporary combination with field crops. In this system, cultivators are allowed to grow agricultural crops in

the inerspaces of forest plantation far three to five years i.e. till the canopy of trees start closing. The main idea is to manage forest plantations.

- **Forest**: Ecologically, the tern forest signifies a complex organism, composed of distinct biological units called forest communities that have come into being by a combined action, reaction, and co-action of a variety of organisms with complex factor of the habitat that themselves changed both in space and time.
- **Agroforestry**: Agroforestry is any sustainable land-use system that maintains or increases total yields by combining food crops (annuals) with tree crops (perennials) and /or livestock on the same unit of land, either alternately or at the same time using management practices that suit the social and cultural characteristics of the local people and the economic and ecological conditions of the area. It is also called the ' multiple land use' system.

  The different agroforestry systems are:

  i) Agri-silviculture system (i.e., agricultural crops intercropped with tree crops),

  ii) Silvi-pastoral system (i.e., raisining of fodder grass and legume mixture in the inter space of trees.) and,

  iii) Silvi-horti-pastoral system (i.e., growing trees along with fruit trees and raising grass and legume mixture in the inter space).
- **Alley cropping**: Alley cropping with the rows aligned along contours is an effective means of erosion control which also imparts stability in production. The hedge rows are cut back at planting of arable crop and kept pruned during the growing season so as to prevent shading and reduce competition with arables. When there is no crop the hedge rows are allowed to grow freely or cut to meet the fodder needs. *Glyricidia, Leucaena* and *Sesbania* are generally preferred for alley cropping.

2.3.3.3 *Animal based component:*

- **Tranhumant pastoralism**: With large seasonal differences in precipitation and temperature the livestock with low input management are moved from one place to another. American ranchers use winter and summer pastures, whereas African pastor use wet-season and dry season pastures. Traditionally, pastoral peoples in West African keep their livestock in arid area during the wet season, where forage quality is relatively high. In dry season, when water becomes scare in the north, they move their animals further south to more humid areas, where the livestock can graze the crop residues in harvested fields and green grass in low-lying areas along streams and rivers. These herds are important source of manure for arable farming, Similar movement of livestock can be found in drought-prone areas of Rajâsthan in dry seasons.

2.4 INTEGRATED FARMING SYSTEM (IFS)

2.4.1 Concept and Definition

Integrated farming system introduces a change in the farming techniques for maximum production in cropping and other components and takes care of optimal utilization of resources. A judicious mix of enterprises like dairy, goat and sheep, poultry, piggery, fishery, sericulture etc., suited to given agro-climatic conditions and socio-economic status of the farmers would bring prosperity in the farming.

The integrated farming system is conceptually a set of elements or components that are interrelated which interact among themselves.

2.4.2 Classification of Farming System Interactions.

The type of interaction with examples are given below (Derek, 1985):

1. Biological interactions between crops
   - (a) Biological interactions in space, e.g., intercropping
   - (b) Biological interactions overtime, e.g.,
     - i. Conflicts in planting a crop in relation to harvest of previous crop.
     - ii. Carry-over of soil structure and crop residues from preceding crops.
     - iii. Carry-over fertility from previous crops.
     - iv. Carry-over and build up of weed seeds and other pest populations from previous crops.

2. Biological interactions between crops and livestock, e.g.,
    i. Use of crops and crop residues for fodder.
    ii. Use of farm yard manure as crop nutrient source.
    iii. Use of animals for draught power.
3. Resource competition and complementarity, e.g.,
    i. Conflict in labour use between enterprises.
    ii. Cash flows from sale.
    iii. Competition for irrigation water between enterprises.
4. Meeting multiple objectives of farm-households, e.g.,
    i. Choice of multiple crop and production practices to manage risk.
    ii Planting and storage of food crops to balance seasonal food needs.

2.3.3 Advantages of integrated farming system (IFS)

The advantages of IFS are listed below:

i. Increased productivity and profitability,
ii. Sustainability in production.
iii. Balanced food,
iv. Prevention of environmental pollution,
v. Recycling of farm waste / organic residues,
vi. Money round the year,
vii Solve energy erisis,
xiii. Solve fodder erisis,
ix. Solve fuel and timber shortage,
x. Avoid degradation of forests,
xi. Provide opportunity for agro-oriented industries,
xii. Employment generation,
xiii. Increases input use efficiency,
xiv. Standard of living of the farmer increases, and
xv. Improves literacy.

The various enterprises that could be included in the farming systems are crops, dairy, poultry, goat and sheep rearing, fishery, sericulture, agrofarestry, horticulture, mushroom cultivation, etc. Thus, IFS deals with whole farm approach to minimise risk and increase the production and profit with better utilizedation of wastes and residues leaving no chance to pollute the environment or to degrade the resource base. To put this concept into practice efficiently it is necessary to study linkages and complementarities of different enterprises in various farming systems. The knowledge of complementarities will help to develop farming systems in which the waste of one enterprise is more efficiently used as inputs in another enterprise within the system.

Farming system may be predominantly crop based, tree based, and/or animal based. However, because of site specific soil and environmental constraints and individual preference and priorities, there are virtually as many systems as farmers.

2.4.4 Aims

The aim of the integrated farming system can be achieved (Rangasamy *et al.*, 2002) by:

(i) efficient recycling of farm and animal wastes;

(ii) minimizing the nutrient losses;

(iii) maximizing the nutrient use efficiencys

(iv) following efficient cropping systems and crop rotations; and,

(v) complementary combination of farm enterprises.

According to Swaminathan (1987), integrated farming system approach is not only a reliable way of obtaining fairly high productivity, but also a concept of ecological soundness leading to sustainable agriculture.

2.4.5 Farming situations

There are three different farming situations as given below (Hosmani *et al.*, 2000):

Systems	Examples
(a) Irrigated/wetland system:	Rice based system + fish culture + duck/poultry.
(b) Rainfed/dryland system:	Grain/fodder crops + sheep/goat rearing + agroforestry (tree/agri-silvi/agri-horticulture.
(c) Garden land system:	Garden land crop with forage + livestock/ poultry (crops with fodder, mulberry + dairy + silkworm rearing).

2.4.6 Integrated Aquaculture - Agriculture Components

The husbandry of plant and animal that live in water is knows as aquacultare. Water resources available to a farm family may include ponds, streams, floodprone and waterlogged land, which serve useful purposes, viz., domestic requirement of water, supplementary irrigation and pisciculture.

2.4.6.1 *Cropping with aquaculture*

The farmer can enhance their income by adopting integrated fish farming methods, such as cropping cum fish culture and livestock cum fish culture. The integration of cropping with fishery includes rice-fish farming, banana -fish farming, coconut -fish farming, cultivation of vegetables and growing of fruit trees on dikes besides pond with fish culture. These are examples of integrated use of natural resources in a sustainable manner. Crop residues crop byproducts, e.g., rice and maize bran, tree leaves, rotten fruits and vegetables and manure fertilize the pond which serve as a feed for fish. Fish converts plant and animal waste into high quality proteins which can be used to revitalize cropland.

2.4.6.2 *Rice-fish farming*

i) **In captural system**, wild fish population reproduced in the flooded rice fields are harvested at the end of rice growing season. In this system, sumps of about 40-50 m^2 size and 2 m deep are dug in the lower region of groups of fields. Species of channa, clarias and anabas are cultivated.

ii) **In cultural system**, fish fingerlings are stocked as in fishponds at a known density depending on the fertility of the system.

   a) In **concurrent culture**, fish and rice are simultaneously grown and harvested. Fish controls weeds, molluscs and harmful insects of rice and thus, helps in higher productivity of rice. However, this system limits the use of modern techniques like use of chemical fertilizers, insecticides, pesticides and herbicides.

   b) **In rotational culture system**, rice and fish are grown alternatively. After one or two weeks of rice harvest, the field is prepared for fish rearing. Bunds of 50 cm height are formed around each field and ditches of 50 cm width and 30 cm depth are dug around the field. Simple bamboo pipe or screens are kept at the inlet and outlet to prevent passage of fish and detritus during the culture period. Common carp is idle for this system.

2.4.6.3 *Rice-fish-azolla farming*

Shanmuga sundaram and Balusamy (1993) reported that in low-lying wetlands, the best farming system was rice-rice-azolla-calotropis +fish farming. Experiment was conducted for two crop seasons planted with rice for a total duration of eight months with one fish crop harvest. *Azolla microphylla* was applied at the rate of 2.0 t per ha as fish feed cum N fixer, five days after transplanting. A wasteland weed named colotropis was incorated at the rate 125 t per ha as green leaf manure, 10 days before transplanting. Recommended dose of fertilizer was applied at the rate of 100 N + 50 P + 50 K kg per ha. Irrigation was scheduled to maintain 5 cm depth of water throughout the crop growth period. With 5 per cent neem seed kernel extract plant protection measures were taken. Fingerlings of catla, rohu and mirgal were stocked in equal ratio with a total of 3000 per ha, 15 days after transplanting. Field trenches were provided with 1.0 m depth and 1.0 m width occupying 10 per cent of rice area to sheller the fish. Banana pesudostem with cowdung (1:1) mixture and rice bran were fed at the rate of 5 per cent of the fish body weight. The results reveled that in low-lying wetlands the most profitable farming systems was rice-rice-azolla- calotropis + fish farming. There was slight decline in rice yield under rice-rice-fish farming due to the presence of fish trenches occupying ten per cent of the area. However, the yield reduction was made to good when azolla and calotropis were introduced.

2.4.6.4 *Rice-fish-vegetable farming*

Experiment conducted at Agricultural Research station, Bhavanisagar, Tamilnadu revealed that the rice-rice-vegetable + fish culture was more remunerative then rice alone or rice + rice + fish farming. The study area was 0.25 ha per system. The field trenches 1 m deep × 1 m wide was provided for shelter of catla, rohu and mirgal stocked 15 days after transplanting at an equal ratio of 3000 fingerlings per ha. Vegetable crops, viz., bitter gouard, ribbed goured, beans etc. were raised on bund of fish trenches.

2.4.6.5 *Horticulture with fish farming*

Fruit trees like papaya, banana, etc., can be raised on the embankments or bunds of the pond for making the integrated farming system economically more viable. The leaves of fruit trees can be utilized as supplementary feed for the grass carp.

2.4.6.6 *Fish-cum-sericulture*

The mulberry trees are grown on the bunds of the fish ponds. The droppings of the silkworm are fed to the fish particularly grass carp. The excreta of the fish decomposes in the bottom of pond into nutrients to be utilized for fertilizing the mullberry trees. The silkworm pupae and their cake after extraction can also be utilized as supplementary feed for the fish. Thus, the fish and plant sustain their yield through mutual benefit.

2.4.7 Integrated Agriculture and livestock Components

2.4.7.1 *Cropping with dairy*

Experiment conducted at farmer's field (Shankaran and Rangaswamy, 1990) in Tanjore delta revealed that an additienal net income obtained from the dairy based mixed farming with 3 number of milch animals over the existing farmer's method (rice-rice-rice-fallow-pulse). Two holdings of one acre each was taken for study. Inclusion of dairy unit generated an additional employment of 190 man-days. Similar findings have been reported by Pracer *et al.* (1983). Rajkumar (1988) reported that the farm yard manure abtained from dairy unit was applied to rice-fallow-cotton in combination with inorganic fertilizer recorded the highest seed cotton yield.

2.4.7.2 *Cropping with goat rearing*

Field trials conducted in Kabisthalam village of Tonjore district of Tamilnadu (Skankaran and Rangaswamy, 1990) to study the feasibility of rearing Tellichery goats under deep litter system (5 does and 1 buck). The conventional method of cropping rice-rice-rice-fallow, blackgram was compared with modified cropping of rice-rice-rice-fallow, cotton, paragrass. The goat manure collected under deep litter system was recycled as a source of nutrient to rice-fallow cotton. The mixed farming generated additional income with additional employment of 109 man-days. For landless labourers rearing of goat and sheep may provide continuous employment (Singh *et al.*, 1986).

2.4.7.3 *Cropping with animal and biogas system*

Mehta *et al.* (1986) reported that in Northern India, with the triple cropping intensity of fodder crops and recycling of feed nutrients in a milch animals and biogas system from one ha of land, it is possible to produce 82.63 GJ of energy and 906.11 kg of protein from milk, 263.93 kg of N; 148.93 kg of P_2O_5 and 90.94 kg of K_20 fertilizer from biogas slurry and 2897 kilowatts of power per annum for farm operation from biogas using Kirloskar biogas engine. Traditional system of wheat and rice cultivation can produce 93.87 GJ of energy and 654.25 kg of protein from food grain per ha per annum. The recycling of feed nurients in the available straws of rice and wheat crop can yield an additional 16.31 GJ of energy and 166.97 kg of protein from milk; 81.23 kg of N; 56.00 kg of P_2O_5.

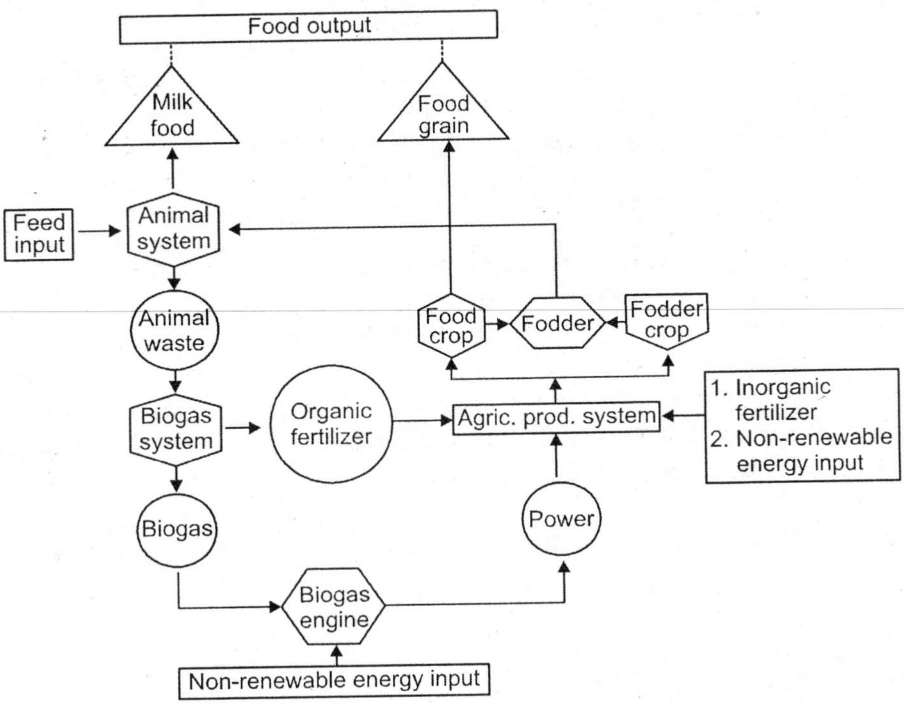

FIG. 8: Modeal for recycling feed nutrient in crop, animal and biogas systems. (Mehla et. al., 1986).

25.14 kg K_2O fertilizer from biogas slurry and 1097 kilowatts of power per ha annually from biogas. A model as constructed by Mehta *et al.* (1986) to integrate the process of recycling of feed nutrients through a crop, animal and biogas system is given in Fig. 8.

2.4.8 Integrated Aquaculture and Animal Components:

2.4.8.1 Fish-cum-pig-farming

On the embankments of the fish pond, the pig house can be constructed. The faeces and urine can be drained into the pond which serve as a rich manure and can be utilized by fish as feed. For a ha of water area of pond 30-40 pigs are sufficient. This integrated farming system reduces the input cost to about 60 per cent and achieves a financial return of 52 per cent (approximately) on the variable cost (Rangasamy *et al.*, 2002)

2.4.8.2 Fish-cum-poultry /duck farming

In case of poultry farming, ponds can be manured with the litter from the poultry. About 500-600 birds can provide adequate quantity of litre for fertilizing a ha of water area of the fish pond. A net financial return on variable cost at the rate of 54 per cent is achieved through the sale of fish, chicken meat and eggs.

Duck-fish culture is highly remunerative. The droppings of the ducks can be recycled far manuring the fish ponds at the rate of 10 to 15 t per ha per year. The droppings can either be collected from the duck house or the duck house can be constructed above the water surface of the pond itself. A total of 200-320 ducks (approximately 1 kg in size) are sufficient to provide menure for fertilizing a hectare of water area of pond (Rangasamy *et al*; 2002)

2.4.9 Integrated Agriculture, Animal and Tree-Components:

One ha of dryland was selected for the study involving crops, goat rearing and silvipasture (Rangasamy *et al*; 1988) as discussed below in section 2.4.9.1 and 2.4.9.2.

2.4.9.1 Cropping with goat and silvi-pasture in rainfed areas

In the integrated approach, the cropping pattern was modified by including both sorgun grain crop (0.2 ha) and sorghum fodder crop (0.2 ha) in place of fodder sorghum raised by farmers. To meet the fodder requirement of goats, subobool and *Cenchrus ciliaris,* a pasture grass as an intercrop was raised in 0.2 ha. Trees like *Acacia senegal* and *Prosopis cineraria* were raised in 0.2 ha. Tillichery goats (1 buck + 5 does) were advantageously integrated into this dry farming situation. Apart from high additional income from grain and fodder, goat meat and manure, this system generated additional 54 man-days employment over the conventional system.

2.4.9.2 *Cropping with dairy, biogas and silviculture in irrigated areas*

A model on integrated farming system in 1.0 ha was developed to suit the farmer of irrigated area in Coimbtore district of Tamil Nadu. The normal cropping pattern of finger millet-cotton-sorghum was slightly modified by inclusion of intercropping, such as, greengram in cotton, cowpea in sorghum and sunflower as border crop in finger millet (0.75 ha). The perennial grass fodder (0.15 ha) and perennial legume (0.05) were also raised in the holding. Three graded Jersey cross-bred milk cow with 2(two) calves were included under dairy component. For effective recycling of farm and animal waste, a biogas unit of 2 m^3 (two cubic metres) capacity was installed for the production of fuel, light and enriched manure. Two-cubic-metres of gas everyday was produced out of sixty kg of cowdung of 5 animals which was equivalent to 1.5 litres of kerosene. Two hundred subabool trees were planted all along the boundary of the field for fodder and timber production. This system facilitated continuous flow of income to the farmer throughout the year besides generating additional employment (Sankaran and Rangasamy, 1990).

2.4.10 *Crops and Cropping /Farming Systems and Prospects of Diversification*

Diversification in agriculture assumes significance with both ecological and balance nutrition viewpoints. The present agriculture scenario is dominated by monoculture of certain crops as more than 80.00 per cent of our food comes from about 10 crop species. Crop diversification may prove to be paramount importance in mitigating the problems arising on account of monoculture. It is also critical to overcome the problem that are currently associated with the most productive cropping system 's sustainability. A more diversified food basket will provide food security and improve the quality of life by adding to mutritional status of our people.

A large number of crops and cropping systems are followed under different soil and climatic regions of India. Major shifts in crops and cropping systems which have taken place in the recent past in some of the agro climatic regions of the country (Chatterjee and Srivastava, 1990) are discussed below:

2.4.10.1 Western Himalayan region

Cultivation of mustard after rice and maize has come up in a big way in Kashmir Valley (Table 10). Besides mustard, in the milkshed areas, oats, berseem and red clover are being intensively cultivated. In Jammu, area under maize has increased, whereas the area under pearl millet has decreased. Area under wheat has increased at the cost of chickpea. Instead of maize-wheat system, the focus has shifted to *kharif* pulses-wheat system or rice-field peas or rice-oats / berseem systems. In Uttar Pradesh hills, wheat is replacing barley. In the tarai and Bhavar regions, area under maize and finger millet has been reduced appreciably, giving way to rice. A large area has been diverted to eucalyptus and tree growing.

2.4.10.2 Eastern Himalayan region

Area under fruit crop (apple, plum, pears) is increasing in Arunachal and adjoining areas. Rice and maiza fallows are being utilised for growing mustard and sunflower (Table 11). There has been a considerable increase in the area under ginger and cardamom. Area under bamboo and pineapple in lower hills is also increasing. In milkshed areas, rice-oats and rice-berseem sequences have been adopted extensively.

2.4.10.3 Indo-gangetic (lower, middle, upper and trans gangetic) plains region

With the setting up of sugar mills, area under cereal crops has been diverted to sugarcane in Haryana. Uttar Pradesh and Punjab (Table 12). Area under soybean and the other newly introduced crop like gobhi-sarson (*Brassica napus*) has recently increased in Haryana, Punjab and Uttar Pradesh. Rajmah (*Phaseolus vulgaris*) is coming up big way in the plains of Uttar Pradesh. Area under winter maize has increased appreciably in this region. Area under cotton in trans-gangetic plains has also increased at the cost of pearl millet, sorghum and barley. Rice-potato-sunflower, rice-toria-onion / sunflower, cotton-wheat, clusterbean-wheat, rice-mustard are gaining ground as against rice-wheat system. With the introduction of new high yielding varieties of pigeonpea and groundnut, short duration pigeonpea-wheat sequence is becoming popular. These crops are also becoming lucrative to the farmers. In West Bengal, area under wheat has decreased appreciably and has given way to rice-mustard, rice-potato-sesame / groundnut system. Provision of loans on easy terms in coastal areas has helped the farmers to shape their land for fish-cum-rice culture with wide and raised bunds for vegetable and fruit (banana and coconut) cultivation. In milkshed areas, rice-oats / berseem system has come up in a big way. Agro-forestry with poplar, as a tree component, as against, eucalyptus, is being taken up by many farmers in Punjab, Haryana and Western Uttar Pradesh. farmers are planning to increase their wood production per unit area and time.

2.4.10.4 Plateau (Eastern, Western, Southern and Central) Region

Cultivation of groundnut followed by *rabi* sorghum with limited or no irrigation is coming up in a big way in Sangli, Satara and Kolhapur districts in Maharashtra and adjoining areas (Table 13). Farmers are preferring chickpea after sorghum / pearl millet as against wheat in black soils. With the introduction of the short duration pigeonpea variety, a new system of pigeonpea-wheat is coming up. Area under sunflower as sole crop or as an intercrop with groundnut or finger millet has increased appreciably in Karanataka. Because of rise in the price of cotton, hybrid cotton under limited irrigation through dug wells or under adequate irrigation facility in canal areas, is becoming popular. In the canal command area, it is very unfortunate that the rice crop, which shows a poor water use efficiency, is being extensively cultivated, even at 200 per cent intensity. The reason ascribed for this is that it is easy to store and sell as compared to other commodities. Coconut plantation has increased appreciably in Karnataka (even with limited amount of irrigation water in dug wells). Area under mulberry for silk -worm raising is also increasing at a fast rate. Area under onion, both direct seeded as well as transplanted, has increased in recent years, In Orissa Plateau and in Jharkhand and West Bengal (district of Purulia), cultivation of groundnut is gaining ground. Soybean is becoming an important crop in Koraput district of Orissa.

TABLE 10: Crops and Cropping System in the Western Himalayan Region

	Existing		Prospect of diversification			
Region (1)	**Crops**	**Cropping / farming systems**	**Crops**	**Cropping / farming systems**	**Resource constraints**	**Remarks**
Western Himalayan comprising of J & K, H. P. Divn. Himalayan U.P. Divn (land use planning based on the concept that land up to 3.0% slope is suitable for agriculture on terraces 3.0-5.0% slope for horticulture and silvi-pastoral programs and above 5.0% slopes for forestry).	Rice, maize, finger-millet, pearl millet wheat, barley, limited area under pulses; oilseeds temperate fruits, vegetables; spices, medicinal plants, forage crops like oats, berseem, red clover etc. in milkshed areas, potato for table & seed: seed production of vegetables.	Cereal based rice-mustard, rice-wheat, maize-wheat / mustard, maize-potato, maize-peas, maize in upland & rice in medium / low land. Maize-toria-wheat, maize-peas / gram, rice-linseed: rice-berseem: rice-oats, maize-soybean-wheat, specialised farming if high valued crops (e.g. saffron etc.).	Improved varieties of maize, fingermillet rice, vegetables, forage crops mustard, sunflower, and soybean, ornamental flowers of high value; spiçes; temperate fruits; forage crops and natural grasses for dairy development; mushroom cultivation.	Inter cropping in orchards; relay seeding in kharif crops (e.g. berseem) in autumn in rice fallows to get forages in early winter and spring; maize-soybean / pulses inter-cropping; reduction in barley area, better grassland management for animal production.	Irrigation water at many places.	There is need to develop appropriate minor & medium irrigation devices through water harvesting techniques, processing of farm produce into value added products, processing of mushrooms.

TABLE 11: Crops and cropping System in the Eastern Himalayan Region

Region (2)	Existing		Prospect of diversification			
	Crops	**cropping / farming systems**	**Crops**	**Cropping / farming systems**	**Resource constraints**	**Remarks**
Eastern Himalayan Region comprising 8 States (Assam. Arunachal Pradesh, Meghalaya, Manipur, Mizoram Tripura, Nagaland, Sikkim) and 3 districts of W.B (Darjeeling, Jalpaiguri, Cooch Bihar) and 49 distruicts are hilly and 10 comprise of plains.	16% area under crops: maize, rice, jute, pulses, wheat. Oilseeds, fruit production- citrus, pineapple, orange, banana, cardamom, ginger.	In Sikkim maize based and rice based systems, fruit growing, livestock farming.	Cereals, pulses and oilseed (mustard, soybean, sunflower), citrus, pineapple, Temperate fruits (apple, plum, pears); bamboo, ginger, cardamom, flower (orchids & bulbous flower), tea, forage crops and improvement of grassland in milkshed areas; silk rearing. Alder trees to improve fertility in *jhum* lands.	Formulations to be based on flood-free, flood-prone and chronically flood affected pockets. Animal Husbandry, sericulture & fishery related systems.	Erosion controlling,; communication, marketing, processing storage; both irrigation and drainage.	Creation of additional employment, erosion control, small scale water harvesting structu res. fruit processing units; custom-hire services (to be organised), processing of farm produce into value added products.

TABLE 12: Crops and Cropping System in the Indo – Gangatic Plain Region

Region (3,4,5 &6)	Existing		Prospect of diversification			
	Crops	**Cropping/ farming systems**	**Crops**	**Cropping/farming systems**	**Resource constraints**	**Remarks**
Indo-Gangetic plains comprising of lower (3) middle (4), upper (5) and trans-Gangetic (6) plains; Delhi ridge dividing eastern and western plains.	Cereals, pulses, oilseed, sugarcane, tobacco, chillies, potato; cotton in (6), jute in (3), fruits (mango, litchi, pineapple, plum, coconut, banana, vegetable and flowers.	Rice -wheat, maize- wheat; jute-wheat / oilseeds / pulses; rice -pulses / oilseeds; cotton-greengram, rice / maize / jute-potato; rice / maize-toria-wheat; rice-sugarcane-ratoon-wheat / fallow; rice-maize-oats / berseem in milkshed areas.	Soybean, mustard sunflower, groundnut, maize, *(rabi)*, Rajmah a new pulse crop; intercropping of maize with pulses or soybean; substitution of upland rice with groundnut, maize & millets or pulses, cotton; sugarcane + potato, vegetables, onion & garlic; alternative cropping of 'makhana', 'sin-ghara', 'mat-sticks' in low-lying areas.	Rice-maize *(rabi);* rice-berseem; rice-mustard; rice/ maize-groundnut (summer), rice/ maize-potato-groundnut; Paira cropping of linseed/ peas/ lentil or khesari for fodder; maize-sweet potato/ yam bean-onion in N. Bihar; inter -cropping of sugar-cane with potato/ wheat/ mentha/ mustard/ maize for fodder, rice-mustard-rice; rice-green manuring-rice inter cropping of cotton with green/ black gram;, inter cropping of pigeonpea/ maize/ sorghum with pulses/ soybean.	Irrigation in (3) (4) & (5), drainage in (3) & (4) salinity in (3), alkalinity; in (5) & (6). Proper deve-lopment of *'diara'* & 'tal' areas sub-jected to season bound water logging. Limitation in efficienct water use; S, Zn, B, MO deficiency; Fe toxicity. Land shaping for fish culture (rice-cum-fish culture), animal husbandry in milkshed areas; bee-keeping,; sericulture.	Employment generation for wasteland development, land shaping, improvement of field irrigation channels, agro-based industries, processing of fruits and other agri-linked activities, animal husbandry, aquaculture. Discourage growing of paddy in summer to substitute other water use efficient crops.

TABLE 13: Crops and cropping System in the Plateau Region

Region (7, 8, 9 & 10)	Existing		Prospect of diversification			
	Crops	**Cropping / farming systems**	**Crops**	**Cropping / farming systems**	**Resource constraints**	**Remarks**
Plateau region comprising of Eastern (4), Central (8), Western (9) & Southern plateau & hills (10), covering 35 districts of AP, Karnataka, TN and Jharkhand, MP, UP, Rajasthan & Manarasntra. 80% of the area is rainfed, rainfall at many places bi-modal, a good number of dams have come up. Wide diameter dug wells present in graden land areas.	Maize, sorghum, pearl millet, finger-millet, rice (in irrigated areas) *Kharif* pulses (pigeonpea, greengram, black-gram) and chickpea *Dolichos*, safflower, sunflower, groundnut, sesame; cotton; forage crops: nybrid-napier, oats, multi-cut sorghum, lucerne etc.; milkshed areas; sugarcane (irrigated), cotton sole or inter cropped.	Usually mono-cropping; inter-cropping of cereals with pulses or oilseeds, sorghum / groundnut, pigeonpea.	Upland crops; dry-land horticulture; guava, tamarind, pomegranate, ber, jaman, sapota jackfruit, cashew, castard apple, citrus, sun ɔwer, soybean, groundnut, maize hybrid sorghum are coming up in big way: *Khanf* `and *rabi* onions direct seeded as well as transplanted; high value horticultural crop-grapes., banana, turmeric, lemon, flower, etc.	Due to mild tem-perature conditions crop can be grown throughout the year, paired row planting of crops show advantage ; set row planting and fertilizing in groundnut in shallow soils; cross-bred animal in milkshed areas, sheep and goat rearing. Farming in big way: fishery in inland water bodies; sericulture; agro-forestry, etc.	Water harvesting and *in-situ* moisture conservation; shallow soil; poor water use efficiency in canal areas; widespread deforestation.	Water use efficiency can be improved by (1) discouraging crop requiring large quantites of water as rice in summer, (2) optimum sowing dates of crops, (3) developing cropping plans. Scope of in - creasing ground water utilisation coupled with conjunctive use of rain water; holistic development of animal husbandry by raising fodder crops, improvement in veterinary services, nutrition and breeds of livestock. Agri-linked activities to be augmented by improving rural employment potential.

TABLE 14: Crops and Cropping System in the Coastal Region

	Existing		Prospect of diversification			
Region (11 & 12)	**Crops**	**Cropping / farming systems**	**Crops**	**Cropping / farming systems**	**Resource constraints**	**Remarks**
Coastal plains comprising coastal districts of Orissa, AP, TN & Pondichery covering 19.6 m ha in the east coast (11) and the plains, coastal districts of Maharashtra, Goa, Karnataka, Kerala and TN covering 7.3 m ha in west coast (12). Rainfall in east (78-130 cm). and west coast (200-400 cm). Irrigation in west coast is low (15%) and have considerable low-lying areas.	Rice, pulses and groundnut (more in east then in west coast), sorghum, pearl millet are grown in east coast, finger millet in both coasts; coconut, sapota, cashew fruit trees.	Rice, mono-cropping; rice-rice with supplemental irrigation; rice-black gram, green-gram or chickpea,; rice-relay cropping with sunnhemp; sorghum-puless; sorghum-groundnut. In Goa and Konkan region in the western ghat 10,000 ha of *rabi* rice and 5000 ha sweet potato and vegetables are the double crops in rice lands.	Rice and other crops; soybean, mustard, oil palm, vegetables. Kerala provides scope to grow rubber, oil, palm, coffee cashew, spices & condiments; In Konkan region there is scope to increase area under banana and sugarcane; under pepper, banana, pineapple in coconut groves, under tapioca, pulses and grasses in cashew plantations.	Mostly rice-based cropping system; there are avenues of growing profitable alternative crops like mat-sticks, water chestnuts; fish-cum-rice culture; water loving forage crops like Para grass, *Coix lacrym-jobii* in milkshed areas; mishroom growing; fishery development.	Drainage; acid sulphate soils; nutritional disorders due to heavy leaching and water logging; BPH and GM incidences in rice; Wilt diseases in coconut. Bunchy top in banana ; Erosion in some areas.	Efficiency in irrigation water use need to be improved through promoting conjunctive use of underground and surface water.

TABLE 15: Crops and cropping System in the Western Dry Region

	Existing		Prospect of diversification			
Region (13 & 14)	**Crops**	**Cropping / farming systems**	**Crops**	**Cropping / farming systems**	**Resource constraints**	**Remarks**
Western dry region comprising of Gujarat plains and hills, Dadar & Nagar Haveli, Daman & Diu, Saurashtra plains and Rajasthan dry region comprising 9 districts (Barmer, Bikaner, Churu, Jaisalmer, Jalore, Jhum, Jhunu, Jodhpur, Nagaur & Sikar: scanty & erratic rainfall; scorching heat.	In light soils pearl millet, cluster bean *(guar)* and moth *(Phaseolus aconitifolius)*, commercial crops in dry region: cumin, fennel + isabgol; in Gujarat in black soil regions sorghum, groundnut, sesame, castor and cotton are taken; where there is irrigation facility cotton, sugarcane, rice, *mustard* tobacco, *rabi* groundnut, potato, banana, are grown; chickpea grows well in black soils. A number of forage crops do well : Napier, sorghum, lucerne, oats, berseem with irrigation.	Usually sole crops are taken but inter-cropping of groundnut, pearl millet and other pulses are also taken; cotton is grown usually as a sole crop and also with green or black gram.	In Gujarat, under irrigated condition a good profit is made by the farmer in seed production of hybrid sorghum, peart millet and cotton. Groundnut is the most important crop in the region; a good amount of seed of berseem and luceme is also raised; early varieties of pigeonpeas; banana; arid fruits are taken.	Livestock aspects are of great significance in the area; natural grasses provide a good amount of fodder and so is the case with *khejri* tree, Groundnut haulms are also used as cattle feed; only mono-cropping is done under rainfed condition with certain amount of risk; scope of growing tobacco, chickpea, onion, rapeseed and mustard, sunflower, safflower in system with limited number of irrigation; rice-whear or rice-groundnut with irrigation; potato gives high yield.	Water, high temperature; less number of trees; dry winds. Lack of contingent plains to meet drought: the choice of crops, crop varieties, time of sowing, soil and fertilizer management, etc.	Emphasis on conjunctive use of ground and surface water. Deep black soils have got enormous potentiality to hold moisture in the soil profile and there is scope to effectively utilise it for crop production.

TABLE 16: Crops and cropping System in Island Region

	Existing		**Prospect of diversification**			
Region (15)	**Crops**	**Cropping / farming systems**	**Crops**	**Cropping / farming systems**	**Resource constraints**	**Remarks**
Island Region-the region consists of Andaman & Nicobar and Lakshadwip divison, High rainfall (200-300 cm) and humid areas.	Coconut and rice are the most important crops, with limited irrigation facility: 2 rice crops can be grown per year; colocasia. pineappale, spices and condiments are important crops.	Mono-cropping of rice; double cropping of rice; small amount of pulses (green and black gram) are grown.	Spicies and condiments, turmeric, clove, Saffron, coconut, oil palm, & rubber, Double cropping of rice; banana, pepper and pineapple in coconut and arecanut orchards.	Multistoried cropping in coconut orchards: garden crops, fruits and vegetables; fish-cum-rice culture: fishery, rain water management through dam and tanks to give supplemental irrigation if necessary.	Leached soil; salinity; erosion.	Runoff water management; silvi-pisci-culture in mangrove vegeta-tional areas.

2.4.10.5 *Coastal (eastern and western) region*

Due to heavy rainfall, rice in the plain lands and coconut in the undulated area are extensively cultivated. Plantation of cashewnut is also coming up in a big way. In the thickly populated state of Kerala, homested farming has become the most important system (Table 14). In addition to this rice, banana-cassava, coconut and areca nut-based cropping system have become popular all over the state. In the districts of ldukki, Kozhikode, Wynad and Cannore, pepper-based cropping system has come up in big way. In the districts of Alleppey, Kottayam and Ernaculam, animal component (fish & prawns) is also involved with farming system. A wide range of food, fruits, vegetables, spices and condiments, lemon grass, rubber, oilseed (sesame), betel vine are extensively grown in this area to utilise every bit of their physical resources. It provides an unique example of different possible diversifications of crops in areas receiving high rainfall along with moderately high temperature.

2.4.10.6 *Western dry region*

Soils are usually deep or medium black. Groundnut, mainly as a sole crop, is grown in *kharif*, with limited irrigation facility through ground water (Table 15). Cotton, due to its high price and demand, is also being extensively cultivated. In canal irrigated areas, rice with poor water use efficiency is cultivated. There is a scope of diverting rice area to high value commodities (cumin, fennel, isabgol, onion, tobacco, mustard, banana, etc). In milkshed areas, cultivation of napier-pearl millet hybrid, multi-cut sorghum, African giant variety of maize and lucerne are being extensively grown. The farmers of Gujarat, specialising in raising hybrid cotton, sorghum and sunflower, are earning a good amount of money and opened new avenues for employment. There is also a great scope of improving grassland productivity in wasteland, pasture and forest land for increasing animal production in addition to arable farming which may help in sustaining both soil fertility and the environment on long term basis.

2.4.10.7 *Island region*

In Andaman Nicobar and Lakshadwip with high rainfall and temperature and undulating topography, there is a wide scope of crop diversification as has been achieved in Kerala (Table 16)

2.4.10.8 *Diversification of area for vegetable, aromatic and medicinal plants and floriculture*

The net product value per unit investment in vegetable cultivation is high, particularly near urban areas and where transport and marketing are better organised. Similarly, floriculture around religious places and towns fetches more profit than any other type of crop production. Mint crops, viz., Japanese mint (*Mentha arvensis* L.) is an essential oil bearing plant. The oil or their active constituents are used in toothpastes, candies, chewing gums, cosmetic and medicinal preparations. This is being cultivated in Uttar Pradesh (*terai* region) as a sole or intercrop with sugarcane. Rice transplanted after the mint crop, makes luxuriant growth. The promising mint-based systems that have come up well are mint-rice-potato / field peas and mint-maize-potato. A number of small scale industries have come up with intensive cultivation of these commercial crops (mentha, vetivars, lemon grass, palmarosa, etc.) This would increase the employment opportunities in rural areas and improve the standard of living in villages. Simultaneously, they will build up soil fertility for future exploitation.

2.5 HARMONIC FARMING SYSTEM

With its diversity of agro-climatic features India abounds in many traditional farming systems that have the virtues of compatibility and harmony with the immediate environment. The need for improved farming systems with improvement in production mechanism of traditional farming systems provides further opportunities and responsibilities. Scientists have considerable success in developing cropping system that can modify natural environment and improve the potential output of a farming system. For example, scientists have been able to alter the genetic structure for the desired characteristics (yield, leaf elongation, disease resistance), enhance water available and soil fertility. Crops, cropping patterns and production will have to be tailored to individual farming situations providing the farmers with some options to suit any contingency.

The rising costs of chemical fertilizers compel attention to improving fertilizer use efficiency by crop plants by minimizing the losses of nitrate from the soil and hence, more nutrients within the rooting zone. The many success stories about organic farming from around the world have reopened the topic of organic versus inorganic merry-go-round. With the inception of green revolution in India, the application of organic manure is tended to decline which is suspected by many farmers as the cause of general yield decline in absence of inorganic fertilizer.

On-farm energy need is an important component of a farming system. By and large, animals are the main source of energy in India particularly by small and marginal farmers constituting 80 per cent of farming community. Hence, prospects of large scale mechanization seem remote. In Indian context, there is need to improve energy efficiency. However, Ward *et al.* (1980) is of the opinion that low energy, use efficiency crop production in India not withstanding when the potential of draft animal is fully used, can provide the vital energy supplement for farming and transport besides dairy products and the means of year-round income to the farmer.

Harmonic farming system is not a romantic motions but is the challenge. In the ultimate analysis, it appears imperative that farming system be developed to fit into harmonic matrix (Table 17) as suggested by Swaminathan (1988).

TABLE 17: Symphonic agriculture matrix.

Needs	Technology	Action points services	Government policies
Ecology	i) Conservation and enrichment of environmental assets ii) Farmer participation in location specific technology development.	Associations for sustainable development.	Agro-ecological rehabilitation. Monsoon management long term commitment.
Economics	Cost reduction without yield reduction	Timely and adequate supplies of inputs, management tools.	Input and output pricing policies. Home and International trade.
Employment	Farming system research to generate favourable growth linkage.	Knowledge, skill input transfer services for on-farm and off-farm employments.	Political will and action for ensuring rural development and gainful self-employment.

2.6 DETERMINANTS OF FARMING SYSTEM

Determinants are environmental factors that influence the performance of farming system, whereas environment is aggregate affecting the life and development of an organism.

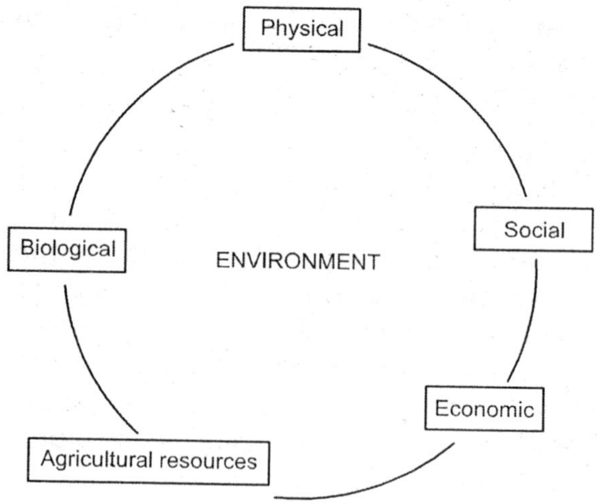

FIG: 9: Environmental Factors

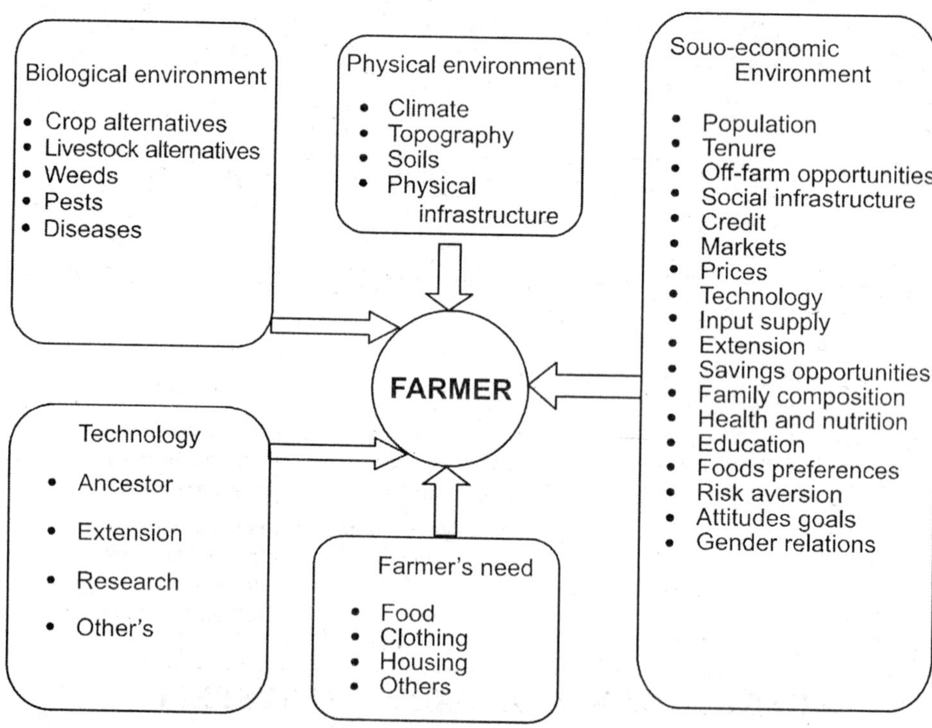

FIG.10: Environment of a farm

A farming system results from a complex interaction of inter-dependent components - soil, water, crops, livestock, labour and other resources within an environment setting that bear upon the agricultural enterprise of a rural household or a farmer. At the center of the interaction is the farm household represented by the farmer, exercising control and choice regarding the types and results of the environmental interactions (Fig. 10.)

The unit allocates different quantities and qualities land, labour, capital and management to crop, livestock and off-farm enterprises in a manner which given the knowledge the household possesses will maximize the attaining of the family goal (s). The household/farmer ultimately decides on the farming system that will emerge, on whether, or not to adopt the proposed intervention and how to assign resources to support it given the constraints imposed by the technical element and exogenous factors (Norman, 1980).

The farm is viewed in a holistic manner. The farmers are subjected to many socioeconomic, biophysical, institutional, administrative and technological constraints. The farming system conceptually is set of elements or components that inter-related which interact among themselves.

The total farm environment in which rural household operate can be divided into natural environment and human environment.

(a) **Natural environment**: Natural environment endowment at any location determines the farming system. Scientists had considerable success in development system that can modify natural environment and improve the potential output of a farming system. Natural environment establishes certain (i) physical and (ii) biological determinants.

(i) **The physical determinants**: The physical determinants (Fig.11) consist of edaphic, aquatic and climatic factors e.g., unfavourable water regimes (erratic distribution of rainfall may cause flood or droughts), poor soils (calcareous with micronutrient deficiencies), adverse topography (basin type dead course of river which causes water logging during monsoon), etc.

(ii) **Biological determinants**: These are pests, disease, weeds, livestock alternatives, crop alternatives, etc. The biological characteristics of agriculture allow and require flexibility in the production decisions of the farmers in his continuous struggle with his natural environment.

(b) **Human environment**: It provides the sufficient condition for development and utilization of a particular system. It takes into account two sets of social, cultural and economic factors:

(i) **Exogenous**: These factors are largely outside the farming household's control, like community structures, external institutions, etc.

PHYSICAL DETERMINANTS			
EDAPHIC		AQUATIC (FISHERIES)	CLIMATIC
land Type Soil Type Hydrology	1 Soil texture 2 Soil structure 3 Available plant Nutrients 4 Soil reaction 5 Soil organic matter 6 Soil micro organism 7 Soil moisture 8 Soil temperature 9 Soil atmosphere	1 Physical (Temp, space, turbidity) 2 Chemical (O_2, H_2S, CO_2, salinity, pH) 3 Biological (Success of spawning, vanishing species, competition, predator, parasitism, disease algal bloom)	1 Rainfall 2 Temperature 3 Light 4 Humidity 5 Wind 6 Air pressure 7 Attitude

FIG.11: Physical determinants of farming system

(ii) **Endogenous**: These factors, on the other hand, are those over which the individual farming household has some control, like allocating labour, capital and development of management skills.

Human environment establishes socioeconomic and agricultural resource determinants:

(i) **Socioeconomic determinants**: These include market, power, tenure, population, credit, tradition, social infra-structure, needs or incentives, infra-structural support and others.

(ii) **Agricultural resource determinants**: These are land, labour, capital, management capability, seeds, fertilizers, animals, machines, and energy.

In a farming system, all the activities-decisions, management, input/output, purchase/sale, resource (s) utilized make the matrix of farming system which are bound to interact with socio-economic and biological environments. The farm activities interact with market forces (socio-economic) and ecosystem (bio-physical) for purchasing input and disposing output by utilizing natural resources (land, water, air, sunshine, etc.) and degrading it.

Sustainability is goal of the farming system where production process is optimized through efficient utilization of inputs without impairing the quality of environment with which it interacts. To achieve this, collaboration between scientists as well as between scientists and farmers is essential for rapidly developing technologies that are appropriate to farmers' circumstances and that help to meet national goals.

2.7 TECHNOLOGY

A technology is a component combination of all management practices for the component of a systems, each practice has to be defined by the timing, amount and type of various technological components, such as varieties, land preparation, fertilizer or weeding. A subsistence farmer who purchases no input is nevertheless using a technology-sometimes quite a complex one. A farming system obvious is very complex and that is why any agricultural technology well suited for a particular agro-ecological situation and socio-economic environment may not be adopted by other farmers. Unless both natural and human environments are considered, agricultural research will not result in relevant agricultural technology.

Technology developed should be appropriate to the circumstances of target groups of farmers. Farmers' circumstances are all the factors that influence farmers' decision about a crop technology-the natural environment (such as, rainfall), the economic environment (such as, product market) and the farmers' goals, preferences and resource constraints. If technologies are appropriate to farmers circumstances, the recommendation will be rapidly adopted by the farmers. Circumstance affecting farmer's choice of a crop technology is depicted in Fig.12

2.8 RECOMMENDATION DOMAIN

It is a solution proposed for group of farmers that is based on their problems and their circumstances. A group of roughly homogenous farmers with similar circumstances for whom more or less the same recommendation can be made is known as recommendation domain.

The proper number of recommendation domains depends on the amount of variation in farmers' circumstances (the more variation, the more domains needed). They may be determined by variation in the natural circumstances of the farmer, such as rainfall, soil or diseases. A region may contain many agro-climatic environments. In an agro-climatic environment a crop exhibits rather uniform biological expression, so that varietal or fertilizer responses would be similar, everything else being equal. Within an agro-climatic environment, however, there may be groups or regions with different socio-economic circumstances that require different recommendation domains. For example, close to a large town, maize may be grown for sale as fresh cobs/ears, while further away it is grown as a subsistence grain. Such differences may impose modifications on varietal selection and planting date. More commonly, even among farmers who are in the same agro-climatic environment, the different resource endowments of farmers may lead to different technological needs. For example, small farmers who have scarce capital relative to labour and who place more emphasis on food security may follow quite different cropping patterns and practices from those of large farmers in the same agro-climatic environment.

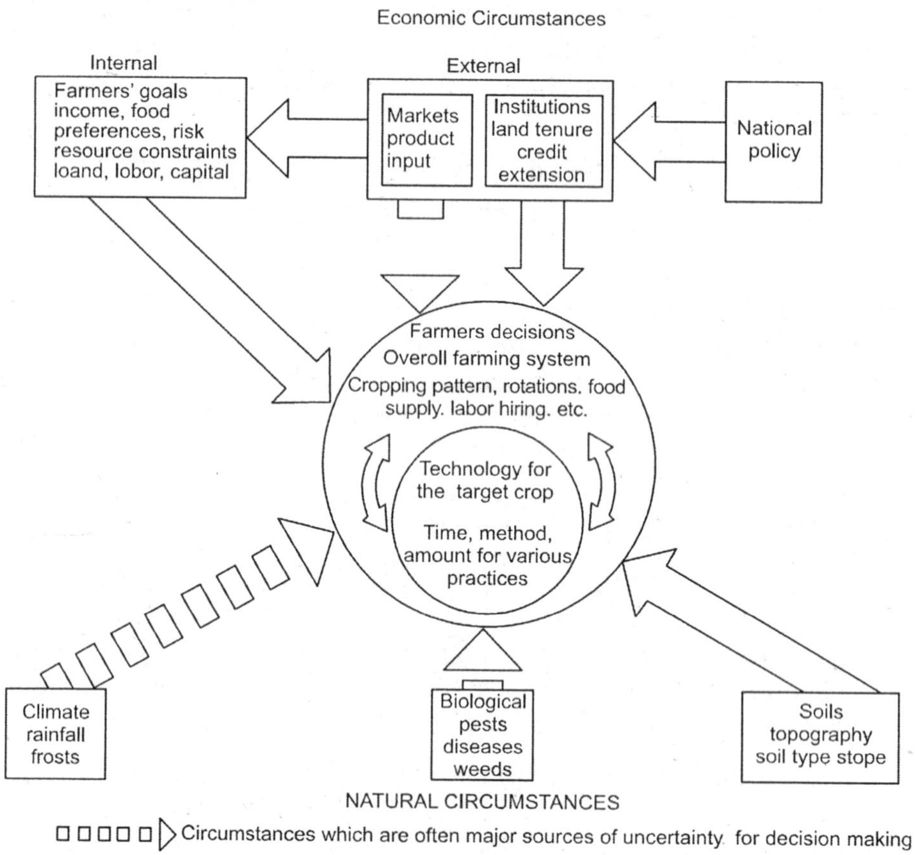

FIG. 12: Circumstances affecting farmers' choice of a crop technology.

At times, a recommendation domain may result from a complex interaction of agro-climatic and socio-economic factors. For example, within an agro-climatic environment for maize, there may be different disease incidences for beans, which cause farmer to plant beans early, therefore delaying maize planting. In this case recommendation domains may result from natural circumstances, diseases, which affect bean production and an economic circumstance, labour scarcity, which conveys this effect to maize practices.

Recommendation domains are not necessarily continuous geographical areas. For example, two neighboring farmers may be in different recommendation domains because of large differences in available resources. Even within a farm there may be different recommendation domains due to variation in soil type or topography. Hence, it is clear that knowledge of farmers' circumstances and how they effect crop technologies are necessary for defining recommendation domain.

2.9 FARMING SYSTEM RESEARCH (FSR)

2.9.1 Introduction

Since the beginning of the century the population through out the world in general and in South Asia in particular has increased at a dramatic rate, but increase in food production has not been substantial to cope with the food requirement of the growing population. Hence, a large section of rural population still live below the poverty line who cannot afford two square meals a day. According to Global Hunger Project based in New York "A country may produce food, if the people do not have money to buy the food, hunger problem persists." Today, there are more hungary people in the world than any time before. Africa has the severest hunger problem in the world. To meet the situation with more pragmatic approach, the possibility of increasing the net crop area is considered to be minimum while the opportunities for raising more food lies in increasing yields per unit area, increasing cropping intensity and increasing both yields and cropping intensity.

2.9.2 Why the Technologies Generated Become Inappropriate?

On-station experiments on national and International research institutes have demonstrated that yield of crops could be raised several fold over the farmer's existing levels provided right technology with right management is adopted. However, a wide gap exists between the yield achieved at the experimental station and farmer's yield.

Farmer very often reject technology because he can not manage to buy the inputs, and he does not have accessibility to credit sources required to follow the recommendations. He is not adequately trained also to on the use of technologies. In other words, the technologies recommended are not appropriate to the farmer's resource base and the skills of the farmers. Farmers would like to see that the technologies are compatible with his resources-physical, climatic, biological and socio-economic conditions. It has to meet the family goals, lower the risk and raise the income. It the recommended technology does not satisfy these conditions, the farmers are not going to adopt it.

Thus, the technologies generated become inappropriate due to following reasons:

(i) The physical environment of a research station is quite different than that of the small farms.

(ii) The level of crop management and input use as well as the degree of control irrigation and pests on experiment stations are high.

(iii) The general recommendation are given for large geographical areas or regions, but large differences exist in the agro-climatic and socio economic situations of farmers within small areas even within a village.

(iv) The recommendation suited for some farmers are not suited for others.

(v) The recommendations are made without a clear understanding of the causes of problems faced by farmers in increasing production and net income.

2.9.3 Limitations of Present Research Thrust

According to Mahapatra (1992), the existing agricultural research is a component research and there is no system based programme. The response of a component in isolation dose not necessarily fit into a system perspective. The individual programme ignores the socio-economic and biophysical aspects of the farming community. The limitations of conventional model of agricultural research and extension in dealing with interactive matrix are well recognised. The reductionist component research approach curtails system perspective resulting in wide gap in technology development and utilization particularly for the resource poor farmers. System research take into consideration appropriate perspective in handling a complex undertaking such as farm enterprises diversification and their interaction on farm productivity.

2.9.4 Concept, Definition and Approach of FSR

The farming system, as a concept, takes into account, the component soil, water, crops, livestock, labour, capital, energy and other resources with farm family at centre managing agriculture and related activities. The farmers for their survival have been developing technologies / systems through experience gained in their environment since the dawn of civilization by utilizing their inharent knowledge and wisdom. From the last four decades a great emphasis has been made on the utilization of farmer's technical knowledge and wisdom to the formal research system through their participation for the improvement of farms in a holistic way. This approach is termed as farming system research (FSR). Shaner *et al.* (1982) have described this as an approach to agricultural research and development that views the whole farm as a system and focuses on (i) the interdependencies between the components under the control of the farm household and (ii) how those components interact with the physical, biological, and social-economic factors not under household control. The approach involves selecting target areas and farmers, identifying problems and opportunities, designing and excuting on farm resarch and evaluating and implementing the results. In the process, opportunities for improving public policies and support systrms affecting the target farmers are also considered.

Farming system research (FSR) addresses itself to each of the farm's enterprises, and to the inter-relationship among these enterprises and between the farm and its environment. It employs information about the farm's various production and consumption systems and about farm environment (physical, institutional, social and economic) to increase the efficiency with which the farm utilizes its resources (Zandstra, *et al.*, 1981).

According to CGIAR TAC Review (1978), farming system research is:

(a) conducted with a recognition of and focus towards the interdependencies and inter-relationship that exist among elements of the farm system, and between these elements and the farm environment, and

(b) aimed at enhancing the efficiency of farming systems through the better focusing of agricultural research so as to facilitate the generation and testing of improved technology.

The basic elements of FSR approach include (i) problem solving, (ii) holistic, (iii) farmer's participation, (iv) location specificity of technological solutions, (v) bottom up research strategy, (vi) interdiscipline, and (vii) sustainability.

2.9.5 Farming System Research versus Conventional Agricultural.Research

Agricultural research is search for additional knowledge, practice and not yet discovered knowledge on the behaviour of species, varieties and combination of practices in any given agro-climatic zone. Farming system research informs about the existing situations and improves prognosis. One of its major task in the testing of innovations in the system and thus, contributes to the relevancy of biological science research work.

Agricultural research is usually commodity or disciplined oriented. Solutions with the highest technical efficiency may not be optimal solutions in economic term, and optimal solutions for a given crop may not be optimal. It is the test of farming system research to assess the worthwhileness of innovations in the context of the total farm unit. Farming system research is, thus, a major tool to close the 'gap' between agricultural research and performance in the actual farm (Ruthenberg, 1980).

Conventional approuch of research and extension in agricultural has been suffering from several lacunae. The characteristic features of conventional approach have been: (a) Single comoddity based (b) Single discipline orinented (c) Top-down planning (d) Resource exploitative and, (e) Input internsive. This apprach has led the Indian agviculture to slugish growth in productivity after initial remarkable success, large scale resource degradation, poor rate of technology transfer, decline in productivity factors and decreasing benefit: cost ratio.

Difterent components of agricultural system must, therefore, be analysed in view of identifying farmer's problems through system approach otherwise the following complications may arise:

(i) The development and research priorities may be established in isolation from the system context and are sometimes inappropriate.

(ii) Poor communication among researchers, development workers and farmers.

(iii) Plans, programmes and projects may tend to concentrate on productivity issues while neglecting issues of stability, sustainability and equity.

(iv) Poor knowledge may be used to analyze the systems and carry out inappropriate research and development interventions.

2.9.6 Goal of FSR

The ultimate goal of farming system research is to integrate the existing sub-systems on a farm to harness maximum efficiencies and develop sustainable resource systems which will optimize their use, minimize degradation with consideration to regenerative capacity and increase income and employment for farm families and promote quality of life and environment. This implies that development of farming system requires proper inter-linkages among various sub-systems if impact is to be realized in terms of income generation, energy efficiency and employment generation.

2.9.7 Principles involved in FSR

The farming systems research approach includes (a) viewing the farm as a whole, (b) identifying the farming systems, the interacting components and delineating boundaries, (c) systematically investing the nature and extent of interdependence among the enterprises and identifying the constraints, (d) applying the modern technical know how to the system so as to make it yield the optimum results, (e) studying the equity gender income, employment and resource use efficiency, and (f) dealing with the issues at integration level through analysis and solution of problems towards sustainable farming system development.

2.9.8 Scope of FSR

Farming enterprises include crop, livestock, poultry, fish, tree crop, plantation crops, sericulture, etc. A combination of one or more enterprises with cropping, when carefully chosen, planned and executed, gives greater dividends than single enterprise specially for small and marginal farmers. Farm as a unit is to be considered and planned for effective integration of the enterprises to be combined with crop production activity. The integration of farm enterprise depends on many factors, such as:

- Soil and climatic features of the selected area,
- Availability of the resources land, labour and capital,
- Present level of utilization of resources,
- Returns from existing farming system,
- Economics of proposed integrated farming system, and
- Managerial skill of the farmer

2.9.9 Objectives of FSR

Any farming system research programs most help the decision maker to understand physical and socio-economic environment within which farm business is to be conducted. It should help to evaluate the existing resource base, its allocation and potential to meet the objective of farming. While adopting the farming system approach, the following important objectives have to be kept in view:

(i) Identify constraints in increasing farm productivity.

(ii) Provide technological intervention options for improving farming systems with a given resource base through On-Farm Adoptive Research (OFAR).

(iii) Refine and develop appropriate farm technology so as to provide stability in production, employment throughout the year, income at periodic intervals, improved management skill of the farmers and decent living.

(iv) Develop methods for measurment and monitoring of sustainability of farming system.

2.9.10 Classification of Farming System Research (FSR)

Simmonds (1984) and Spedding (1988) identified three categories of farming system research:

(1) **FSR sensu-stricto or Farming system analysis (FSA)**: It is the study of farming system *per se*, as they exist. Typically, the analysis goes deep (technically and socio-economically) and the object is more academic or scholarly rather than practical; the view taken is nominally "holistic" and numerical system modelling is a fairly natural outcome if a "holistic" approach is claimed.

(2) **Farming system component research (FSCR) or On-farm research with a farming systems perspective (OFR&FSP)**: It is a practical adjustment to agricultural research which starts from the precept that only farmer's experience can reveal to the researcher what farmers really need. Typically, the on-farm research with farming system perspective process isolates a sub-system or the management of a specific resource of the whole farm, studies it in just sufficient depth to gain the necessary farming system perspective and proceeds as quickly as possible to experiment on-farm with farmer's collaboration. There is an implicit assumption that stepwise change in an economically favourable direction is possible and worth seeking.

(3) **New farming system development (NFSD)**: It starts with the view point that many tropical farming systems are already so stressed that radical restructuring rather than stepwise change is necessary; the invention, testing and exploitation of new system is, therefore the object. While on-farm research with a farming systems perspective (OFR&FSP) seeks to adopt technology to the farmer's economics, new farming system development (NFSD) must usually imply government intervention and the adoption of economics to technology. NFSD is fundamentally a 'top-down' approach to agriculture research and technology which seeks revolutionary rather than evolutionary change. It requires multiple and complex changes instituted simultaneously with ultimate product being a radically new farming system (Simmonds,1986) In addition to mentioned categories of FSR, the other types of FSR described are mentioned below:

(4) **Farming system adoptive research (FSAR)**: It is interdisciplinary on farm adoptive reseanoh targeting a defined group of relatively homogeneous farming systems characterised by similar constraints. It draws on the existing 'body of knowledge' generated by strategic and applied research and on farmers knowledge of their environment to develop and fine-tune short term location specific technologies for the designated farming system. According to Collinson (1982), the objective of FSAR is to increase productivity of the farming system through the development of relevant technology, which conforms to farmer's circumstances. This means that technology must be possible and attractive for the

farmers to adopt which confirms to farmer's determination of their need.

(5) **Farming systems bare-line data analysis (FSBDA)**: It refers to collection of basic data on an agro-climatic zone and analysis of data after a large scale survey with the objectives of diagnosing major constraints and developing a classification of major types of farming systems. The greater emphasis is placed on biological and physical, rather than socio-economic variables.

(6) **Farming system research and agricultural development (FSRAD**: It refer to farming system programme which integrates agricultural research and development strategies. The objective of FSRAD is the long-term transformation and development of agriculture in a designated sub-region through both technological development for major farming system and institutional and economic reforms in the environment. It, thus, combines mainstream agricultural research, FSCR, FSAR and sometimes NFSD with the development of agrarian infrastructure and services as a means of increasing overall agricultural productivity of the sub-region.

The differences, as well as similarities among the six types of research (FSA, FSAR, FSCR, FSBDA, NFSD, and FSRAD) exist when they are compared in terms of the degree to which they incorporate the key concept of FSR as originally conceived.

The difference form of research are given below:

(i) **Basic research** is designed to generate new understanding.

(ii) **Applied research** is designed to create new technology.

(iii) **Adoptive research** is designed to adjust technology to the specific needs of a particular set of environmental conditions.

(iv) **Strategic research** is designed for the solution of specific research problems.

2.9.11 Characteristics of Farming System Research (FSR)

1. **It is farmer-oriented**: FSR targets small farmers as the clients of agricultural research and technology development. Farmers are integrated into the research process. The existing farming system is studied before proposed improvement-social scientists collaborate with technical scientists on the analysis

of existing system, problem diagnosis and development of alternate technologies.

2. **It is system-oriented**: FSR is essentially operational research which first identifies technical, biological, and socio-economic constraints at the farm level for major type of farming systems. It then endeavors to develop technologies, which are feasible for the targeted farming households to adopt and meant to alleviate those constraints. The research process is both iterative and dynamic with adjustments being made in technologies design as understanding of and communication with small farmers.

3. **Interdisciplinary and multidisciplinary**. Collaboration with agricultural scientists of various disciplines and social scientists is needed to understand the condition under which small farmers operate to accurately diagnose constraints, and to develop improved technologies suitable to those conditions. Each programme should determine the composition of its FSR team. Generally it should consists of an agronomist, a socio-economist, and local agriculturist, backed up by extension personnel and biological and social scientists from the region or a nearby experiment station.

4. **Complements mainstream commodity and disciplinary research, it does not replace it**: FSR draws knowledge of technologies and management strategies generated by discipline and commodity research and adapts them to the specific environments and socio-economic circumstances of targeted group of relatively homogenous farmers. Problem found on the farm which cannot be solved are fed back to the discipline and commodity researchers.

5. **Tests technology on-farm**: On-farm experimentation establishes the context of collaboration between farmers and researchers and fosters a deeper understanding of the farming system among researchers. It also provides for true evaluation of technologies under the environmental and management conditions in which it will be used. On-farm research is central to FSR approach.

6. **Provides feedback from farmers**: FSR channels feedback on farmer's goals, needs, priorities, and criteria for evaluating technologies. Farmers give opinions on proposed new technology, help plan on-farm trials, implement trials under

the supervision and express opinion on results of trials. Farmers reaction and adoption are closely monitored.

7. **FSR is a problem-solving approach**: FSR is operational research, which first identifies technical, biological and socio-economic constraints to improve production in a farming system and then endeavors to develop solutions, which are appropriate for the management conditions of that system.

8. **FSR views** the farm or production unit and the rural household or consumption unit, which in the case of small farmers are ofter synonymous, in a comprehensive manner.

9. **FSR recognizes** the interdependencies and interrelationship between the natural and human environments. Research on a sub-system can be considered part of the FSR process if the connection with other sub-systems are recognized and accounted for.

10. FSR does not require in-depth research, but a broad range of investigations directed at an integration of the relationship among the salient features of the systems.

2.9.12.1 Methodology

Since farming system research programme implies the concept of total environment consisting of both natural and human elements, the methodological complexities may be great. In practice, instead of assuming all factors as potential variables subject to manipulation, some are treated as promoters not subject to manipulation. The whole research process is recognized as dynamic and iterative with both 'backward' and 'forward' linkages between farmers and research-workers. The "feed in" from the farmer helps to set experimental priorities and problems; the "feed back" from the farmer lets the scientist know, if they are providing useful results. In farming system research the farmers by participating in research process help in identification of research problems as well as take part in testing the possible solution and realistic result can be obtained through farmer managed trials.

Thus, a farming system research approach includes agro-climatic, technological and sociological aspects of the problem. That is, the technical, institutional and behavioral dimension of, say, a particular system and the approach focuses on developing a new technology for that particular system, through farmer's participatory mode taking into account availability of land, labour, capital, energy and other resources while keeping farm family at centre. To achieve this, goal should be to evaluate and refine / develop client oriented need based research under bio-physical and socio-economic conditions of resource poor farmers (Fig. 13).

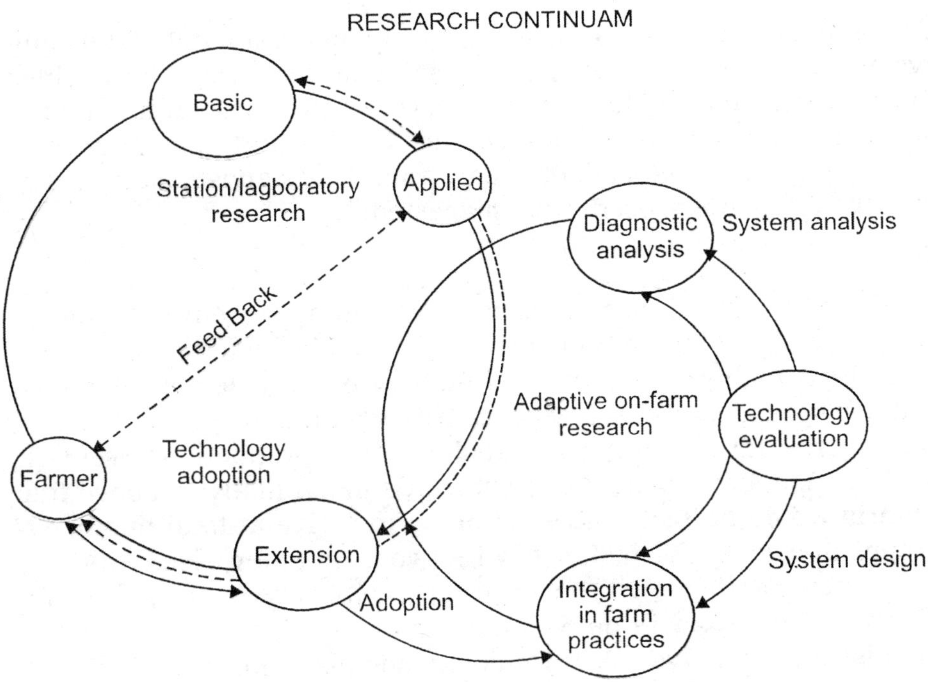

FIG. 13: From technology generation to adoption.

Methodological issues aim to:

- minimize the time required to move through the existing research stage. The method, in addition to ensure fast turn-around, needs to be practical, replicable and inexpensive,
- maximize the return by making results more widely applicable, and
- use of second best solution or the best of those readily available.

The methodology of farming system research (FSR) consists of five stages multi-disciplinary efforts: (a) Descriptive and diagnostic, (b) Design, (c) Testing, (d) Evaluation, and (e) Diffusion and extension. Multidisciplinary efforts of a team of scientists combining agronomists, social scientists, economists and specialists in livestock, fishery, forestry, tree, plantation crops and farm management are essential to carry out the FSR programme at all the stages. These stages are briefly described (Mahapatra, 1992; Ghildyal, 1985) below:

(a) Descriptive and diagnostic stage

The aim of the diagnostic survey is to understand the farming systems that are practiced in the selected target areas. This enables the scientists and the farmers to determine the constraints that the decision making households face and the flexibility in the current farm enterprises could adopt to carry out the diagnostic survey. For this, the following procedure is suggested:

Step I

All the districts falling in a particular farming system zone should be grouped based on four criteria, viz., (i) high productivity and low spread (ii) high productivity and high spread (iii) low productivity and low spread, and (iv) low productivity and high spread. Further, each of the farm enterprises, such as crop production, livestock, poultry, fish, etc. having been subjected individually to above four criteria would be factorially combined to arrive at total number of groups in a zone. For instance, with two farm enterprises (say crop and livestock) there will be 16 groups, with three enterprises, the number of groups will be 64, and so on. After having classified all the districts as above select randomly adequate number of districts from each group.

Step II

In the district chosen for the study, prepare a list of all the blocks / *taluks* and select few blocks randomly for the study

Step III

From each of the selected block / *taluk*, select adequate number of villages, say, ten randomly from the totality of the villages in a block / *taluk*

Step IV

Carryout the diagnostic survey by rapid rural appraisal for each of the ten selected villages and prepare an inventory of farm resources, production constraints and support services. Using the information so obtained, the entire population of farm holding may be grouped, based on the four criteria: (i) large (ii) medium (iii) small and (iv) marginal holdings. Further, each of the enterprises, such as, crop, livestock, poultry, fish, etc, having subjected individually to above criteria, would be factorically combined to arrive at the number strata as has been done above. For two enterprises the number of strata will be 16, for three, 64 and so on.

From each of the stratum so obtained, select randomly adequate number of farm holding, say six. Four out of this, will be the sites which FSR will be conducted while the remaining two would serve as control.

(b) Design stage

In this stage, a suitable action plan for the selected farmer is formulated. The improved technologies relevant to overcoming or avoiding constraints identified in the first stage are specified. The cumulative store of knowledge at the research station is the best source of ideas for potential improved technologies that might be appropriate Equipped with this knowledge and that obtained from the diagnostic survey, a farm management design for the selected farmer, taking into consideration his preference, resources and overall production benefits is to be prepared. All intervention plan taking into consideration farmer's preference, risk involved, labour resources, gender distribution of labour and family age structure etc. ensures that the intervention fits in the farmer's existing skill as far as possible.

(c) Testing stage

The objective of the testing stage is to evaluate a few of the more promising technologies arising from the design stage on the farmer's farms. First part consists of trials at the farmer's level that use farmer's land, labour and other resources but the research team provides the managerial input skill. The second part involves farmer's testing providing their own land, labour and other resources. The second part on-farm experimentation under farmer-managed trial technique will provide the potential for substantial research worker / farmer interaction and result in assessing the technologies under conditions as close to reality as possible. In this testing stage, the involvement of extension personnel can particularly strengthen the programme.

Since no two farmers have identical circumstances, the farmers are classified with similar circumstances into recommendation domains-groups of farmers for whom more or less the same recommendations can be made or for homogenous population on the basis of their farming system. They are determined by variation in the natural (agroclimatic) or socio-economic circumstances or from a complex interaction of agroclimatic and socio-economic factors.

(d) Evaluation stage

The research team by visiting the farmers periodically will collect the necessary data for analysis and drawing conclusions. The data needed would consist of (i) input use (both available on the farm and purchased), (ii) labour utilization patterrn, (iii) output of crops, animal, birds, etc., (iv) production of byproducts, like crop residue and manures, (v) soil fertility in case of crops, and (vi) income from farm enterprises and the build up of assets.

The data would then be subjected to economic analysis with a view to workout the profitability, technical feasibility, social and cultural acceptability and sustainability of intervention with existing resource base and infrastructure. An index of acceptability is calculated and if acceptability is low, the research team considers the benefits of the improved technology and may even drop it. The technology should not simply be designed, introduced and abandoned. Information needs to be collected on the technologies acceptance or rejection by the farmer, the final judge to the appropriateness of the proposed technology. Until this point, all scientific evaluations remain at the level of hypothesis. Unless the circle can be completed, unless research results reach the farmer, prior efforts may have been fruitless and research findings are shelved in the experiment station archives. And if the technology is rejected by the farmer, the research process can be repeated to determine the reasons and find ways to overcome them.

(e) Diffusion and extension stage

The acceptable new technology is promoted on collaboration with State Department of Agriculture and other non-government voluntary agencies. To achieve this, it is necessary to bridge the gap between the knowledge of the farmer and the know-how of the technology he is going to adopt. Since all identified problems are not research issues but on a whole farm basis, in terms of cash, labour and land allocation, imparting training to the farmer is essential before technology is promoted. So also is the availability of critical inputs of farming by increasing the infrastructure facilities and by associating the different services and agencies involved.

2.9.13 Steps followed for implementation of a project

(i) **Participatory rural appraisal or PRA:** It is adopted for site selection after a village selection by a multi-disciplinary team meeting with target group of farmers.

(ii) **Site characterization:** It is done for its demographic features, institutional support, infrastructure, topographic features, soil climate and existing farming systems.

(iii) **Problem identification and prioritization:** Problem faced by target groups are listed and then prioritized as per merit based on severity, extent, economic importance and farmer's perception.

(iv) **Indigenous technology knowledge (ITK):** ITK of farmers are gathered and documentated for technology refinement for finalization of intervention options.

(v) **Intervention options identified:** Relevant technologies generated are assessed through a number of on-farm trials. Same are refined according to local needs or omitted and resources are give due consideration.

(vi) **Intervention execution:** The identified interventions in different production system are executed with close participation of farmers.

(vii) **Impact analysis:** Assessment of the impact of different interventions are undertaken on production, productivity and benefit derived or supposed to be derived. Detailed economic analyses is done.

2.9.14 Strategy

The strategy is to develop a science based agriculture through balancing the inherent soil resources and crop requirement with innovative soil management techniques and judicious use of inputs. Some appropriate and scientifically proven components are listed in Table. 18.

TABLE 18: Scientifically proven components for developing sustainable farming systems

Objective	Recommended component / sub-systems
A. Controlling crosion	Conservation tillage, strip cropping, cover crops, mixed cropping and agro- forestry, mulch farming, terracing tied ridges, shelter belts, windbreaks.
B. Conserving water	Mulch farming, eco-fallow, weed Control, tied ridges, irrigation
C. Alleviating compaction	Sowing deep-rooted legumes, chisel Ploughing, reducing vehicular traffic, guided traffic.
D. Enhancing soil fertility	Legume based rotations, cover crop, organic amendments, chemical fertilizers
E. Improving soil fauna,	Mulch farming, crop diversification, reduced tillage, organic

Objective	Recommended component / sub-systems
maintaining Soil tilth	amendment, planted fallows
F. Pest management	Rotations, crop diversification, residue management, pesticide use based on threshold needs.

In view of serious limitation on horizontal expansion of land for agriculture, only alternative left is for vertical expansion through various farm enterprises requiring less space and time but giving high productivity and ensuring periodic income especially for the small marginal farmers belonging to handicapped zones (rainfed areas, drylands, arid zone, hilly areas, tribal belts and problem soils). For this, several farm enterprises could combine, viz., crop + livestock, crop + poultry, crop + dairy, agri-horticulture system, crop + horticulture + sericulture system, agro-forestry, silvi-pastoral, rice + fish, rice + fish + mushroom, flouriculture + apiculture, fish + duck + poultry, etc.

For meaningful execution of integrated farm enterprises, the following activities should be undertaken by a multidisciplinary team of scientists with farmer's participation and involvement at all stages.

(i) Thorough understanding of existing farming systems and their components.

(ii) Assessment of resource availability in the farm environment and identification of bio-physical, socio-economic, institutional, administrative, and technological constraints.

(iii) Development of economically viable and efficient integrated farming systems suitable for various domains.

(iv) Diffusion of improved technologies and receiving “feed back” for further improvement of the system as a whole.

(v) Continuous improvement in components technology to fit into a given farming system.

(vi) Improvement in quality of farming system research.

(vii) National and international linkages,

2.10 SHIFT IN FUTURE STRATEGIES OF RESEARCH

(a) **Programme mode of research**: The first and formost Change is needed in the way research is done. A shift from ‘Project mode’ to ‘Programme approach’ is needed which cuts across disciplines, commodities, division and institution. This will require prioritization of national and institutional programme and strengthening of inter-disciplinary and inter-institutional linkages. The programme approach will also ensure teamwork culture, the most crucial element for research.

(b) **Eco-regional approach:** More than half of required growth rate in yields has to come from research efforts by developing location specific low input use and eco-friendly technologies. The research activities have to be upscaled within the eco-regions and by dovetailing research and development (R&D) priorities between and within eco-regions. Success of the approach would necessarily depend on harmony between priority setting at the eco-region level and effective execution at the local levels.

Land use planning involving GIS and simulation modeling techniques have to be blended with this approach which will offer agro-technologies tailored to match the resources and needs of the regions with a commitment to improve the productivity of scarce resources, while protecting the quality of soil and water for posterity (Paroda, 1998).

(c) **Precision Farming (PF):** Every shrinking resource base, increased cost of production and ecological concerns necessitate for precision farming (PF) which refers to most efficient use of inputs, use of inputs as per requirement and agronomic practices. In this context, it is pertinent to mention that application of fertilizer nutrients on the basis of soil test value will not only economise on the cost of fertilizer application but will also minimise soil degradation. Land leveling and root zone wetting through micro irrigation system might lead to efficient use of scarce water resources. Also use of sprinkler and drip irrigation systems would not only enhance water use efficiency but would also enable effective use of nutrients for higher crop productivity.

The Precision farming (PF) sometimes also referred to as site specific farming (SSF) is an emerging technology that allows farmers to adjust inputs duly considering the in-field soil variability, crop condition and yield resulting in the reduction of cost of production.

The precision farming (PF) technology uses Global Positioning System (GPS), which comprises satellites that transmit the signals picked up by user receivers to define the receivers location. With this information and on-board sensors and farm equipment like crop monitors and yield monitors, application of crop inputs like fertilizers and plant protection chemicals can be guided thereby reducing the cost of production. Integration of information on resources with socio-economic and meteorological inputs have a tremenolous scope for optimum crop management strategies (Rao, 1999).

REFERENCES

Ahmad, S and Srivastava, V.C. 1993. Legume + cereal and legume + legume Intercropping system., I. Assessment of crop competition and productivity. Journal of Research (BAU), 5(1): 51-54

Aiyer, A.K.Y.N. 1949. Mixed cropping in India. Indian Joural of Agricultural Sciences,19:439-543.

Balasubramaniun, P. and Palaniappan, S.P. 2002. Principles and Practices of Agronomy. Published by Agrobios (India), Jodhpur, pp:511-512.

Bateman, R.J. and Rowling, R.W. 1980. Conservation cropping-a new way of farming for the south Bernitt. Queensland Agriculture Journal, 106(3):13-16.

CGIAR. 1976. Consultative Group on International Agricultural Research (CGIAR), New York, USA.

CGIAR. 1978. Farming systems research at the International Agricultural Research Centre. Consultative Group on International Agricultural Research (CGIAR), New York, USA. State University International Development, Michigan, USA.

Chatterjee, B.N. and Srivastava, V.C. 1990. Diversification of crops and cropping system. Agronomic Research Towards Sustainable Agriculture. (eds. Singh, K.N. and Singh, R.P.). Indian Society of Agronomy, Division of Agronomy, IARI, New Delhi-12. Ch. VII, pp:81-95.

Collinson, M.P. 1982. Farming system research in East Africa: The experience of CIMMYT and some National Agricultural Research Services, 1976-1981. Michiqan State University International Development Paper 7, East Lansing, Michgan, USA.

Committee on Fodder and Grasses, 1987. Report on National Wasteland Development Board, New Delhi, India.

De candolle, M.A.P. 1982. Physiologie veqetale. TOME III. Becht Jeune, Parris, pp: 1474-1475.

Derek Byerlee.1985. Developing crop technologies within the context of multiple cropping schemes of South-Asia. Paper presented at the International Conference on Multiple Cropping Systems Research, Nanjing, China, October 8-12,1985.

Donald, C.M. 1963. Competition among crop and pasture plants. Advances in Agronomy, 15:1-118.

FAO. 1988. Animal production year Book. Food and Agriculture Organization, Rome, Italy. Fresco, L.O. and Westphal, E. 1988. A hierachical classification of farm system. Experimental Agriculture, 24: 399-419.

Ghildyal, B.P. 1985. Farming system research for improved technology. National symposium of Indian society of Agronomy, CSSRI, Karnal, Haryana, India, April 3-5, 1985. Cropping systems, India Society of Agronomy. New Delhi,: pp 95-101

Hall, A.E., Carnell, G.H. and Howton, H.W. 1979. Agriculture in semi-arid environment. Ecological Studies.34., Springer Verlog, Berlin Heidilberg, New York, USA, pp:200- 223,

Hans Molisch. 1937. Der Einfluss einer Pflanze auf die andere. Allelopathic Fischer, Jena.

Hosmani, M.M., Chittapur, B.M. and Babalad, H.B. 2000. Farm Productivity- New century, New challenges. Publ.5910, Wood Ridge Hill San Antanio, Texas 78249, USA.

Hunsigi, G. and Krishna K.R. 1989. Science of Field Crop Production. Oxford IBM Publication Co. Pvt. Ltd., New Delhi. Calcutta. pp: 23-24.

Kanwar, J.S., Virmani, S.M and Das, S.K. 1992. Farming systems research in India: A historical perspective. Experimental Agriculture, 28: 1-17.

Krantz, B.A., Singh, S., Sharma, S.K., Singh, P., Sahrawat, K.L. and Nigam, S.N. 1975. Cropping patterns for increasing and stabilizing agricultural production in semi-arid tropics. *In* Proceedingss of International Workshop on Farming Systems held on November 18-21, ICRISAT, Hyderabad, India, pp: 217-248.

Khoshoo, T.N. 1983. Energy form Biomass. Indian Farming, 33(3): 10-13.

Kulkarni, K.R., Devi, L.S., Jagannath, M.K. and Setty, T.K. 1686, Studies on rice farming systems in India. Report of the Asian Rice Farming System Workshop Group, IRRI, Phillippines. Manila.

Lal, R and Miller, F.P. 1990. Sustainable farming for tropics. *In*: Sustainable Agriculture: Issues, Perspectives and Prospects in Semi- Ariol Tropics (*ed.* R.P. Singh). Indian society of Agromomy, IARI, New Delhi, Vol.1: 69-89.

Ludwig, W.1959. Die selection theorie. *In*. G. Herberer (ed.) Die evaluationen der organismen. Fischer, Stuttgart, pp:662-712.

Mahapatra, I.C. 1992. Farming Systems Research- Challenges and Opportunities. Orissa University of Agriculture and Technology, Bhubneswar, India, pp: 1-10.

Maxwell, S. 1984. Farming systems research: Hitting a moving target. Discussion Paper No. 199. Sussex Institute of Development Studies.

Mehta, R.K., Srivastavaa, A., Manik, R.S. and Mudgal, V.D. 1986. Agricultural Systems, 21:659-669.

Nair, P.K.R. 1979. Intensive multiple cropping with coconuts in India-principles, programmes and prospets, Verlag Paul Parey, Berlin, pp:147.

Norman, D.W. 1980. The farming system approach: Relevency for the small farmers. Michigan State University Development paper 8, East Lansing, Michigan, USA.

Ofori, F. and Stern, W.R. 1987. Cereal-legume intercropping systems. Advances in Agronomy, 41:41-90.

Pal, M., Singh, K.A. and Ahlwat, I.P.S. 1985. Cropping system research. I. Concept, needs and directions. Indians Society of Agronomy National Symposium, CSSRI, Karnal, Haryana held on April 3-5, 1985, pp: 1-20.

Palaniappan, S.P., Balasubraminiam, A. and Ramaswami, C. 1976. Intercropping of sorghum with legume for saving nitrogen. Sorghum News letter, 19:86.

Palaniappan, S.P., and Sivaraman, K. 1994. Cropping Systems in the Tropies-Prinaples and Management (Second edition). New Age International (P) Limited, Publishers, Chennai.

Prasad, K, and Srivastava, V.C. 1991. Weed management in pure and mixed crops of pigeonpea (*Cajans cajan)* and soybean (*Glycine max*). Indian Journal of Agriculturl Sciences, 61: 374-378.

Parcer, R.D., Choudhary, S.K. and Gautam, V.1983. Income and employment effects ofmilch animals on different farm size in Karnal district of Haryana. Paper presented in seninar on "Challenge for Rural Development in North-Western Region in India." held in HUA, Hissar, Haryana.

Paroda, R.S. 1998. Natural resource management for sustainable agriculture: a new paradigm. Pleanary lecture delivered on 24 Nov., 1998, pp:(i) – (xxi). *In*: Proceedings of First International Agronomy congress, Indian Society of Agronomy, ICAR, New Delhi, India.

Radke, J.K. and Hagstorm, R.T. 1976, Strip-intercropping for wind protection,: pp. 201-222. *In*. R.I Papendick P.A. Sanchez and G.B. Triplett (eds.). Multiple cropping ASA Spl. Publ. No. 27. Madison, Wisconsin, USA.

Rajkumar, S. 1988. Feasibility of introducing dairy as a complementary enterprise for small farmers in Tanjavur delta. M.Sc. Ag. Thesis, TNAU, Coimbtore, India.

Rangasmy, A., Samkaran, S. and Subharayalu, M.1988. Integrating farming system-a viable approach. Research Bulletion, TNAU, Coimbtore, India.

Rangasmy, A., Annadurai, K., Subbian, P. and Chinnsamy, J 2002. Farming System in the Tropics. Kakyani Publishers, Ludhiana. New Delhi.

Rao, D.P. 1999. Remote sensing-valuable information aid. Survey of Indian Agriculture, The Hindu. National Press, Kasturi Buildings, Chennai, India.

Rice, E.l. 1974. Allelopathy, Academic Press, New York, USA.

Rice, F.L. 1984. Allelopathy (2nd Edition). Academic Press, New York, USA.

Ruthenberg, H.1980. Farming Systems in the Tropics. Third Edition. Clarendon Press, Oxford, U.K.

Sankaran, S. and Rangasamy, A., 1990. Farming System Research, pp:69-80. *In* Agronomic Resarch Towards Sustrainable Agriculture, Indian Society of Agronomy, IARI, New Delhi, India.

Shanmugasundran, V.S. and Balusamy. 1993. Rice-fish-azolla integrated farming system in lowlying wetlands. Farming Systems, 9:105-107.

Shanner, W.W., Philipp, P.F. and chemeht W.R. 1982. Farming systems research and developments. Guidelines for Developing Countries. Westview Press, Boulder, Colorado.

Simmonds, N.W., 1984. The state of the art of farming systems research. World Bank Agricultural Symposium, 1984, Washington.

Simmonds, N.W. 1986. A short review of farming system research in the tropics. Experimental Agriculture,. 22:1-13.

Singh, N., Kairon, M.S. and Malik, B.S. 1986. Better farming systems for rural development. Bulletin published by HAU, Hissar, India.

Singh, R.P., Das, S.K., Rao U.M.B. and Reddy, M.N. 1990. Towards sustainable dryland agriculture practices. Central Research Institute for Dryland Agriculture (CRIDA), Hyderabad, India.

Singlachar, M.A. 1987. Farming system in 21st century: A rambling in agrarian structure and strategies for agricultural development for the 21st century. USA Technical Series No. 27, vol. 2: 13-16.

Spedding, C.R.W. 1988 An introduction to agricultural systems. Elsevier Applied Science. London. -New York (Second edition).

Srivastava, V.C. 1990. Farming system a resource management strategies for resource poor farms in India. Paper presented in SAARC Programme held at Gazipur (Bangladesh), Aug 25, 1990.

Srivastava, G.P. and Srivastava, V.C. 1993. Effect of winter legumes on nitrogen economy of succeeding rice (*Oryza sative)*. Indian Journal of Agricultural Sciences, 63(8): 501-503.

Srivastava, V.C. 1995. Crops and Cropping systems. *In* Acid Soil Management Kalyani Publication, Ludhiana,: pp. 359-382.

Stewart, J.I. 1987. Proceealings INRAN-ICRISAT-TROP SOILS- USAID-USDA workshop on Soil, Crop and Water Management Systems for Rainfed Agriculture in Sudano- Sheliian Zone, Niamey, Niger, 11-16, pp:224-264.

Stewart, J.I. 1988. Development of management strategies for minimizing the impect of seasonal rainfall variations. *In* Bidinger, F.R. and Johnson, C.(eds). Drought Research Priorities for the dryland tropics. ICRISAT, Hyderabad, India.

Swaminathan, M.S. 1988. Sustainable food and nutrition security for the 1990's. Agricultural situation in India: 363-367.

Swaminathan, M.S. 1991. The environment scene-maladies and remedies. Survey of the Environment. The Hindu, pp:4-7.

Trenbath, B.R. 1976. Plant interaction in mixed crop communities, pp: 129-169. *In* R.I. Pependick, P.A. Sanchez and G.B. Triplett. (eds). Multiple Eropping Systems, Macmillan Publishing Company, New York.

Verma, M.K. and Rao, M.S.S. 1969. Selection of intercrop for cotton in India. Experimental Agriculture, 5:233-230.

Verma, R.S. 1995. Ratoon decline in sugarcane. Cooperative Sugar, 26:349-351.

Warld, G.M., Sutherland, T.M. and Sutherland, J.M. 1980. Animal as energy source in third world. Agriculture Science, 208:507-574.

Willey, R.W. 1979. Intercropping-its important and research needs. Part -I. Competition and yield advantages. Part-2. Agronomy and research approaches. Field crop abstracts, 32:1-10 and 73-85.

Young A. 1989. Agroforestry for soil conservation. CAB International. Walling Ford, U.K.

Zandstra, H.G. Price, E.C., Litsinger, J.A. and Morris, R.A. 1981. A methodology for on-farm cropping system research, IRRI, Los Banos, Philipines.

CHAPTER III

AGROFORESTRY

3.1 INTRODUCTION

The whole world, particularly developing countries are now facing serious challenges of the conflict between rapid growth of population and limited available resources. People's increasing demands lead to the irrational exploitation of natural resources. Indiscriminate extension of agriculture by large scale cultivation of land for food production causes destruction of original forests since the carrying capacity of land in many developing countries is overstretched and they do not have sufficient land resources to meet the food requirements of their increasing population. Gradually the irrational land management brings serious water and soil erosion, soil degradation, extension of desertification and environmental pollution.

India too has reached the physical frontiers with regard to cultivated area but its population of human beings and animals is rising in geometrical progression. With the increase in population, more and more of marginal and submarginal land, slopy barren land and other lands not suitable for arable farming are put to cultivation. It is doing more harm than good to cultivators. There is increasing siltation of the dams and waterways, leading to floods and loss of production besides decline in soil fertility. The disturbing feature is that not only the food, fuel and fodder requirements are increasing, but the timber needs are also increasing rapidly. The soil erosion and degradation are increasing and good quality land is shrinking at an alarming rate. According to the 2001 report of forest survey of India, the forest cover in the country is 6,75,538 km^2 constituting 20.55 percent of the total geographical area (329 million ha). Out of this, dense forest constitutes 12.68 per cent and open forest 7.87 per cent. As against the national goal of keeping one-third of land (33 per cent) under forest cover for maintaining ecological balance, it has nearly 12.5 per cent less forest coverage and hardly half of total forest cover reported can be considered really good. The area under forest is further declining rapidly because of the poor men's needs and rich men's greed. Under Tenth Five Year Plan, the Government of India has set a target to increase the country's forest cover up to 25 per cent.

The integration of farming with forestry originates from the realisation that trees play a vital role in safeguarding the long range intrests of agriculture including animal husbandry. The tree based system or agroforestry, an age old practice, has been recognized as a land use system which is capable of yielding wood, food, fuel, forage and medicine and at the same time conserving and rehabilitating the ecosystem vital for sustainable agriculture and environment. The National Agricultural Policy (2000) emphasized the role of agroforestry for efficient nutrient cycling, nitrogen fixation, organic matter addition and for improving drainage and underlined the need for diversification by promoting integrated and holistic development of rainfed areas on watershed basis through involvement of community to augument biomass production through agroforestry and farm forestry.

3.1.1 Historical Background

Historically trees, crop and animals co-existed well in ancient times when forests were in plenty and agriculture flourished. It is an ancient practice of growing trees and agricultural crops in intimate combination with one onother for the welfare of farm family. It has been in use for atleast 1300 years (Brookfield and Padoch, 1994). In about 700 BC, during the process of human evolution from the forest when man learnt the art of domesticating plants and animals and thus changed from a system of hunting and food gathering to food production. The origin of agroforestry practices, i.e. growing trees with food crops and grasses, is believed to have been during Vedic era (ancient period, 1000 BC).

In an account of the origin of life, the bible (Gen. 2: 8-9) described gardens where all kind of trees grew for both beauty and food. Brownrigg (1985) reviewed literature that describes paintings, illustrations and texural description of home gardens in the Near East dating back to 3000 BC. Between 400 to 2000 BC the agroforestry existed in India in its primitive form. The early men cleared forests and planted fruit trees, flowering trees and shrubs and allowed natural grasses to come up. Then they grew minor millets and crops that could add to hermitage's puritan's life. These dwellings were called *ashrams*. Kings and scholars sent their sons and daughters to learn in these abodes. Shifting cultivation in India is prehistoric originated during the Neolithic period (Swaminathan, 1987) and partly in a response to agro-ecological conditions in the region. Horticulture as co-existent with agriculture is found to have been prevalent in India from early historic period as evident from references occuring in different texts of the Vedic literature. During the reign of Ashoka, Samudragupta, Harshwardhana from 300 BC to 800 AD, trees were planted on road-sides as grooves all over the country for shelter, food, fuel, fodder, medicine, etc., for the welfare of people. This continued in the Cholas, Chalukyas, Pandyas, Cheras who gave prominence to trees in various places. During the Moguls and pre-moguls in 1200 to 1700 AD, the fruit trees like mangoes were popular. Trees with good foliage, flowers, fruit, fodder, timber were introduced in parks and cultivated lands as agroforestry. Nizam during his reign protected and advocated formation of natural silvi-pasture. Kanchas with *Hardwickia binata* with *Schima nervosum* which is still found in degraded condition in Telangana, Hyderabad, Andhra Pradesh (Chinnamani, 1969). Tipu Sultan had declared sandal a royal tree protecting it in agricultural land. Duting British regime, *Eucalytus* was introduced for fuel in Nilgiris and other places.

In Europe until middle ages, it was a practice to clear fell trees, slash and burn, cultivate agricultural crops for varying periods, and plant tree species after the sowing of agricultural crops. Such system was practiced in Finland upto the end of nineteenth century and in few areas of Germany as late as 1920s. In Central America, the farmers have long initiated the structure and species diversity of tropical forests by cultivating variety of crops with different growth habits (Gliessman, 1981). Forde (1937) reported that in southern Nigeria, yams, maize, pumpkins and beans were typically grown together under a cover of scattered trees.

In Asia, the Hanunoo of the Philippines practised a complex and somewhat sophisticated type of shifting cultivation. In clearing the forest for agricultural use, they deliberately left certain selected trees which, by the end of the rice-growing season, would "provide a partial canopy of new foliage" to prevent excessive exposure to the sun at a time when moisture is more important. In Hanunoo farming system, trees were either planted or conserved to provide food, medicines, construction wood and cosmetics in addition to their protective services (Conklin, 1957). In 1806, U Pan Hle, a Karen in the Tonze forests of Tharawaddy Division of Burma (Myanmar), established a plantation of teak through the use of what he called the "taungya" method where agroforestry was utilized as a system of land management (Blanford, 1958). From these beginings, the practice became more and more widespread. It was introduced into South Africa as early as 1887 (Hailey, 1957) and was taken from Myanmar to Chittagong area of India (now Bangladesh) in 1890 and to Bengal in 1896. Regular taungya cultivation was used in India for raising plantation of *Shorea robusta* and *Tectona grandis* during 1912. In 1920 it was adopted in Travancore (Kerala), in 1923 in the United Province (now Uttar Pradesh), and in 1925 in the Central Province, now Madhya Pradesh (Raghavan, 1960).

Agroforestry, a combined agricultural and forest production system, has been recognized since time immemorial as an important system of cultivation in India. It is followed traditionally in different parts of country particularly in North-Eastern and Southern regions by the name of shifting cultivation and taungya systems, etc. The multitier tree-crop combinations in the home gardens of humid low lands, particularly in Tamilnadu and Kerala, generate cash income and meet household requirements. Similarly, *Khejri* (*Prosopis cineraria*) combined with agricultural crops in the hot arid region meets the requirement of fodder, small timber and food. Combination of *Alnus nepalensis* and *Amomum subulantum* in the humid sub-temprate regions of Nepal, Bhutan and Sikkim is an example of the commercial, but traditional agroforestry system.

Agroforestry as a science has become of paramount importance due to acute shortage of wood and tree based products. Deforestation of catchment area has resulted in severe erosion and caused floods of unprecedented nature. It has now been well recognized that agroforestry can address some of the major land-use problems of rainfed and irrigated farming systems in India and that a great deal can be accomplished by improving indigenous systems.

3.1.2 Concept of Agroforestry

Agroforestry is "a land use system that involves deliberate retention, introduction of trees or other woody perennials in crop/animal production to benefit from the resultant ecological and economic interactions" (Nair, 1993). The basic ideas emerging from this are given below:

(i) Agroforestry is a land-use system including combination of agriculture, agrostology, forestry, horticulture and animal husbandry, etc.

(ii) Integration of trees / woody perennial for reducing risk and increasing total productivity.

(iii) Ideal agroforestry systems are stable and sustainable.

(iv) Integration of trees with pastures/crops may result in more efficient use of sunlight, moisture and plant nutrients than is generally not possible by mono-cropping of either agriculture or forestry crops. There may be competition between trees and crops for these resources but net effect is positive, if such systems are designed and managed scientifically (Mac Dicken and Vergara, 1990).

Agroforestry combines production of multiple outputs with protection of resource base. It places emphasis on the use of multiple indigenous trees and shrubs and is particularly suitable for low-input conditions, resource poor farmers, and fragile environments. It involves the interplay of socio-cultural values more than in most other land-use systems.

The **main objectives** of agroforestry are (a) maximizing productivity, involving a multiplicity of outputs, (b) ensuring sustainability, which implies the conservation, or even improvement of soils, so as to broaden future land use options and adaptability for the intended users.

In India, because of various factors, such as competition for land due to intense population pressure, rapidly depleting fuel wood resources, time-scale for orchard operations, dispersed distribution of benefits from forestry and seasonal shortage of labour, the growing trees did not find a suitable place in rural economy in the past. However, due to ecological and social cost and land degradation besides highly adverse land-man and land-livestock ratio that has placed enormous pressure on non-renewable land resource, emphasis is now shifting towards agroforestry based land-use systems (Singh and Roy, 1998).

3.1.3 Definitions

Agroforestry is a new name for a set of old practices. As a science it has been introduced only recently. Hence, the agroforestry systems have been the target of scientific enquiry and analysis and thus, have been defined by many in different ways:

- The word 'agroforestry' was actually coined in 1977. It means practice of Agriculture and forestry on the same piece of land. Bene *et al.* (1977) defined "agroforestry as sustainable management system for land that increases overall production, combines agricultural crop, tree crops and forest plants and/or animals simultaneously or sequentially and applies management practices that are compitable with socio-cultural patterns of local population."
- According to King and Chandler (1978) " Agroforestry means a sustainable land management system which increases the overall yield of the land, combines the production of crops (including tree crops) and forest plants and/or on the same unit of land and applies management practices that are compatible with the cultural practices of the local population. Hence, agroforestry is a complex land use technology intimately associated with factors like structural description (trees, crops and/or animal combination), production objectives (sustained yield) and socio-economic objectives (cultural practices of local community) which encompasses myriads of problems and needs of the local community."
- Nair (1979) defines "agroforestry as a land use system that integrates trees, crops and animal in a way that is scientifically sound, ecologically desirable, practically feasible and socially acceptable to the farmers."
- According to Lundgren and Raintree (1982): "Agroforestry is a collective name for land use systems and technologies in which woody perennials (trees, shrubs, palms, bamboos etc.) are deliberately combined on the same land management unit with herbaceous crops and/or animal, either in some form of spatial arrangement or temporal sequence. In agroforestry systems, there are both ecological and economic interactions amongst the different components."
- Another widely used definition given by the International centre for Research in agroforestry (ICRAF), as reported by Nair (1989), states that "agroforestry is a collective name for all land use systems and practices where woody perennials are deliberately grown on the same land management unit as agroforestry crops or animals in some form of spatial arrangement or temporal sequence."
- Melvin G.R. of United Kingdom (quoted by Tewary, 1995) describes "agroforestry a land use system (a) in which woody

perennials and herbaceous crops are grown together in mixture, zonally and / or sequentially, with or without animals, and (b) which provide greater benefits for the land use than agriculture or forestry alone, including one or more of the following: sustained soil fertility, soil conservation, increased yield, diminished risk of crop failure, ease of management, pest and disease control and / or greater fulfilment of the socio-economic needs of the local population."

- "Agroforestry is the growing of trees in combination or in sequence with agricultural crops and / or pastures on small holdings or large estates. It is not synonymous with community forestry but is often an appropriate means of implementing a community forestry project (Laurence Rode of United Kingdom as quoted by Tewary, 1995)
- "Agroforestry is a form of multiple cropping which satisfies basic conditions- (a) there exists at least two plant species that interact biologically, (b) at least one of the plant species is a woody perennial, and (c) at least one of the plant species is managed for forage, annual or perennial crop production" (Pathak *et al.*, 2006).

A few other definitions of agroforestry are as under:

- Agroforestry is a system of combining agricultural and tree crops of various longevity (ranging from annual through biannual and perennial plants) arranged either temporally (crop rotation) or spatially (intercropping) to maximize and sustain agricultural production."
- "Agroforestry in its applied aspects is a set of management techniques combining elements of agronomy, animal production and silviculture with those arising from the interaction of plant and animal species of different sizes, growth patterns and requirements."
- "Agroforestry is a dynamic, ecologically based, natural resource management practice that, through the integration of trees on farms and in agricultural landscape, diversifies and sustains production for increased social, economic and environmental benfits."
- "Agroforestry is a land-use that involves delibrate retention, introduction of mixture trees or other woody perennials in crop/animal production field to benefit from the resultant ecological and economic interactions."

Both ecological and socio-economic interactions between different components in agroforestry system implies (Lundgren and Raintree, 1982) that

(i) agroforestry normally involves two or more species of plants (or plants and animals), at least one of which is a woody perennial;

(ii) an agroforestry system always has two or more outputs;

(iii) the cycle of an agroforestry system is always more than one year; and

(iv) even the simplest agroforestry system is structurally, functionally, and socio-economically more complex than a monocropping system.

3.1.4 Other Related Terms

- **Social forestry** (Growing trees on community lands / Government lands / wastelands): "It is transfermation of unproductive or unutilized waste lands into abundant reservoir of solar energy through production of high yielding, rapidly growing biomass with involvement of society of all level for its own benefits. "In other words," it is the creation of sustained forest resources of the people, by the people and for the people." Social forestry programme aims at establishing multipurpose tree plantation that supply fuel, fodder, fruits and small timbers to meet the basic requirement of the rural communities besides ensuring environmental stability.

- **Farm forestry** (Growing trees on farmers field): "It is a practice of raising small woods on the privately owned farms in addition to normal cultivation to derive indirect benefits, like protection of crops against high winds and control of erosion besides meeting demands of fuel, small timber, grazing, fodder and leaf manure."

  "It is an integration of farming with forestry to promote commercial tree growing by farmers on their own lands." It does not meet forest tree culture at the cost of agriculture."

- **Community forestry**: "It is a form of social forestry which refers to tree plantation activities undertaken by a community (local people) on community land or *panchayt* land and processing or utilizing the tree products locally in a systematic way."

- **Urban forestry**: " It is the planting of trees on public and privately owned land in and adjuscent to urban centres." In urban forestry trees with more aesthetic value are planted.
- **Silviculture**: Trees and shrubs with or without animal.
- **Agri-silviculture / Silvi-agriculture:** Growing trees with annual or perennial agricultural crops.
- **Silvi-pastoral:** Growing trees with grasses for rearing cattle.
- **Silvi-horticulture:** Growing trees with fruit crops.
- **Silvi-horti-pasture:** Growing of trees, fruits and grasses together.
- **Agri-horticulture**: Agricultural crops with fruit trees.
- **Silvi-horti-floticulture**: Growing trees, fruit trees and flowers together.
- Alley cropping: Growing of shrubs or trees alternated with crops with fixed width.
- **Multistory**: Trees, shrubs and agricultural crops are combined to exploit height advantages of the associated species.
- **Boundary plantation:** Trees are raised on the boundaries of agricultural fields.
- **Block plantation:** Planting sole trees as a block in a given area.
- **Energy plantation:** Planting fast growing tree species at closer spacing for higher biomass production.
- **Aquaforestry:** Combined cultivation of trees and fish.
- **Silvi-sericulture:** Practicing sericulture on trees
- **Live fence/biofence plantation:** Planting fast growing thorny trees and bushes at a very close spacing on the boundary of agricultural fields as a guard against grazing by wild animals.
- **Shelter belt plantation:** Planting few rows of trees and bushes on field boundary depending upon wind direction and speed to moderate high speed winds in arid regions.

3.1.5. Scope and Need

Pressure on the land resources on our planet will increase significantly in the next few decades. The amount of land available for food supply and other human needs will continue to decline in per capita terms. The FAO study says that about 80 million ha of culturable wastelands in developing countries would have to be brought under cultivation / plantation in the 15 years. The study also stressed that the trend of stripping forest eover to meet food, fodder and fuel need must be controlled and efforts be made to increase productivity of agricultural land by advocating agroforestry as one of the most practical ways of meeting the rising demand or food, fodder, and fuel and finally the requirements of timber for multifarious purposes. Thus, the increasing pressure on land as well as on forest reinforced as never before the absolute and inseparable nexus between agriculture and forestry which is covered under Agroforestry to ensure optimum use of available land resources.

Agroforestry has great technical and economic potential to overcome in the way of higher and sustained crop production and farm income. This system fulfils both production and service role (Gill and Bisaria, 1997,1999). It can play a very effective role in the utilization of all the natural resources for sustained crop production.

The need for agroforestry is given below (Singh *et al.*, 2000):

- To create an alternate food source in ecologically disadvantaged situations, such as dry/arid lands;
- To utilize the available farm resources efficiently and judiciously;
- To maximize the per unit production for food, folder, fuel, livestock and other forest products, Besides, it should be raised with a view to optimizing the productivity of biological and physical resources, *viz.* land, labour, livestock, soil moisture, soloar radiation, etc.,
- To reduce pressure on protective and productive forests for meeting the local demands of fuelwood, fodder, building material, industrial timber, etc., so that the existing forests can be fully spared for their protective and productive roles;
- To maintain the ecological balance by mitigating green house effect and sequestering atmospheric carbon dioxide;
- To check the soil erosion, conserve soil moisture and increase the soil fertility, and
- Alternative option for reclaimation of degraded lands, such as salt-affected, mine spoils, etc.

3.1.6. Benefits of Agroforestry

The unique feature of agroforestry which distinguishes it from all other farming systems is the delibrate introduction of multipurpose tree species (MPT) into the landscape (Steppler and Lundgren, 1988). The agroforestry helps to achieve simultaneously one or more of the following categories of benefits:

A. Economic

(i) Agroforestry systems are often claimed to be superior to agricultural systems because of their ability to contribute to sustained production of crops, trees and livestock especially on fragile lands which prevent low external input agriculture.

(ii) Sustainability of cropping systems is enhanced through soil and water conservation by plantation of trees which control run-off and erosion. Tree crowns protect the soil from the impact of raindrops and create specific microclimate.

(iii) Food and feed production from crops associated with trees are increased through nitrogen fixation, better access to nutrients brought to surface layer from deeper layer of soil by tree roots, improved availability of nutrients due to high cation-exchange capacity of the soil and its organic matter and mycorrhizal association.

(iv) It provicles food for human beings and produce feed for livestock. Many tree species (e.g., *Leucaena leucocephala*) supply nutritious and often protein rich fodder.

(v) It meets the demand of energy by producing fuel-wood for direct combustion; pyrolytic-conversion products such as charcoal, oil and gas; latex, oils and other combustible saps and resins. Ethnol is produced from fermentation of high carbohydrate fruits.

(vi) Agroforestry systems provide shade trees for people, livestock and shade-loving crops; building materials for construction, wind breaks and selter belts for protection of settlements, crop lands, pasture and roadways; and live fences and fence posts for protection.

(vii) Agroforestry is capable of meeting the demands of raw materials of several agriculture and forest based industries, e.g., raw materials for paper and pulp mills; wood for agricultural implements, sport goods, furniture and sawmills;

tannins, essential oils and medicinal ingradients; and fibre for weaving.

(viii) Agroforestry systems increase employment opportunities. Most of the forestry activities are labour intensive and considerable alternative employment opportunities are generated. Only plantation including nursery operation generate employment of about 200 to 500 man-days per ha (NCA, 1976). Subsequent care and maintenance also provide employment to the tune of 50 to 75 man-days per ha per annum. Harvesting of trees is onother major activity which provides about 10 to 15 man-days per cubic meter of wood harvested.

(ix) Agroforestry is an insurance against total crop failure common to monocropping in rainfed areas. Cultivation at single agricultural crops in the arid areas is highly prone to fruequent risks posed by drought.

(x) Farmers can get direct cash benefit from sale of products. Excepting food, fodder and fuclwood most of other tree products cannot be consumed by the producers. The level of farm income is increased due to improved and sustained productivity of the agroforestry systems.

B. Enviornmental

(i) Depletion of ozone layer, increased build up of carbon-di-oxide (CO_2) in the atmosphere, green-house effect and global warming are some of the serious consequences of environmental degradation caused by excessive deforestation and heavy pressure of increasing population. In general large scale deforestation is blamed for reduction of rainfall (its total quantity, spread of number of raing days and wet months). It is also one of the major causes for increase in CO_2 in the atmosphere (at the rate of 0.3 per sent per year since 1950s), which is potentially serious threat to agriculture and environment.

Trees protect us from different kinds of pollutions. They protect us from dust, dirt, noise pollutions and other physical air pollutants. One hectare of a close forest can filter about 50 tonnes of dust and durt (Dirk, 1966). Some of the chemical air pollutants are also absorbed by the trees protectintg human beings form their adverse effect.

(ii) Trees help to protect environment and improve microclimate not only by reducing soil erosion but also reduce effect and repeated occurance of droughts and floods, modeerate temprature and increase preceptitation and humidity. Trees due to their canopy architecture, shade and density help in reducing temprature extremes of environment and surface soil due to reduction in radiation.

(iii) Agroforestry has potential to conserve and utilize biodiversity. Pathak an Hegde (1996) reported that single species plantations though provide wood at short rotation, have been found to deplete biodiversity compared to agroforestry including home-gardens in tropical humid climate.

(iv) Tree components of agroforestry improve soil-moisture retention in rainfed crop lands and pastures through improved soil structure by constant addition of organic matter from decomposed litter.

(v) Trees with high water requirements improve drainage from waterlogged or saline soils.

(vi) Nutrient release through litter deposition and fine root decay of trees keep the soil more fertile assuring sustainable supply of nutrients. The leguminous trees have a higher capacity of mineralization which is governed by precipitation. The accumulation of soil organic matter under trees is most commonly reported effect of trees on soil which bring important change in physical and chemical properties, e.g., reduction in soil bulk density and increases in soil aeration and soil hydraulic conductivity.

Trees, belonging to family leguminaceae, can fix nitrogen in soil through symbiotic association of *Rhizobium* or *Frankia*.

(vii) Trees reduces surface run-off, nutrient leaching through impeding effect of tree roots. They also reduces evaporation of soil moisture through a combination of mulching (by litter falls) and shading. It has been reported that association of multipurpose tree species (MPTS) with crops could control 78.3 per cent runoff and 91.2 per cent soil losses compared to the bare plot under Jhansi conditions (Singh, 1994). Similarly, Prasad *et al.* (1986) advocated 99.7 per cent reduction in run-off when bamboo was grown with grass as compared to shifting cultivation in high rainfall areas.

C. Social

(i) The social benefits from agroforestry is the result of increase in the yield of crop and tree products and in their sustainability by stabilisation and improvement of communities.

(ii) Agroforestry is instrumental in improvement in nutrion and health of society due to increased quality and diversity of food products and environments.

(iii) It improves living standard of rural people through sustained employment and generation of higher income.

Some of the advantages of agroforestry summarised by Singh, *et al.* (2000) are:

- It yields extended range of products, viz. food, fodder, fuel, fibre, fruits, manures, cosmetics, medicines, oils, resins and gums.
- It helps in utilization of off-season precipitation. Most of the drylands are cropped for a single season but if a perennial tree component is there in the system, it would make use of the same efficiently. A dryland ecosystem without perennial component can never be a sustainable system.
- Sustainable option for reclaimation and management of wastelands such as sand dunes, salty lands, revines, mine spoils, etc.
- Helps in maintaining hydrological balance of ground water. Integration of trees having higher transpiration rate with annual crops helps in lowering water table in regions having rising water trends.
- Agroforestry in dryland helps soil and water conservation. The perennials component of the agroforestry system either on contours or in some other geometry helps in reducing soil and water losses.
- It imparts stability and results in risk reduction of the two or more components of the system, even if one fails, the other would give a harvest of some economic value.
- A viable option for building organic matter in the soil by moderating temprature under tropical conditions.
- Presence of trees in sgroforestry system gives extended management options for the farmers, i.e., the trees can be

harvested either for food or fuel or pole or small timber or for nitrogen as per farmers requirements and market demands.

- Disposal option for poor quality waters, such as domestric and industrial effluents including heavy metal loaded waters.
- Agroforestry system checks the rise in soil temprature, especially during summer months. Thus, it protects the soil microorganism, which would be of great benfit in crop production. Trees also provide as shade for cattle.
- It helps in recycling of nutrients from lower soil horizons. The deep rooted nature of most trees results in tapping of nutrients from deeper soil layers and returning them to plough layer through leaf drop and litter.
- Mitigation of green house effect, depletion of ozone layer and pollution check.
- It helps in off-season family labour utilization. During off-season there would be no field work for family labour in dryland areas. Time can, thus, be effectively used for harvesting, pruning and marketing of the tree components in agroforestry systems.
- Tree component of agroforestry system results in improvement in the micro-climate of the whole system.
- Agroforestry results in improving and stabilizing the income of farmers.

3.1.7 Limitation of agroforestry

Inspite of significant impact of agroforestry on counry's economy, improvement of natural resources and environment, it has not become as popular with the farmers as it should have been. There are certain inheretent limitations with this land use system. Some of these are listed below:

(i) Competition of trees with crops in close proximity for sunlight may reduce crop yield due to shading by trees and effect of changes in spectral quality of light on the growth of crops. Root and shade of trees affect cultivated crops even at the distance of 20 m in case of babool (*A. nilotica*). as reported by Osman *et al.* (1989).

(ii) Fast growing species of trees may place a heavy demand on soil moisture, and unless properly managed, this can lead to adverse effects on crops, especially in drier environments.

(iii) Nutrient depletion by trees may temporarily deprive adjuscent crops of nutrients, although litter fall, root biobass, etc., may compensate for this to some extent in long run.

(iv) Nutrient losses from whole tree harvesting may be excessive especially in plantation forestry. Food crops may also be damaged during tree harvest operations.

(v) Adverse chemical or biological effects may result from certain tree species leading to allelopathy, acidification, accumulation of taxic exudates, provision of alternate hosts of pests and pathogens, etc., and thus, reducing growth and yield and sometimes complete failure of crops.

(vi) A badly managed tree stand, as with any vegetation, can cause accelerated erosion.

(vii) Prolific seeding habit of *L. Leucocephala* resuting in weedy growth depress the yield of arable crops.

(viii) This system harbour birds, pests and diseases. *Leucaena* psyllids (*Heteropsylla cubana)* in near future may become the pest of cultivated crops.

(ix) Agroforestey systems are labour intensive and reduces the scope of farm mechanisation.

(x) Allelopathic effects of trees in crops my occur, e.g. *Eucalyptus spp.*

(xi) Farmers, in general, are reluctant to grow trees on their farm lands owing to long gestation period in getting returns from trees.

(xii) Proven agroforestry technologies are available only for limited situations / locations / regions.

(xiii) Lack of quality seed / planting material sources of promising agroforestry species.

(xiv) Rigid legal laws restricting harvesting, transporting and sale of trees.

(xv) Lack of trained manpower and infrastructure for agroforestry.

In order to avoid adverse effects of tree, the following management strategies may be adopted:

(a) Trees like legume trees with small and light crowns may be selected so that sufficent sunlight may be available to reduce competion for light.

(b) Trees selected for agroforestry should be deep-rooted so that they will absorb moisture and nutrients from the deeper subsoil while crops receive moisture and nutrient from the upper layers.

(c) Trees in agroforestry should be spaced farther apart to reduce their competition effects on crops.

3.2 CLASSIFICATION OF AGROFORESTRY SYSTEMS

3.2.1. Criteria of Classification

According to Nair (1985), classification of agroforestry systems is based on the following sets of criteria:

(i) **Structural basis**: considers composition of the components including spatial admixture of the woody component, vertical stratification, and temporal arrangement of different components.

(ii) **Functional basis**: refers to the major function or role of the system, mainly the woody components (e.g., windbreak, shelterbelt, soil conservation and so on).

(iii) **Ecological basis**: considers environmental condition on the assumption that certain types of systems can be more appropriate and sustainable for certain ecological conditions. For example, there may be a set of agroforestry systems for arid and semi-arid lands, tropical highlands, lowland humid tropics, etc.

(iv) **Socioeconomic basis**: takes into account the level of input management (e.g., low goals subsistence, commercial, intermediate, etc.).

3.2.2. Structural Classification

The structure of agroforestry system can be defined in terms of its components (constituents) and the expected role or the function of each (manifested in terms of output). The nature of the component is not the only important aspect but also their arrengement. Hence, on the basis of the structure, the agroforestry systems can be grouped into (A) Nature of components (B) Arrangement of components, and (C) Stratification of components.

In addition to above, Dwivedi (1992) suggested classification based on three more sets of criteria, i.e. (a) Physiognomy (characters of veqetation, such as (i) xeromorphic, (ii) mesomerphic, (iii) hydromorphic, etc.; (b) Floristic characters, and (c) History.

A. Classification based on nature of components

Based on the nature of components, agroforestry system can be classified into the following major systems and each system can be further grouped into subsystem / practices.

(1) **Agri-silviculture system:** (Crops and trees including shrubs/vines): This system integrates agricultural crops and trees and involves systematic and deliberate use of land for the concurrent production of agricultural crops (food and/or fodder crops) in the available space between trees of either forest (with a productive and/or service role) or woody perennials. Compatibility between different components of this system depends on growth stage due to varying shade effect and root competition.

In the plantation of *Leucaena leucocephala*, field crops such as pigeonpea + blackgram/greengram/direct seeded rice can be grown successfully with least reduction in fodder yield of tree component. The nositive association effect of *Leucaena leucocephala* and *Sesbania aegyptica* has been found in forage crops like hybrid napier, lucerne, oats and several other cereal forages (Singh, 1988). Agri-silviculture involving poplars and berseem in *tarai* belt of Uttar Pradesh under irrigated condition gave a fodder production of 5 tons per ha in 5 to 6 months (Chaturvedi, 1983). Intercropping with sorghum (*Kharif*), taramira (*rabi*), cowpea found to be compatible with *Eucalyptus teretictonis.* Wood yield of *Eucalyptus* was found to increase in association with the intercrops of cassava + groundnut in Kerala (India) in sloping laterite soils. *Dalbergia sissoo* grown with rice + pigeonea was found to be the best under rainfed condition of Orissa (India). In the hilli region, tree species such as *Grevillea robusta*, Albizia lebbeck, *Ailanthes excelsa, Populus sp* and *Eucalyptus globus* are grown in association with wheat, peas. potaroes, cauliflower, mustard, etc.

Subsystems / practices of agri-silviculture. Based on the nature of components, the agri-silviculture system can be groped into the following forms:

(i) **Improved fallow in shifting cultivation:** Improvement in shifting cultivation practices in several areas of north-east India refers to land left without crops for period ranging from one season to several years to recover depleted soil nutrients.

(ii) **The Taungya system:** Forest plantations are established and agricultural crops are allowed to be grown until canopy closes. This practice is popular in several states of India, e.g., U.P., West Bengal, Jharkhand, Chattisgarh M.P., etc. It is also termed as lease system.

(iii) **Tree gardens:** Fruit trees in combination with trees yielding timber, firewood, fodder, etc., locally termed as orchards. The major function of this system is production of food, fodder and wood products for home consumption and sale for cash.

(iv) **Multipurpose tree species and shrub on farm lands:** In this sub system, several forms of plantation are done e.g., single row or multiple rows on fields on one side or sides, scattered trees or according to some systematic patterns on bunds, terraces or plot/ field boundaries. The role of this system is production of various tree products and protective function is fencing, social values and plot demarcation. Examples of multipurpose trees planted in this subsystem of agroforestry are *Acacia albida*, *Cassia siamea, Azadirachta indica, Acacia senegal, Leucaena leucocephala,* etc.

(v) **Hedgerow cropping or alley cropping:** In alley cropping, fast growing N fixing shrubs are planted in hedge rows. Food crops are grown in alleys formed by hedge row of trees (usually legumimous) and shrubs. It is also known as hedge row intercropping system. Based on different objectives/purposes for which the alleys are raised, alley cropping may be grouped into: (a) Forage alley cropping, (b) Green manure cum mulch alley cropping (c) Forage cum mulch alley cropping, and (d) Forage cum green manure alley cropping.

Trees legumes like *Sesbania grandiflora*, *Leucaena leucocephala*, *Prosopis sp* and trees like *Cassia siamea*, *Gliricidia sepium* etc. are used in this system. *Leucaena* performs better than any other perennial herbaceous legume due to its easy management on heavy soils with less than 800 mm of rainfall. Wilson and King (1981) reported that the alley system based on Leucaena-maize was promising with no additional N input on a sandy soil. Leucaena can be grown with sorghum + pigeonpea (1:1) for higher biomass and net return under rainfed condition in entisols with 3 per cent slope. *Brassica compestris var.* yellow sanson grown within the alleys of Eucalyptus hybrids and *Dalbergia sissoo* gave fairly good yield at Jhansi (UP), India.

(vi) **Trees with plantation crop:** Plantation of several multipurpose trees such as *Albizia lebbeck*, *Terminalia myriocarpa* and *Gliricidia spp*, in coffee and tea plantation in South India and Assam are examples of combined production systems of agroforestry. Crops, such as coconut and cocoa are also combined as intercrops with perennial trees as (a) integrated multistorey mixture of plantation crops; (b) mixture of plantation crops in alternate or other regular arrangements; (c) shade trees for plantation crops, and (d) intercropping with other agricultural crops.

(vii) **Homestead plantations or agroforestry:** In north India, the trees of different species, e.g., *Acacia nilotica*, *Azadirachta indica*, *Dalbergia sissoo*, *Moringa oleifera*, mango, peach, bamboos, etc. are grown around houses, cattle-sheds, wells, tanks, etc. to supply small timbers, firewood, fruits and fodder and also combine vegetable gardening.

(viii) **Shelterbelt / windbreak and plantation:** Shelter-belt are belts or blocks consisting of several rows of trees established at right angles to the prevailing wind with the sole purpose of deflecting air current, reducing the velocity of prevailing winds, providing protection against effect of wind erosion and protecting leeward areas from the desicating effects of hot wind.

Wind-breaks are strip of trees and/or shrubs planted to protect fields, homes, canals or other areas from wind and blowing soil or sand. Wind-breaks are planted to redue or prevent soil erosion, to improve the microclimate for growing crops and to shelter people and livestock, to protect crops and pastures from hot and drying winds or cold winds, to reduce evaporation from farmlands, to provide habital for wildlife, and for fencing and boundary demarcation.

Shelterbelts and windbreaks of *Acacia species, Prosopis species, Eucalyptus species, Dalbergia species, Albizia species*, etc. in arid areas and *Casuarina equisetifolia* in coastal areas are popular in India. In general, trees with narrow and vertical growth are ideal for to wind-breaks to minimise their adverse effect for crop production.

(ix) **Agroforestry for soil-conservation:** Trees can be on bunds, terraces, slopes, etc. planted to benefit soil conservation by stablising the structure and to make the productive use of the land. The suitable tree species for soil conservation are *Acacia catechu, Pinus roxburghii, Leucaena leucocephala, Prosopis juliflora,* etc. Trees may be combined with mixed and staple food crops. Eruit trees planted along the edges of terraces are *ber (Zizyphus)*, papaya, citrus, mango, guava, etc.

(x) There are many common agri-silviculture systems followed in differen agroecological regions of India, e.g., agri-silvi-horticulture (trees + fruit trees + crops), energy plantation (trees + crop during initial years), horti-olericulture (fruit trees + vegetables), etc.

(2) **Silvi-pastoral system (trees + pasture and/or animals):** Such systems integrate pasture and/or animals with trees. Here, woody perennials, both timber yielding trees and fodder yielding trees plus grases are introduced deliberately and systematically and managed scientifically. Thus, silvi-pastoral practice including management of fodder grasses in natural forests or in plantations or in grasslands. Under this system, grazing cattle is permitted and stall feeding is also practiced.

Under poor soil, water and nutrient situations where cropping is not possible, such system can serve the twin purpose of forage and firewood production and ecosystem conservation. In India such systems have great potential on wastelands, especially in the arid and semi-arid regions (Singh and Roy, 1993)

Subsystems/practices of silvi-pastoral System: Silvi-pastoral system (trees + pasture and/or animals) is further classified into following subsystems or practices:

(i) **Protein bank or cut-and-carry fodder production:** In this silvi-pastoral sub system, various protein rich multipurpose trees are planted on or around farmlands and rangelands for cut-and-carry fodder production to meet the feed requirements of livestock during the fodder-deficit period in winter. Suitable tree species for arid and semi-arid regions are *Acacia nilotica, Opuntia ficus, Prosopis species, Rhus sinuta*, etc. For humid and sub-humid regions, *Artocarpus species, Bombax malabaricum, Cordia dichtotoma, Dalbergia jambolana, Zizyphus jujuba*, etc are sutable tree species.

(ii) **Living fence of fodder trees and hedges:** In this sub system, verious fodder trees and hedges are planted as live-fences to protect the property from stray animals or other biotic influences. Several multipurpose trees and shrubs, *Sesbania spp., Leucaena leucocephala, Acacia nilotica, Gliricidia sepium, Erythrina abyssinica*, etc, are planted on the boundary of the field.

(iii) **Trees and shrubs in pastures:** In this subsystem, various trees and shrubs are scattered or arranged systematically to supplement forage production. Several tree species, eg. *Prosopis cineraria* and *Acacia spp.* in arid areas; *Derris indica, Emblica officianalis, Tamarindus indica*, etc, in humid and sub-humid regions; *Acacia nilotica, Dalbergia sissoo, Pongamia pinnate*, etc, in plains and oaks, celtis, *etc.* in the hills are widely used.

(3) **Agri-silvi-pastoral system** (trees + crops + pasture/animals): This system consists of following subsystems/practices:

(i) **Woody hedges for mulch, browse, green manure, soil conservation etc:** In this subsystem, various woody hedges, especially fast-growing and coppicing fodder shrubs and trees are planted for mulch, green manure, soil conservation, browse and production of food, fodder and fuelwood. *Sesbania spp., Leucaena leucocephala, Erythrina spp*, etc. are generally used in several forms.

(ii) **Home gardens** (multiple combination of trees, fruit trees, vegetables etc.): This subsystem is common in almost all areas. Many species of trees, bushes, vegetables and other herbaceous plants are grown in dense and apprantely random arrangements, although some rational control over choice plants and their spatial and temporal arrangements may be exercised. Most home gardens also support a variety of animals (cow, buffalo, bullock, goat, pig, sheep, etc) and birds (chicken, duck, etc). Some of the most frequently used species for home gardens are: (i) woody species like *Anacardium occidentale* (cashew), *Psidium guajava* (guava), *Mangifera indica* (mango), *Azadirachta indica* (neem), *Cocos nucifera* (coconut) and (ii) Herbaceous species, like *Abelmoschus esculentus* (Okra), *Allium* spp (onion), *Brassica oleracea* (Cabbage), Cucurbita spp. (pumpkin), *Ipomea batatas* (sweet potato), *Lycopersicon esculentum* (tomato), *Musa* spp. (banana), *Phaseolus* spp. (beans), etc.

(4) **Other systems:** The following subsystems/proctices can be included:

(i) **Multipurpose woodlots:** Location specific multipurpose tree species are groun mixed or separately in areas where acute shortage of fuelwood, fodder and small timber is felt. This is also practised on wasteland both for production and reclaimation etc.

(ii) **Agri-silvi-fishery or Aqua-forestry** (trees + fishes): In agri-silvi-fishery subsystem is being practised in Nagaland where fish ponds are made in paddy fields. In aqua-forestry, various trees and shrubs preferred by fish are planted on the boundary and around fish ponds for fish production and bund stabilisation. Trees leaves are used to feed fishes.

(iii) **Apiculture with trees:** In this subsystem, various trees species frequently visited by honeybees are planted for the main purpose of honey production.

(iv) **Agriculture with sericulture:** This is practised in tribal belts of Jharkhand, Orissa, Chattisgarh and Madhya Pradesh.

(v) **Shifting cultivation:** Common in North-Eastren States of India.

B. Classification based on arrangement of components

Nair (1989) classified such plant arranggements as multispecies combinations which can involve the diamensions of space and time.

(1) **Spatial arrangement:** Spatial arrangement of plants in an agroforestry can result in (a) mixed dense stands as in home

gardens (b) sparse mixed stands as scattered trees in pastures and agricultural fields and (c) trees in strips or trees in rows on field boundaries or row plantation grown on agriculture fields and allowing intercropping. The species (or species mixture) may be laid out in zones or strips of varying width. There may be several forms of such zones, varying from microzonal arrangements (such as alternating rows) to microzonal ones. A common example of zonal pattern is hedge-row intercropping (alley cropping). An extreme form of the zonal planting is boundary planting of trees on edges of plots or fields for a variety of purposes (fruit, fodder, fuel-wood, fencing, soil conservation or wind-break). The extreme forms of microzonal arrangements can lead to sole-cropping systems, but interactive association of different components can be used as criteria to decide limits between zonal agroforestry and sole crop (component) plots.

(2) Temporal arrangement: On the basis of arrangement of components in respect of time, the agroforestry systems are classified into:

(i) **Concident:** When different crops occupy the land together as coffee under shade trees, and pasture under trees.

(ii) **Concomitant:** When components stay together for some part of life. For example, agricultural crops are allowed to be grown only for a few years in taungya plantation.

(iii) **Intermittent:** (space dominated): When annual crops are grown with perennial crops, e.g. paddy with coconut.

(iv) **Interpolated:** (space and time dominated): When different eomponents occupy space during different times as in home gardens.

(v) **Separate:** When components occupy space during separte times as improved fallow in shifting cultivation. An extreme example is the conventional shifting cultivation cycle; involving 2-4 years of cropping and more than 15 years of fallow cycle. Similarly, some silvi-pasrural systems may involve grass alleys in rotations with some species of grasses remaining on the land for several years.

C. Classification based on stratification of components

On the basis of vertical stratifications agroforestry systems are classified (Dwivedi, 1992) into:

(1) **Single layered:** The crops are usually is one layer or storey, as in hedge row intercropping, multipurpose tree cropping, tree gardens, etc.

(2) **Double layered:** The crops are usually in two layers as in agri-silviculture, wheat cultivation under poplar, etc. (Common in Haryana, Uttar Pradesh and other States of India), plantation crops under shade trees, silvipasture, etc.

(3) **Multi-layered:** The crops form several layers in some systems in agrisilviculture, silvipasture and agrisilvipasture. For example, *Coffee arabica*, *Piper nigrum* under *Terminalia myriocarpa* or *Albizia spp*., home gardens where large trees, small trees, shrubs, fruits, vegetables, etc, are grown together.

3.2.3 Functional Classification of Agroforestry Systems

Agroforestry systems perform two important functions, viz., (a) *Production* (meeting basic needs for fuel, timber, fodder, fruit, fibre and several other products) and (b) *Protection* (Protecting and maintaining the production system by moderating microclimate, improving soil, conserving soil and mointure and mainting biological diversity).

On the basis of functional classification, agroforestry systems may be catagorised as:

(1) **Productive agroforestry system** includes shifting cultivation, agri-silviculture, home gardens, woodlots etc.

(2) **Protective agroforestry system** includes windbreak and shelterbelts, alley cropping, tree plantation grown for soil conservation, wasleland reclaimation, sand dune stabilization and shade (for crop, animal and man)

(3) **Multipurpose agroforestry system** covers both functions of production and protection.

3.2.4 Socioeconomic Classification

On the basis of socioeconomic criteria as scale of production, level of technology input and management, Lundgren and Raintree (1982) grouped agroforestry systems into three categories:

(1) **Commercial agroforestry system:** In this case the scale of production of output (usually a single commodity) is the major consideration of the system and the scale of operations are generally medium to large, the land ownership is either government, corporate or private, the engagement of labour is either on daily wages or on contract. In this system, the commodities like rubber, oil palms and coconut with intercropping of food crops or pasture/animals; commercial production of shade tolerating plantation crops, such as coffee, tea and cacao; rotational timber or food crops system in which a short phase of food-crop production is used as silvi-cultural method to establishment of timber species (various forms of taungya); commercial grazing and ranching under large scale timber and pulp plantations, etc.

(2) **Intermediate agroforestry system:** The system is intermediate between commercial and subsistence scale of production and management. This system is undertaken mainly on medium to small sized farms wherein the food/cash crops are grown where food crops satisfy food needs of family and the cash crops cater for the cash needs. Usually farmers who either own the land or have long-term tenancy right to land, reside and work themselves on the land, supplemented by paid temporary labour. In India, for example, intercultivation of crops in orchards, short rotation plantation of Poplar, Eucalyptus, Casurina for meeting row materials requirement of paper pulp, matchwood, energy needs, etc.

(3) **Subsistance agroforestry system:** Subsistence agroforestry system are those practised for satisfying the basic needs and are mostly managed by the owner and his family without any outside labour and with low investments. The examples are shifting cultivation, scattered trees on farm lands, home gardens, etc.

3.2.5 Ecological Basis of Classification

Most of the agroforestry practices/systems have been developed based on specific ecological situations and different geographical regions. This specifically denotes to climatic conditions i.e. (i) Tropical, (ii) Sub-tropical (iii) Dry tropical (iv) Temparate (v) Alpine and (vi) Sub-alpine zones. Each zone may be further sub-divided on the basis of wet, moist and dry conditions, i.e. (a) humid/subhumid lowlands (b) semi-arid/arid lands and (c) highlands.

(a) **Humid/sub-humid region** is characterised by hot humid climate for all and most of the year and an evergreen and semi-evergreen vegetation. Because of climatic condition that favour rapid growth of a large number of plant species, various types of agroforestry system can be found in areas with a high human population, e.g. different forms of home gardens, plantation crops and multilayer tree gardens. In areas of low population density, trees on rangelands and pastures, silvipastoral systems, improved fallow in shifting cultivation and multipurpose tree wood lots, etc. are the major agroforestry systems.

(b) **Semi-arid and arid lands** is characterised by rainfalls confined to 9-12 days in July-September, 2 to 4½ wet months, various pressure deficit ranging from 9 mb in January to 30 mb in April-May, solar radiation incidence (400-500 cal / cm^2/ day), high wind velocity (20 km / hour), high potential evapotranspiration (6 mm / day) and high mean aridity index (70-74.8%). The main agroforestry system in this zone are various forms to silvipastoral systems, wind-breaks and shelter-belts and multilayer, multipurpose trees on croplands

(c) **Tropical highland** is characterised by uneven topography, varying levels of rainfall, degraded and shallow soils at high altitude to deep rich soils in valleys and great climatic variations. The Himalyan region in an example of this type of area. The area has rich biological diversity and significant agroforestry potential particularly in humid and sub-humid tropics. The main agroforestry systems are production systems involving plantation crops, such as coffee and tea, use of woody perennials in soil-conservations, and silvipastoral combinations.

3.3 AGROFORESTRY SYSTEM

An agroforestry system is a specific local example of practice, characterized by environment, plant species and their arrangement, and socio-economic functioning. An agroforestry practice denotes a distinctive arrangement of the components in space and time. Agroforestry systems are decision making units designed to meet farmers objectives by manipulating biological processes and that produce crops through interactions among their soil and wrody components with either herbaceous or livestock components, or both. Such systems are considered to be more efficient in utilization of solar energy and available nutrients and water. The agroforestry include all activities of tree planting and forestry where farmers are involved, derive benefits and/ or depend on them for daily domestic needs in prevailing land uses in the ambit of agroforestry helping agricultural economy and livelihood need.

Integration of trees in agricultural system is possible and suitable to farming systems in India to take large scale plantation to restore forest coverage and achieve the objective of sustainability. The sustainability objective must be of conservation oriented agroforestry practices designed to increase the long-term productivity of the production system.

In agroforestry the words like "systems", "sub-system" and practice are often used synonymously. In land- use term, a ***system*** refers to a type of land-use specific to an area and described according to its biotechnical composition, arrangement and level of techanical management of socio-economic features. Examples of these systems are subsistence agricultural systems, plantation crop system, livestock production system, commercial timber production of specific commodity. ***Sub-system*** refers to a part of system with a more restricted role, content and complexity than the system itself. For example, an agrisilviculture system can have several types of sub-system according to the type and arrangement of its constituent components. ***Practice*** in agroforestry denotes a specific land management operation of an agroforeatry nature on a farm or other management unit. Examples of agroforestry practices are alley cropping (hedge-row intercropping), boundary planting of trees, and shrubs as shelter-belts and windbreaks, use of woody perennials in soil conservation, tree gardens, woodlots on agricultural land, etc.

In India, most commonly found agroforestry systems are: (a) agrisilvicultural (b) silvipastoral (c) agrisilvopastoral and (d) other combination of various components. The major agroforestry practices and their charactistics as reported by Nair (1993) are summarized in Table 19. The essential characteristics of different regions in India and the major agroforestry emphasis in each are summarized in Table 20 (Pathak *et al.*, 2006).

TABLE 19: Major agroforestry practices and their main characteristics

	Agroforestry paractics	Brief description (of arrangement of components)	Mojor groups of components (w=woody, h=herbacious f=fodder for grazing, a=animal)	Agroecological adaptability
a. Agrisilvicultural systems (crops- including shrub/vine/ tree crops-and trees)				
(1)	Improved fallow	Woody species planted and left to grow during the 'fallow phase'	w: fast- growing preferably leguminous h: common agricultural crops	In shifting cultivation areas
(2)	Taungya	Combined stand of woody and agricultural apecies during early stages of establishment of plantations.	w: usually plantation forestry spp. h: common agricultural crops	All ecological regions (where taungya is in practices): several improvements possible
(3)	Alley cropping (hedge-row intercropping)	Woody species in hedges: agricultural species in alleys in between hedges; microzonal or strip arrangement	w: fast-growing leguminous, that coppice vigorously. h: common agricultural crops	Subhumid to humid area with high human population pressure and fragile (productive but easily degradable) soils
(4)	Multilayer tree gardens	Multispecies, multilayer dense plant associations with no organized planting arrangements	w: different woody components of varying form and growth habits h: ususlly absent; shade tolerant ones sometimes present	Areas with fertile soils, good availability of labour, and high human population pressure
(5)	Multipurpose tree on crop lands	Trees scattered haphazardly or according to some systematic patterns on bunds, terraces or plot/ field boundaries	w: multipurpose trees and other fruit trees h: common agricultural crops	In all ecogical regions esp. in subsistence farming; also commonly integrated with animal
(6)	Plantation crop combinations	(i) Integrated multistory (mixed, dense) mixtures of plantation crops (ii) Mixtures of plantation crops in alternate or other regular arrangement (iii) Shade trees for plantation crops;	w: plantation crops like coffee, cacao, coconut, etc. and fruit trees, esp. in (i) fuelwood/ fodder spp. esp. in (iii) h: usually present in (iv) and to some	In humid lowlands or tropical humid/ subhumid highlands (depending on the plantation crops concerned); usually in smallholder subsistence system

	Agroforestry paractics	Brief description (of arrangement of components)	Mojor groups of components (w=woody, h=herbacious f=fodder for grazing, a=animal)	Agroecological adaptability
		shade trees scattered (iv) Intercropping with agricultural crops	extent in (i); shade-tolerant species	
(7)	Home gardens	Intimate, multistory combination of various trees and crops around homesteads	w: fruit trees predominate; also other woody species, vines, etc. h: shade tolerant agricultural species	In all ecological regions, esp. in areas of high population density
(8)	Trees in soil conservation and reclamation	Trees on bunds, terraces, raisers, etc. with or without grass strips; trees for soil reclamation	w: multipurpose and / or fruit trees h: common agricultural species	In sloping areas, esp. in highlands, reclaimation of degraded, acid, alkali soils and sand dune stablization.
(9)	Shelterbelts and windbreaks, live hedges	Trees around farmland/ plots	w: combination of tall-growing spreading trees h: agricultural crops of the locality	In wind-prone area
(10)	Fuelwood production	Interplanting firewood species on or around agril lands	W: firewood species h: agril crops	In all ecological regions
b. Silvopastoral systems (trees + pastutre and/or animals)				
(11)	Trees on rangeland or pastures	Trees scattered irregularly or arranged according to some systematic pattern	w: multipurpose; of fodder value f: present a: present	Extensive grazing areas
(12)	Protein banks	Production of protein rich tree fodder on farm/rangelands for cut-and-carry fodder production	w: leguminous fodder trees. h: present	Usually in areas with high person: land ratio
(13)	Plantation crops with pastures and animals	Cattle under coconut in south-east Asia and the south pacific	w: plantation crops. f: present. a: present	In areas with less pressure on plantation crop lands

	Agroforestry paractics	Brief description (of arrangement of components)	Mojor groups of components (w=woody, h=herbacious f=fodder for grazing, a=animal)	Agroecological adaptability
c. Agroslvipastoral systems (trees+crops+ pasture/animals)				
(14)	Homegardens involving animals	Intimate multistory combination of various trees and crops and animals around homesteads	w: fruit trees predominate; also other woody species. a: present	In all ecological regions with high density of human population
(15)	Multipurpose woody hedgerows	Woody hedges for browse, mulch, green manure, soil conservation, etc.	w: fast-growing and coppicing fodder shrubs and trees h: (similar to alley cropping and soil conservation)	Humid to subhumid areas with hilly and sloping terrain
d.	**Others**			
(16)	Apiculture with trees	Trees for honey production	w: honey producing (other components may be present)	Depending on the feasibility of apiculture
(17)	Aquaforestry	Trees for fish ponds, tree leaves being used as 'forage' for fish	w: trees and shrubs preferred by fish (other components may be present)	Lowlands
(18)	Multipurpose woodlots	For various purposes (wood, fodder, soil, protection, etc.)	w: multipurpose species: special location-specific species (other components may be present)	Various

Source: Nair, 1993

TABLE 20: Main characteristics and agroforestry emphasis in the major agroecological regions of India

Chamatetistics	Himalyan region	Indo- Gangetic alluvial plains	Arid and semi-arid region	Humid and sub- humid region	Tropical coastal and Island region
Climate	Varies from temperate cold alpine to sub- tropical; rainfall 8 to 350 cm in West and 200-400 cm in east	Very hot summer and cold winter; rainfall 30 to 200 cm; ground frost common in winter	Dry-humid to very hot-sub-humid; rainfall in arid 40-65 cm, in semi-arid 70-125 cm	Humid and sub-humid; hot; rainfall 100 to 40 m, monsoonic	Tropical humid, lowlands; rainfall 60-310cm
Geographical spread	Extreme northeastern to northwestern regions covering parts of Sikkim, Arunachal Pradesh, Assam, Nagaland, Manipur, Maghalaya, Mizoram, Uttar Pradesh, Uttranchal, Himachal Pradesh, Jammu and Kashmir; elevation 1,000-3,000 m	Below foot hills of Himalaya covering about 47 million ha of Punjab, Haryana, Delhi, Jharkhand Uttar Pradesh, Bihar, parts of West Bengal; elevation 150 to 600 m	Spreads over Rajasthan, Gujarat, Punjab, Haryana; parts of Uttar Pradesh, Madhya Pradesh, Andhra Pradesh, Maharashtra, Karnataka and Tamil Nadu	Spreads over parts of Assam, Meghalaya, Mizoram, Tripura; West Bengal, Orissa, Madhya Pradesh, Chhattisgarh, and the Southern States; elevation 150 to 1,500 m	Coastal regions of West Bengal, Orissa, Andhra Pradesh, Tamil Nadu, Kerala, Karnataka, Maharashtra and the Islands
Soils and vegetation	Light to heavy textured, immature, mostly Inceptisols. Alpine scrub sub-alpine coniferous, tropical evergreen, semi-evergreen forests, alpine grasslands	Mostly alluvial with patches of saline and alkali soils. Plant woodlots of Eucalyptus, Populus, Acacia, etc., xerophytic trees and shrubs occur naturally in degraded soils	Sandy to clay loam; xerophytic thorny scrub forests and woodlands and psamophytic scrub vegetation, grazing lands	Alluvial to clay loam; Oxisols, Alfiols, Evergreen, semi-evergreen and deciduous forests	Entisols, Oxisols, Inceptisols, Alfisols. Evergrren, semi-evergreen, moist deciduous, littoral and mangrove forests.
Main land-use systems	Grazing lands, forestry, horticultural and forestry species with agricultural crops, shifting cultivation in NE states	Agricultural crops (Wheat, pulses, sugarcane, oilseeds), commercial wood lots, MPTS with crops and	Trees and shrubs in crop fields and along boundaries permanent grazing lands having trees and shrubs with	Cash crops (tea and large cardamom), agricultural crops, shifting cultivation, forests,	Plantation crops and multistoreyed cropping, field crops, fish culture, forestry

Chamatetistics	Himalyan region	Indo- Gangetic alluvial plains	Arid and semi-arid region	Humid and sub- humid region	Tropical coastal and Island region
		along boundaries, degraded grazing lands	stunt growth		
Main land-use and ecological problems	Excessive deforestation, degraded lands, over-grazing, land sliding and shortage of fuel and fodder	Soil salinity, water logging, decline in soil fertility due to over-cultivation, fuel and fodder shortage	Drought, overgrazing, salinity, extension of sand dunes, low water table, soil erosion, degraded lands	Deforestation, soil erosion, soil acidity and consequent problems, decline in soil fertility, shortage of fuel and fodder, shortening of fallows	Deforestation and degradation of environment, coastal erosion, acid-soils and related soil problems soil erosion along slopes, shortage of fodder
Maior agroforestry emphasis	Planting woodlots on hills, contour farming blended with MPTs management of pastures, raising orchards	Soil reclamation using MPTs, NFTs, suitable grasses; growing NFTs/ shrubs with crops along field boundary, raising woodlots on degraded lands	Raising shelter belts, wind breaks; growing NFTs in field, raising woodlots (NFTs) on degraded lands	Intercropping with nitrogen fixing trees, alley cropping, improved fallows	Afforestation of coastal regions with littoral and mangrove species, aquaculture with mangroves, multistoreyed cropping system; alley cropping

Source: Pathak *et al., 2006*

3.3.1 Traditional Agroforestory Systems in India

The agroforestry systems practised in India have been developed through ages. These are widespread in all ecological and geographical regions of India ranging from simple forms of shifting cultivation to complex home gardens (Table 21) The systems vary in their productive and protective attributes, and their socio-economic dimensions.

3.3.1.1 Forest-based agroforestry systems or Swidden cultivation

(i) **Shifting cultivation:** It is a pattern of land use and oldest system of cultivation of crop under which natural vegetation is cleared by slash and burn method using axe and fire, cropped with common arable crops for a few years and then left unattened allowing the land to rest longer than the period of cultivation when natural vegetation regenerates within the general patterns of alternating follow and cropping cycles, the nature of shifting cultivation varies from place to place. Due to the increasing population pressure, the fallow period is drasticallty reduced and system has degenerated causing serious soil erosion depleting soil fertility and resulting low productivity. The system is practiced extensively in the north –eastern hill region and to some extent Abndra Pradesh, Madhya Pradesh, Orissa and Karnataka states. It is mostly practiced in sub-humid and tropical uplands except in Andra Pradesh and Karnataka States where the rainfall is relatively lower. It is called 'Jhum' in the north-eastern hill region 'Podu' in Andhra Pradesh, 'Dahiya' or 'podu' in Orissa, Chattisgarha and Madhya Pradesh,'*Khallu*' or '*Kurwa*' in Jharkhand and Bihar.

The crops, such as potato, rice, maize and ginger are grown in monoculture or mixed culture along with *Pinus kesiya.* The tree species which may be considered sultable for afforestation of fallow areas or to intercrop with arable crops must be fast growing, preferably nitrogen-fixing and must efficiently recycle available nutrients within the system to restore fertility. These may include *Acacia, Albizia, Casuarina, Erythrina, Gliricidia, Leucaena, Prosopis,* etc.

(ii) **Taungya cutivation:** The taungya (taung = hill, ya = cultivation) is a burmes word coined in Muynmar (Burma). It was introduced by Sir Dietrich Brandis to Chittagong and Bengal areas in colonial India in 1890. Later it had spread thoughout Asia, Africa and Latin America. It is practiced in the state of Ketala, West Bengal, Uttar Pradesh, Madhya Pradesh and Maharashtra, north -east hill region and to a limited extent in Tamilnadu, Andhra Predesh, Orissa and Karnataka for regenerating and/or establishing forest plantations. Generally, it is practiced in area with an assured annual rainfall of 1200-1500 mm.

TABLE 21: Major types of agroforestry systems and the nature of their benefits in different regions of India

Major types of existing AF systems	Major types of benefits/social attributes of systems
Himalayan Regions	
Shifting cultivation	Major requirements of food for the family, leads to soil deterioration
Taungya	Food for family, soil conservation, timber
Fruit trees in combination with agricultural crops and fodder trees	Production of fruits, food crops, fodder and fuel wood
Fodder trees with pastures	Fodder, fuel, cattle rearing
Seasonal grazing in forest	Fodder, cattle rearing
Trees and grasses for soil conservation	Soil conservation, fuel, timber, fodder
Fruits tree in combination with pastures and fodder trees.	Fruit, fodder, fuel
Cash crops in forests / shades of trees	Cash, spices, fuel, fodder
Gangaetic Plains	
Trees for rehabilitation of degraded lands (saline/ sodic/erode)	Reclamation of dgraded lands, fuel, fodder
Fodder trees in degraded grazing land	Fodder, fuelwood, soil conservation
Commercial trees on slope land with commercial or forage grasses.	Cash, soil conservation, forge, fuel
Trees on boundaries of agricultural fields	Food crops, fuel timber, cash protection
Fodders banks	Fodder, soil conservation, fuel
Block plantation	Cash, timber, fuel
Industrial plantation with crops	Cash, food, crops
Fruits Orchards and multi purpose trees	Cash, fruit, fuel, fodder
Fruit or shade-trees on agricultural farms	Food crops, shade, fruits
Trees along canals, root sides	Control on water logging and seepage avenue fuel, forage
Trees on community grazing land	Shade, fuel, forage, cattle rearing
Arid and Semi-arid Regions	
MPTs on agricultural fields	Fodder, fuelwood, cash, crops, minor products
Trees for soil reclamation/sand dune stabilization	Reclamation of soil/stabilization of sand dunes, fuel, fodder, minor products

Major types of existing AF systems	Major types of benefits/social attributes of systems
Wind breaks	Soil conservation, fodder, fuel
Fruit trees with MPTs	Cash, fruits, forage, fuel, minor products
Trees on pasture/grazing lands	Cash, fruits, forage, fuel, minor products Forage, fuel, shade, cattle rearing
Humid and Sub-humid Regions	
Home gardens	Production of multiple outputs
Plantation-crop combination	Cash, multiple products
MPTs on agricultural fields	Minor products, crops
Trees and grasses for soil conservation	Soil conservation, fodder, fuel
Coastal and Island Regions	
Plantation crop combination	Cash, multiple products
Multistoreyed cropping	Cash, multiple products
Home gardens/steads	Cash, multiple products
MPTs with aquaculture	Fish, fuel, fodder, minor products
Trees on pasture lands	Fodder, soil conservation
Grasses under plantation crops	Forage, cattle rearing, multiple products
Shelter belts and with breaks ♦	Share/beach stabilization
Boundary trees on sgricultural crops	Fodder, fuel, minor products
Mangrove plantation	Shore protection, fuel, fodder, coastline protection
Mangrove with fish, prawn, pearl culture, etc.	Shore protection, fish/prawn, fuel, fodder, etc.

Source: Pathal *et. al.*, 2006

Taungya cultivation is scientifically managed shifting cultivation in which the labour is permitted to raise crops in an area but only side by side with the forest species planted. The practice consists of land preparation, tree planting, growing agricultural crops for 1-3 years until shade becomes to dense and then moving on to repeat cycle in a different area. The labour is responsible for the upkeep of a plantation. The owner of the land, and hence the trees, is the state Forest Department. Establishment of teak (*Tectona grandis*) plantation is the most rumemerative taungya practice for forest departments.

Taungya cultivation are of three types:

(a) **Under Departmental Taungya**, agricultural crops and plantation are raised by the forest on daily wayges. The main aim of raising crops along with plantation is to keep down weed growth.

(b) **In Leased Taungya,** plantation land is given on lease to the person who offers the highest money for raising crops for a specified number of years and ensure care of tree plantation.

(c) **In Village Taungya,** crops are raided by the people who have settled down in a village inside the forest for this purpose. Usually each family has about 0.8 to 1.7 ha of land to raise trees and cultivate crops for 3 to 4 years.

Although Taungya cultivation helps considerably in the initial establishment and growth of forest species, mininise the cost of maintenance of the trees and in some cases earns revenue to the forest department on the one hand and provides employment to the rural poor on the other. However, the difficulty faced by the forest officers in getting the land vacated is said to be the main deterrent factor.

3.3.1.2 Farm- based agrorestry practices

(i) **Home–gardens/homesteads:** A home garden is an operational farm unit involving deliberate management of multipurpose trees and shrubs in close association with annual and perennial agricultural crops and, invariably livestock, poultry and /or fish production, managed by family labour, for the purpose of meeting the routine basic needs of mainly the farmer. It is one of the oldest agroforestry practices particularly in Kerala, Tamilnadu, Karnataka, North-eastern states, parts of West Bengal and Andamans and Nicobar Islands.

In home-gardens, there is wide variatrion in the intensity of trees, species and crops based on size of holdings, need of people residing in homesteads and micro-climate. Guava, mango, citrus, jackfruit, coconut, arecnut, papaya, banana, moringa, sesbania, custardapple and many multipurpose trees found grown in home-gardens. Magroves form an essential part of homesteasds of backwater areas in the lowland and costal tracks which form an ideal habital for fish farming. Fish and prawns are also cultivated in fárm ponds, rice fields and some home-gardens also maintain bee hives. Most home-gardens also support a variety of animals (cow, buffalo, bullock, goat, sheep, pigs and birds (chicken, duck). Fodder and legumes are widely grown to meet the daily fodder requirement of cattle. The waste materials from crop and homes are used as fodder/feed for animals / birds and barn wastes are used as manure for crops.

Many multipurpose trees having productive and protective functions are integrated into the system in different spatial and temporal arrangements Teak (*Teaktona grandis*), a commercially valuable timber species is common in the home-gardens. Most of the trees closely interact with agricultural crops when grown. These trees include jack-trees *(Artocarpus spp.)*, *Casuarina equisetifolia*, *Mangifera indica*, *Leucaena leucocephala*, *Bambusa arundinacea*, *Erythrina variegata* and *Gliricidia sepium* (both good support to black pepper). Home-gardens are highly productive, extremely sustainable (through self fertilization effect of abundant litter) and very practicable.

(ii) **Arid regions:** The arid regions of India cover 3,17,909 km^2 and spread over seven states, mostly in north- western part of the country. Inspite of hostile environmental conditions, people in these parts of the country have been growing arable crops along with trees. Through integrated with agriculture, agroforestry is a well defined part of land use systems of western Rajasthan. *Khejri (Prosopis cineraria)* and Bordi *(Ziziphus nummularia)* are the two most important multipurpose woody components of traditional agroforestry systems in the parts of Rajasthan, Gujarat, Haryana and Punjab. In order to reduce negative effect of shading on grain production of moong-bean and cluster-bean, it is a common practice in traditional land use system to frequently harvest folige of *Khejri* amd *Bordi* for getting fodder as well as higher grain yield.

Horticultural fruit trees combined with silvicultural trees to form silvihorticultural system are *Zizyphus mauritiana*, *Punica granatum*, *Phoenix dactylifera*, *Feronia limonia*, *Tamarindus indica, etc.* Agricultural crops which combine with trees for arid areas are pearl-millet, cowpea, sorghum, cluster-bean, black-gram, green-gram, moth, chickpea etc.

In central India, *Dalbergia sissoo*, *Acacia nilotica*, *A. leucopholea*, *Albizia lebbeek*, *Azadirachta indica*, *Pongamia pinnata*, etc. form importan constituent of silvi-pastoral system. On black cotton soil region tall grasses supporting this system include *Hetropogan contortus*, *Dichanthium annulatum*, etc. *D. annulatum* has also maximum soil conservation value.

Tree species recommended for sandy plains are *Acacia tortilis*, *Prosopis juliflora*, *Prosopis cineraria*, *Albizia lebbek*, *Azadirachta indica*, *Acacia senegal*, *Cassia siamea*,

Eucalyptus spp, *Leucaena leucocephala*, etc. Grass species recommended for sandy plains are *Cenchrus ciliaris*, and *Cenchrus setigerus* Suitable tree species for sandstone rocky sites are *Prosopis juliflora*, *Albizia lebbek*, *Acacia senegal*, *Cassia auriculata*. For saline areas, suitable tree species recommended are *Salvadora aleoidea*, *Tamarix auriculata*, *Prosopis juliflora*, etc. Suitable grass species for saline arid regions are *Cenchrus ciliaris*, *Cenchrus setigerus*, *Sacharum munga*, etc.

(iii) **Semi-arid tropics:** Semi-arid tropics encompasses nearly 58 per cent cultivated area in India. The soil type vary form very shallow to deep red and black soils with highly erratic rainfall. The most extensively grown tree species are *Acacia* and Eucalyptus. Other tree species reported are *Borassus flabellifer*, *Tamarindus indica* and *Ceiba pentandra* in Tamil Nadu. Crop production in semi-arid tropics is highly dependent on monsoon and sometimes uneconomical. The deficit of fodder and fuel is alarming. Agroforestry involving perennial component with diversified end products meet several needs but still trees are seldom considered as companion crop. Farmers do raise trees on field boundries but lack adequate information on suitable models in relation to soil type and rainfall.

Food, fodder, fuel and timber are the primary requirements of the farmers, which is set out in Table 22.

Table 22: Standard requirements of food, fodder and fuel

I Food	(i) Grain	400 g / day / adult
		200 g / day / child
	(ii) Pulses	75 g / day / adult
		38 g / day / child
	(iii) Oils	30 g / day / Person
II Fodder		15 kg dry matter / day / animal
		10 kg cereal + 0.5 Kg legume / animal
III Fuel		1.0 t / person / year (Plains)
		1.5 t / person / year (hills)

Tree species suitable for semi-arid tropics having different annual rainfall is presented in Table 23.

TABLE 23: Tree species suitable for semi-arid tropics (SAT) having different annual rainfall

< 500 mm	500-750 mm	>750 mm
Acacia nilotica	*Acacia nilotica*	*Acacia nilotica*
Acacia aneura	*Acacia feruginea*	*Albizia lebbek*
Acacia tortilis	*Albizia lebbek*	*Azadirachta indica*
Acacia albida	*Albizia anara*	*Casuarina equisetifolia*
Prosopis cineraria	*Azadirachta indica*	*Dalbergia sissoo*
Prosopis juliflora	*Casuarina equisetifolia*	*Dalbergia latifolia*
	Dalbergia sissoo	*Ganelina arborea*
	D. latifolia	*Grewia optiva*
	Leucaena leucocephala	*Greville robusta*
	Tamrindus indica	*Hardwickia binata*
		Sesbania sp.

Source: Osman *et el.*(1989)

The purpose of raising trees vary with requirements. *Acacia nilotica* is raised to meet fuel and fodder requirements of sheep and goat, while *neem* (*Azadirachta indica*) is grown mainly for timber purpose. *A. lebbek*, *Gliricidia* and *Pongamia* trees are raised mainly on field bunds for the purpose of green leaf manuring. Palmyra palm (*Borassus flabellifer*), *Tamarindus indica, Acacia leucophplea* and *Aalbizia lebbeck* widely exists in cultivated fields in peninsular India providing food, shelter and livelihood for millions of poor families.

Among several needs of a farmer, food always remains the first priority item. Alley cropping or hedge row farming and intercropping with NFT (nitrogen fixing trees) are components of food based system as briefly described below.

Alley cropping: It is essentially an agroforestry system in which food crops are grown in alleys formed by hedge rows of tree or shrubs. The hedge rows are cut back at planting and kept pruned during the cropping season to prevent shading and to reduce competition with food crops. When there are no crops, the hedge rows are allowed to grow freely. Short duration cereals or millets are better compatible than long duration crops like castor and pigeonpea.

Intercropping with NFTs: Raising of nitrogen fixing trees (NFT) in arable land not only provide rich organic matter and atmospheric nitrogen, but also improves soil structure and prevents land degradation. Studies on intercropping in 4-year old *Leucaena leucocephala* trees planted at 7.5 m x 2 m showed that grain yield of sorghum and fodder yield of pearl millet were high in lopped treatment i.e., 53 per cent of sole (Hocking and Rao, 1985).

Two-tier model of fodder based agroforestry systems are the better options for class III and above land. Top-feed tree species and some of the hardy fruit trees can be integrated with forage crops to support livestock, generate assured income and to prevent land degradation. Pasture management and ley farming are the suitable ways for the land rich, resource poor farmers as these are low-input and risk-free systems. *Styosanthese hamata* is a better option due to its perennial nature in the region of frequent crop failures. Studies on silvi-pastoral system carried out at All India coordinated Research Project in Gujarat and Maharashtra showed that Leucaena + *Stylosanthes hamata* performed better and yielded higher gross return than all other systems.

(iv) **Himalayan region and north-eastern India:** Eastern Himalayan ranges cover Sikkim, part at Arunachal Pradesh and West Bangal in India, Bhutan, Nepal, Tibet and part of china. The region is characterized by soil degradation and gradually shrinking natural resources. Large cardamom (*Amomum subulantum*), a traditional native spice crop grown in a typical agroforestry system from 600 to 2000 meters altitudes, has export potential in Middle East, Europe and East Asia. It is a perennial herbaceous seophyte (shade loving) grown under the shade of tree under moist soil with high organic matter content. Domestication of cardamom plantation under *Alnus nepalensis* is an age old agroforestry practice in eastern himalaya. There are 29 other tree species which support this plantation crop, viz. *Sehima nervosum, Prunus nepalensis, Querens lamellose*, etc. Management practice involve harvesting trees above 16 cm basal diameter to assist natural regeneration of younger tree seedlings and open canopy to regulate light at the ground. This provides continuous supply of fodder and fuel wood.

Amliso (*Thysanolaena agrostis* syn. *T. maxima*) is most promising shade tolerant grass having multiuse value (leaves for fodder, stem for pulp and spike for bloom) and is best suited for silvipastoral systems. Agricultural lands shaded by *Bambusa nutans* can be effectively utilized by growing ginger, turmeric, large cardamom, orchard grass and dinanath grass depending upon the choice of farmers. Further, napier grass, nandi, setaria and stylos are quite promising for establishing terrace risers and their effective utilization. The model for mixed land use system developed at ICAR Research Complex for North – Eastern Hill Region, Shillong, to put 1/3 area of top of hills to bottom under forest with contour bunds, 1/3 under horti-pastoral system with half moon terracing and contour lines, and the lower 1/3 portion put under bench terracing for production of cultivated crops can be extrapolated for effective watershed management of hill or hillocks.

In North-eastern India, alder (*Alnus nepalensis*), a nitrogen fixing tree, is grown in Nagaland to enhance soil fertility for growing maize, millets, potato, chillies, barley, pumpkin etc. It also serves as a shade tree for coffee and cardamom at lower and higher altiude.

In the Himalayan hills of Uttar Pradesh, Jammu and Kashmir, Himachal Pradesh, Sikkim, Meghalaya, Assam and Manipur, there is an acute shortage of fodder. The fodder is obtained from trees of *Oak, Ficus, Albizia, Grewia optiva, Celtis austalis, Subabool* and *Aneguissus latifolia.* Some trees like *Grewia optiva* and *Bauhinia* are grown in the boundries of the farm but they do not meet the heavy demand of fodder. Grazing is adopted by nomads who migrate with their cattles from alpine areas to sub-tropical areas, depending on the change of season and availability of grass fodder for their herds. Fuel timber for packing cases and cash are the needs of the hill people. Fuel is in great demand in winter as it is the only source of energy, whereas forests are fast depleting. Concerted effort are underway for planting trees like *Poplars,* Subabool, *Robiniapseudo acacia, Alnus nepalensis, Bamboo, Dalbergia sissoo, Cryptomeria Japonica,* etc. The species are location specific and depend on precipitation, snow, altitude, soil and other biotic factors. The fruits are apple, pear, peaches, almond, plums, citrus and crops are rice potato, wheat, barley, vegetables and other temprate crops. The study revealed that about 70 to 80 per cent of farmers prefer agri-horticultural system, 40 to 70 per cent agri-silviculture system, 20 to 40 per cent silvi-pasture system and 1 to 20 per cent hort-pastural system (Chinnamani, 1989).

(v) **Gangaetic plains:** Indo-gangetic plains have been formed through sedimentation carried from Himalays by Ganga and Indus rivers. Punjab, Haryana, Uttar Pradesh, Bihar, Part of Jharkhand and Bengal falls under these plains. The region contributes 51.9 per cent of national food grains output. Fuel shortage in the region has resulted in cow dung being used for cooking and as a source of energy. They have switched to gobar gas, kerosene, gas for cooking and are not in favour of fuel wood if the other alternatives are available. The area is rich agricultural belt devoid of forest. The farmers use the agricultural waste as fuel from pigeonpea, cotton, castor, *sesbania, crotolaria* and others. The common fuel wood species are *Eucalyptus tereticomis, E. camaldulensis, Acacia nilotica, Subabool, Casuarina equisetifolia, Albizia lebbeck, Cassia siamea* and *Bauhinia variegata.* The next in order of need is timber, followed by fodder and poles. Small timber is in great demand followed by soft wood which is met by *Poplar,* sissoo, *babool,* siris, teak, Eucalyptus, subabool and bamboo. Farmers prefer energy plantation in small blocks followed by agri-silvicultural practices. The next in order are silvi-pasture, horti-pasture and hort-silvipasture.

Mango (*Mangifera indica*), guava (*Psidium guajava*), *Jamun* (*Syzygium cuminii*), and ber (*Ziziphus mauritiana*) are important fruit yielding trees in Gangatic plains. In Bihar, litchi (*Litchi chinensis*) is important fruit. Species like *Madhuca indica* and *neem* (*Azadirachta indica*) have also shown their preferences in most of the region.

(vi) **Tropical region:** Karnatka, Tamil-Nadu, Kerala, Maharashtra and Andaman and Nicobar have the tropical climate influenced by seacoast. The whole of Kerala (barring few places of rain shadow regions) and small portion of Tamil Nadu fall under humid tropical region in Peninsular India. Ninety-six per cent of the land is under agriculture in this region. Fodder and fuel are scarce and timber is very scarce. *Dicanthus annulatum, Cynodon dactylon, Prosopis,* leaves of *Cordia myxa, neem, babool* are used as fodder. The fuel includes crop residues (cotton, pigeonpea, safflower, and castor stalks) and *Prosopis juliflora* from river bank and tree pruning around farmers' field. Timber is very scarce and *babool* and *neem* are raised as a source of timber and are highly priced (ICRAF,1985 & 86).

In Tamil-Nadu, the need of fodder is acute. Fuel requirement is to be met from *Prosopis juliflora* which grows abundantly on the wasteland. At Thiruvarambakam, Palghat, Kerala, fodder shortage is acute in non-monsoon dry months. The availability of fuel and timber are equally difficult at different seasons due to lack of scientific agroforestry and pressure of land for cash crop, like rubber, spices, etc.

The result of research conducted on gravelly soil in Dharwad, Karnataka indicated that *Casuarina equisetifolia* + rice combination was best followed by *Albizzia molacana.* The combination of groundnut + papaya + teak gave the highest return. In horti-silvicultural studies, sapota + *Albizzia melocana* was found to be the best. Multistorey cropping is highly remunerative. The crops are coconut + coco + pepper + ginger + ragi or rice. Arecanut, cardamom, jackfruit are some of the other crops which are also associated in the multistoried component. It is practiced under rainfed conditions in Kerala, Karnataka, Maharashtra and Tamil Nadu. Other crops are tea + silver oak, coffee + jackfruit, banana + jackfruit, nutmeg + cloves + cinnamom + jackfriut + mangostein cultivated in the hilly tract of Western Ghats. Array of trees, grass and crop combinations are taken for productive agroforestry. In addition, agroforestry with hedge rows alley cropping can be practiced as erosion control. The suitable species are *Subabool, Albizzia, Crotolaria, Arundodanax, Ochlandra,* Bamboo, *Tripsacum laxym,* Napier, etc.

(vii) **Humid and sub-humid regions:** The region is characterized by hot humid climate for all or most of the year and an evergreen and semi-evergreen vegetation. The region consists of Tripura, Orissa, Jharkhand, Chattisgarh and West Bengal. Acute seasonal shortage of fodder, fuel and timber is noticed. Important fuel species are *Acacia auriculiformis, Subabool, Eucalyptus* and *Albizzia* and bamboos. Palm, jackfruit, arecanut, coconut, teak, sal, mango, *mahua* are also common. Interaction of crops and trees are pronounced during dry spells when poor yield of crop occurs. *Acacia auriculiformis* are planted with blackgram, groundnut, maize, rice and sweet potato. Intercrop of *Eucalyptus* is taken with blackgram, horsegram, cowpea, niger and stylosanthes. Agri-horticulture of citrus + cabbage are best in orissa. In Bhubneshwar (Orissa), agri-silviculture of *Cocos nucifera* for boundary plantation, *Casuarina equisetifolia, Anacardium occidentale* for block plantation and homesteads are being practiced. In Tripura, tree components are used along with livestock and poultry components and in irrigated areas, fish component is also incorporated. Agri-silviculture is common in Ranchi

(Jharkhand) and Raipur (Chattisgarh) areas. *Ziziphus mauritiana, Butea monosperma, Aegle marmelos, Mangifera indica, Schleichera oleosa (Kusum)* are grown in Ranchi area. *Acacia nilotica, Terminalia arjuna, Butea monosperma* and *Alhizia spp.* are grown in Raipur area. Homestead agroforestry is also practiced using *Gamelina arborea, Artocarpus heterophyllus, Madhuca indica* and *Ziziphus mauritiana* in these areas. *Shorea robusta* in Jharkhand and *Acacia nilotica* in Chattisgarh are used as timber trees. *Leucaena leucocephala, Pongamia pinnata, Acacia nilotica* and *Dalbergia sissoo* are also used as fodder in these areas. *Mangifera indica, Artocarpus heterophyllus Moringa oleifera, Carcia papaya*, etc. for fruit, *Terminalia arjuna* for raising silk worm and *Butea monosperma, Schleichera oleosa* for *lac* cultivation are being used by farmers in Jharkhand.

(viii) **Coastal and island regions:** The coastal areas swaps region is characterized by low-lying waterlogged marshy areas, flood plains and ill-drained lands. The river banks are occupied by mangroves and associate halophytes. Other natural vegetation include evergreen, semi-evergreen and deciduous forests on uplands and grazing lands in pockets. The main system of this region consists of plantation crops integrated with livestock and poultry and cultivation of rice. The agroforestry practices range from shifting cultivation and farming in forests to hedge-row intercropping systems; sparse stands of trees on farm lands to high density complex multistoreyed home gardens of lowland. In this system, the trees play a predominantly 'service' role (e.g. shelterbelt) and provide the main commercial product (e.g. intercropping with plantation crops). Large areas in coconut plantation are neglected and remain open for grazing. These areas are suitable for multistoreyed cropping systems Spices and cinnamon may be planted as middle storey crops and pineapple or forage grasses like hybrid napier (*Pennisetum purpureum*), guinea (*Panicum maximum*) and legumes like *Stylosanthes guianensis* or industrial grasses like lemon grass (*Cymbopogon fulvus*) may be grown as cover crop. Rubber and red-oil palm plantation may also be integrated with other spice or forage crops particularly the legume covers. In Andamans, even forest-trees are retained in multi storeyed plantations and in almost all plantation including spices, coffee and fruit trees (Pathak *et al.*, 2006)

(ix) **Agroforestry in problematic soils:** Utilization of problematic soils for cultivation of forest trees is a desirable alternative since the crop production on these soils is uneconomical.

(a) **Salt affected soils:** Salt affected soils broadly grouped into saline soils and alkali soils. Saline soils are predominant in arid, semi-arid and coastal areas. The alkali (sodic) soils occur in Indo-Gangatic plain and parts of Madhya Pradesh. In semi-arid tracts of peninsular India, sodic soils occur because of faulty and injudicious irrigation. Classification of salt aftected soils based on United States Salinity Laboratery, United States Department of Agricukture (USDA) is given in Table 24.

TABLE 24: Classification of salt affected soils (USDA)

Class of soil	Electrical Conductivity (EC) mmhos / cm	Exchangeable Na percentage (ECP)	pH
Saline	>4	<15	<8.5
Saline-alkali	>4	>15	>8.5
Non saline-alkali	<4	<15	8.5-10

Sources of excess salt in the soil consist of salt deposit due to impeded drainage, salt containing irrigation water and its continuous use, capillary movement from the ground close to the soil surface, high sub-soil water table, and back water flow intrusion of sea water.

On saline soils furrow method of planting trees have been found successful. *Acacia nilotice, A. farnesiana, A. tortilis, Prosopis juliflora, Parkinsonia aculeata, Casuarina glauca, C. equisetifolia* and *Tamarix troupii* have been found guite successful. On partial reclaimed land Populus deltoides and *Eucalyptus tereticornis* are grows as boundary plantation. On agricultural land, Populus deltoides is preferred. On highly alkaline soil (pH>10), *Prosopis juliflora, Acacia nilotica* and *Tamaris aphylla* can be grown with success. On soil on moderate alkalinity (pH 9-10), *Eucalyptus tereticornis, Parkinsonia aculeata* and *Terminalia arjuna* can be planted. Among fruit trees aonla (*Phylanthus emblica*), guava (*Psidium guajava*), ber (*Ziziphus mauritiana*), Karondaa (*Carissa carandas*), jamun (*Syzygium cuminii*) and pomogranate (*Punica granatum*) on raised bunds may be grown.

Sapota is found to be most promising. Date-palm is highly salt tolerant. These tree species including fruit trees may be blended with a forage grass like Kallar grass (*Leptochloa fusca*), para grass (*Brachiaria mutica*), rhodes grass (*Chloris gayana*), Karnal grass (*Diplachne fusca*), Bermuda grass (*Cynodon dactylon*), Blue panic (*Panicum species*), Guinea grassa (*Panicum maximum*), hybrid napier (*Pennisetum pureum*) and alkali sacaton (*Sporobolus spp.*)

(b) **Acid alfisols on sloppy lands:** According to Dhyani *et al.* (1994) tree species like *Alnus nepalensis, Parika javanica, Michelia oblonga, Prnus cerasoides, Gmelix arborea,* and *Perserianthes falcataria* are recommended for acid alfisols on sloppy lands.

(c) **Gullied and ravenous lands:** Gully is considered as the severest form of erosion. A narrow deep gash formed as a result of localized and concentrated run off by running water is called 'gully'. Very deep (>9 m) and wide gullies with steep sides are known as 'revines'. The revines are classified as shallow,. Medium and deep based on the gully depth, bed width and area.

Ravinous lands are mainly confined to the states of Uttar Pradesh, Madhya Pradesh, Rajasthan and Gujrat along the banks of Yamuna, Chambal, Mahi, Sabarmati and tributes.

For soil conservation in revine lands trees like Acacia nilotica, Prosopis julifiora, Dalbergia sissoo, Azadirachta indica and Pongamia trinnote are most effective in association with forage grasses like *Dichanthium annulatum, Cenchrus ciliaris, Chrysopogon fulvus, Cynodon dactylon, Sehima nervosum* and *Panicum* spp. In shiwaliks, *Eucalyptus* – Bhabar grass (*Eulaliopsis binata*) system have been found quiet efficient, followed by *Acacia catechu* – forage grass, and *Leucaena* – napier grass.

(d) **Minespoils:** Dedhwal and Katiyar (1996) identified suitable species for planting in different kinds of minespoiled lands in India. Prominent species are listed as under:

Minespoil category	Promising species
(i) Bauxite mined areas of MP	*Grevillea pteridifolia, Eucalyptus camaldulensis*
(ii) Coal minespoils of MP	*Eucalyptus species, Acaia species*
(iii) Limestone minespoils of outer Himalays	*Salix, Leucaena, Bauhinia*
(iv) Rock-phosphate minespoils of Mussorrie	*Pennisetum, Saccharum, Vitex, Rumex* etc.
(v) Liginite minespoils of Tamilnadu	*Eucalyptus species, Leucaena, Acacia, Agave*

(e) **Barren rock hill ridges:** The rock is often barren and devoid if soil cover and vegetation. The reason for low canopy coverage of the top of hill is restricted growth in absence of soil in the hills. Easter and western ghats, Deccan plateau, the Aravallis and Vindhyan ranges are characterized by existence of extensive rock outcrops. They are termed as skeletal area. The species suitable for barren rock hill ridges and rock outcrops (Balasubramaniyan and Palaniappan, 2001) are given below:

Tree species	Shrubs	Grasses
A. Area of low and medium rainfall		
Eucalyptus tereticornis	*Carrisa spinarium*	*Cenchrus cilliaris*
Dandocalamus strictus	*Dichlorostachyas cinerea*	*C. setigerus*
Meilia azadarach	*Indigofera articulata*	*Dicanthum annulatum*
Albiza lebbeck	*Inga dulcis*	*Panicum antidotale*
Acacia catechu		*Pennisetum pedicellatum*
Cassia siamea		
Prosopis chilensis		
Dalbergia sissoo		
B. Area of high alavation upto 2000 m		
Robina pseudoacacia		
Populus ciliata, Pinus roxburghii		

(f) **Waterlogged and marshy lands**: Waterlogging may be defined as "an area where the water table rises to an extent that the soil pores in the root-zone of a crop becomes saturated resulting in restriction of normal circulation of air, decline in the level of oxygen and increase in the level of carbon-di-oxide." Stagnation of water in waterlogged areas occurs for most part of the year because of poor drainage. In the coastal regions, waterlogging may be caused by intrusion of sea water in the lands below mean sea level as found in parts of West

Coast. Many fields in canal irrigated areas become unproductive due to waterlogging. Proper drainage is essential to reclaim these areas.

Tree species suitable for waterlogged area are Eucalyptus robusta, Syzygium cumini, Solix spp., Terminalia arjuna, Populus nigra and Acacia nilotica. In marshy places, recommended tree species are Baringtonia spp., Bischofia javania, Eucalyptus robusta and Casuarina equisetifolia.

For surface waterlogged area and riverbeds, the suitable tree opecies are *Dalbergia sissoo, Acacia nilotica, Acacia catechu* and *Leucaena leucocephala.*

(g) **Sand dune management:** In India, about 22.7 per cent of total geographical areas are occupied by sandy soils mainly in Rajasthan, Gujrat, Punjab and Haryana. The region is characterised by sandy to sandy loam soils with excessive permeability, severe wind erosion, high temprature, scanty aned erratic rainfall, high rate of evaporation, devoid of perennial river and sparse vegetation. Most of these areas is covered by drifting or semi-established sand dunes.

A three step technology for the stablization of sand dunes developed by Central Arid Zone Rosearch Institute (CAZRI), Jodhpur includes: (i) protection against biotic interfernce by fencing the area using barbed wire or thorny bushes/live fences.(ii) sand dune fixation to check the movement of sand and (iii) afforestation by grasses and trees. Among the tree species *Prosopis cineraria* is considered as best because of its natural spread in the area, followed by *Dalbergia sissoo* capable of establishing itself in desert ecosystem. The other suitable tree species are *Prosopis chilensis, P. cineraria, Acacia tortilis, Acacia radiana, Zizyphus mauritiana* and *Erianthus munja,* Among the grass species *Cenchrus ciliaris* performs better than others. The other grass species are *Lasiurus sindicus, Panicum turgidum* and *Panicum antidotale.*

(h) **Shelter belts and wind breaks:** A belt of trees or shrubs maintained for the purpose of shelter from wind, sun or snow drift is called a shelterbelt. Any barrier erected to break or slow down the effect of wind is a wind break.

Farmers build shelter-belt by either small dead wood or local vegetation to check wind velocity within safer limits. *Crotalaria burhia, Leptadenia pyrotechnica* and *Aerva pseudomentosa* bushes are planted 20 to 25 m apart rows across the wind direction. Between the lines of these shrubs, grasses such as *Cenchrus ciliaris, C. setigerus, Lasiurus sindicus* are planted on leeward side of each break. This permanet vegetation helps accumulate sand near them, which is again spread in the field. Tall plants of species like *Dalbergia sissoo, Albizia lebbeck, Acacia nilotica,* etc. are used as wind breaks for sand dune management.

(x) Other agroforestry system:

(a) **Apiculture with trees:** In this system farmers have been harvesting honey from bees making hive on trees. The forest trees with high floral diversity are preferred for apiculture

(b) **Aquaforestry:** In this system various trees and shrubs, the litter-feed of which is preferred by fish, are planted on the boundary and around the fish ponds for fish production and bund stabilisation around fish ponds. Farmers of coastal region cultivate fish and prawn in saline water and grow coconut and other trees on bunds of ponds. Fish culture in mangroves is also advocated which forms a rich source of nutrition to aquatic life and breeding ground for juveline fish, prawn and mussels. A well balanced system of animal husbandry including goatry, poultry, duck farming, turtles and fishes in the small ponds in home-gardens make a system of water, energy and nutrient use efficiency per unit area. The leaves of leguminous trees like *Gliricidia sepium, Moringa oleifera, Leucaena leucocephala* have been found to serve as a good fish feed when offered as pellets and thus, improve fish productivity.

(c) **Multipurpose tree species (MPTs):** The areas like rainfed do not support stable crop production and hence, can be utilized effectively for multipurpose tree production system which provide multipurpose uses. The tree species selected for MPTs should have the characteristic features of high survival percentage and easy establishment, fast growing, ability to fix atmospheric N, multiple uses such as firewood, timber, green manure, fodder and raw material for local industry, acceptability by local people, produce nutritious and palatable foliage and high coppicing and pollarding capacity.

- **Fuel:** The species selected for fuel plantation should be amenable for easy management with minimal skill. e.g., *Albizia lebbeck, Acacia albida, A. nilotica, A. senegal, Prosopis juliflora.*
- **Fodder:** Based on palatability rating, trees and shrubs identified suitable for fodder are *Hardwiakia binata, Prosopis cineraria, Ziziphus nummularia, Acacia nilotica, Acacia senegal, Albizia lebbeck, Azadirachta indica.*
- **Green manuring value:** The Foliage of some of the species can be pruned and used as green leaf manure. e.g., *Pungamia glabra, Glyricidia cebium, Azadirachta indica, Erythrina indica, Leucaena leucocephala, Sesbania grandiflora,* etc.
- **N-fixing tree species:** These trees grow rapidly on poor soils because of their ability to fix atmospheric N. The suitable species are *Leucaena leucocephala, Sesbania spp., Gliricidia, etc.* Leucaena is known to fix about 200 kg N per ha. More than 650 tree species are known to fix nitrogen with nitrogen fixing potentials ranging from 20 to 300 kg N per ha. N fixation per annum varies from tree to tree. It is 20 kg N per ha per year for *Acacia albida* (Nair, 1984), 40 to 50 kg for *Prosopis juliflora* (Virginia *et al,* 1986), 58 to 100 kg for *Casuarina equisetifolia* (Dommergues, 1987), and 350 kg for *Sesbania sesban* (Rao, *et al.,* 1990)
- **Wodlots:** These trees are grown on lands which are not suitable for profitable agriculture. It is now being practiced even on good agriculture lands in association with agricultural crops because of attractive return, shortage of fuel-wood and demand of poles or pulp-wood in industry. For example, bamboo poles for orange orchard in Nagpur area and *Eucalyptus* and *Populus* for industries in Punjab, Haryana, Gujrat and UP are in great demand. Wood-lots of *Casuarina,* bamboo,

poplar, *Eucalyptus, Leucaena Leucocephala,* red sandal (*Pterocarpus santalinus*) and *Dalbergia sissoo* have become popular in many parts of India. *Casuarina equisetifolia* is planted in large woodlots in the states of Andhra Pradesh, Tamilnadu, Karnataka and Orissa.

3.4 TREE- CROP INTERACTIONS

Agroforestry is a land-use system involving growing together of woody perennial (trees) and herbaceous plant (crops or pasture) on the same piece of land in a given environmental situation. Various interactions take place between tree and herbaceous components due to difference in growth pattern and resource requriment, which are referred to as tree-crop interaction or interface.(Huxley, 1983). These interactions occur both above and below the ground through microclimate and the media of soil, respectively exerting favourable or adverse effects on herbaceous components (e.g., crops, grasses) and include a complex set of interactions relating to radiation exchange, the water balance, nutrient budgets, cycling and shelter and other environmental modifications. The success of agroforestry system depends on the exploitation of the component interactions. The component interaction refers to inflluence of one component of a systems on the performance of the other component as well as the system as a whole. Example of various tree- crop interactions are listed below:

(i) Crop inhibition (adverse effect) in association with tree is caused by shade effects as well as competition for below ground resources such as nutrients and water. However, provision of shading is beneficial to livestock and species like cocoa. Allelochemicals secreted by some of the tree species cause inhibitory effect on crops.

(ii) In agroforestry system production of associated components increases due to influence of trees because trees are capable of improving productivity of soil, such as (a) nitrogen fixation symbiotically, (b) improvement in organic matter by leaf litters and crop residues, (c) bringing up up nutrients released by weathering in lower soil horizon, and (d) recycling and trapping of nutrients that would otherwise have been lost by leaching.

(iii) Trees can conserve soil-moisture through microclimatic effects of shading and as wind-bresks.

(iv) Trees stabilize soil conservation structures (e.g., grass strips, bunds, terracers) by binding soil particles. Wind breaks serve to control wind erosion and thus, protect crops.

3.4.1 Nature of Interactions

This can be described on the basis of the observable net effect of one component on the other in the system.. In agroforestry system the concepts of observable net effects can be expressed as described below:

(i) **Complementary**: The interaction between the tree and crop components is complementary in nature if the agroforestry system provides a greater yield than the yield of their corresponding sole crops by creating favourable conditions for their growth. The complementary interaction is of two types, viz., spatial and tremporal. Spatial complementary interaction is achieve by varying proportion of desired species. In this, the geometary of the system is altered by changing spacing and density of the components to suit one's requirement without any sacrifice. By more effective use of growth resources the two components complement each other by more effective use of growth resources by utilizing them at specific times due to difference in growth pattern and resource requirement. For example, in poplar and wheat system, when wheat requires more sunlight, poplar is leafless, thus reducing competition between themselves.

(ii) **Supplementary:** The interaction is known as supplementary if the two components interact in such a way that yield of one component exceeds yield corresponding to its sole crop without affecting the yield of the other compoment. In this interaction one component is maximizing its resource utilization without limiting it for other component.

(iii) **Competitive:** The interaction is called competitive when the tree and crop components interact in such a way that increase in the yield of one component leads to decrease in the yield of other component due to competative interaction. In this type of interaction, the components are competing for the same resources (e.g., light, water, nutrients, CO_2). To reduce competition in an agroforestry system, the tree and crop components should be selected in such a manner that they can utilize finite resources at different times and have different growth and developent phases.

3.4.2 Factors affecting interactions

Factors affecting interactive relations among components of agroforestry are given below:

(i) **Species:** Growth pattern and yield varies from species to species and accordingly affects relationship with associated components. Different crops interact differently with the same tree component. For example, studies conducted at the National Research Centre for Agroforestry, Jhansi (India) revealed that reduction in yield of chickpea was more than that of wheat grown in association with different multipurpose tree species. Similarly, reduction in blackgram yield was than that of greengram when grown in association with bamboo. Studies have showed that yield of wheat under *Dalbergia sissoo* was comparable to that of control plot, whereas under *Populus deltoids, Eucalyptus citriodora* and *Bombax ceiba,* it was considerably lower.

(ii) **Age**: Age and growth of the components affect completion by the components. Initially no competition is offered by trees to crops till they attain full size and stature. For example, the clusterbean yield was suppressed by *Faidherbia albida* in the third year of growth, though *Prosopis cineraria* did not affect crop yield eren in the third year.

(iii) **Density**: Density of trees influences the interception of light by canopy covers which ultimately decides the performance of undergrown crops, however the characteristics fo crop specis also deterine impact of canopy cover. Clusterbean yield inproved considerably by coppicing *Holoptelea integrifolia* but yield of greengram did not change significantly. Trees also midify microclimate of crop grown below and improve physical conditions and productivity of the soil. The optimum density in agroforestry system would vary with variations in components. Higher grass yield was observed under widely spaced *Hardwiakia binata* but not in the case of *Acacia tortilis.*

(iv) **Site:** Plant growth, yield and interactive relationship of component species are strongly influenced by climatic, edaphic and physiographic features of an area. In a study, the grass production under *Tecomella undulata* was found to be highest in high rainfall year but in low rainfall year, it was maximum under *Aalbizia lebbeck.*

(v) **Management:** Tree crop interaction is influenced by various management practices. In order to achieve near desired characteristics of trees to favorably influence crop components, tree crowns and roots can be manipulated through management operations, e.g., pruning and thinning. In fodder tree species, lopping is done for harvest of forage periodically, thus reducing shading effect and increasing light availability to undergrown crops.

Management options to achieve increased growth are: (a) microclimatic amelioration, (b) fertilization, (c) application of mulch or manure, (d) irrigation, (e) tillage, (f) adapted species, and (g) supplemental feeding.

Management options for decreased growth are: (a) pruning, (b) pollarding, (c) root pruning, (d) trenching, (e) excessive shading, (f) herbicides, and (g) grazing or browsing.

3.4.3 Types of interactions

The major positive effects that occur at tree-crop interface are shading trees (strees reduction), biomass contribution, water and soil conservation, and that of negative effects are light competition, nutrient competition, water competition and allelopathy (Nair, 1993).

The different types of interaction between tree and crop components classified as positive (beneficial) and negative (harmful) are described below:

(A) **Positive interactions:** The major type of positive intevactions are brefiely described below:

(a) **Increased productivity:** There is no doubt that overall productivity (biomass) of an agroforestry system is generally greater than that of an annual crop. This is due to capture of more growth resources (e.g., light, water, improved soil productivity) and net effect of complex interaction resulting into microlimatic amelioration, thus affecting transpiration rate, photosynthesis and energy balance of associated components.

(b) **Improved soil fertility:** Micro-site enrichment by some trees is an important aspect of agroforestry. For example *Faidherbia albida* in west Africa, Prosopis cinreria in Rajasthan and *Pinus caribea* in Savanna are known to accumulate nutrient through precipitation and nutrient-rich litter. Growing of nitrogen-fixing trees, e.g. *Leucaena leucocephala* and *Gliricidia sepium* in the humid tropics may enrich soil in 2 to 3 years, A large quantity of biomass for green manure is also produced by trees.

(c) **Nutrient cycling:** In agroforestry, trees provide additional nutrient inputs via biological nitrogen fixation and nutrient capture from deeper layer by tree roots, where crop roots are not active.

(d) **Soil conservation**: The woody hedge-rows on contour provide a semi-permeable barrier to surface movement of water, and mulch from trees reduces impact of rain drops on the soil and minimizes splash and sheet erosions through providing continuous leaf cover over the ground surface.

(e) **Water conservation:** A mulch and litter layer under shade trees increases infiltration of rain water and reduces evaporation from soil.

(f) **Weed control:** The effect of shade is more severe for light demanding weeds. More shade under *Senna siamea* is reported to control weeds better than *Gliricidia sepium* or *Flemingia macrophylla.*

(g) **Micro-climate improvement:** Trees for shade, live-fences, wind-breaks and shelberbelts are known to improve micro-climate. Apart from reduction in heat and light, agroforestry systems bring changes in temperature, humidity, movement of air, and moisture of soil. For example, crops that require shading for normal growth are black pepper, turmeric, etc. Example of one crop providing shelter to other crop is growing of tea under the shade of *Albizzia.* Buffering effect of trees on micro-climate beneath them are reported for combination of coconut and cocoa, *Leucaena* and millet, *Hardwickia* and mustard or soybean and *Paulownia* with tea.

(h) **Physical support:** In agroforestry, trees provide physical support to crop components, e.g. growing of betel vine or black pepper vines on the support of mango or coconut and arecanut.

B. **Negative interactions:** Nagative interactions occurring in agrofotetry systems are given below:

(i) **Competition:** The decreasing, effect on growth and yield of particularly crops component in a agroforestry system arise from competion for light, water, nutrient and space. In a fertile soil with adequate moisture, light becomes the most limiting factor. In semi-arid condition, the relative importance of light will decrease. Generally the shade tolerance of crop-plant depends on the photosynthesis pathway and product to be harvested. In dry areas, productivity of agroforestry systems depend largely upon competition for water.

(ii) **Allelopathy:** Inhibition of growth of one plant by chemical compounds released into the soil from the neighbouring plants is termed as allelopathy. Allelopathic effect of tree species on crop plant are presented in table 25.

TABLE 25: Allelopathic effect of tree species on crop plants.

Tree species	Crop plants
Alnus nepalensis	Soybean
Casuarina equisetifolia	Cowpea, sorghum, sunflower
Eucalyptus tereticornis	Cowpea, sorghum, sunflower, potato
Gliricidia sepium	Maize, rice, tropical grasses
Leucaena leucocephala	Maize, rice, cowpea, sorghum, sunflower

(iii) **Pests and diseases:** Shade and more humid environment in a agroforestry system may increase incidence of bacterial and fungal diseases. For example, the incidence of *Phytophthora palmivora* on cocoa increases greatly under conditions of heavy shading.

3.4.4 Methods for Quantifying Interactions

(i) **Land Equivalent Ratio (LER):** Land Equivalent ratio (LER) can be defined as the relative land area required as a sole crops to produce the same yields as intercropping. It is most widely accepted index proposed by Willey (1979) for evaluating the effectiveness of intercropping in terms of yield advantage, which has been extended to agroforestry also. The LER can be expressed as:

$$\text{LER} = L_A + L_B = \frac{Y_A}{S_A} + \frac{Y_B}{S_B}$$

Where, $L_A + L_B$ are the LERs for individual crops or components, and Y_A and Y_B are the individual crop yields in intercropping, and S_A and S_B are their yields as a sole crops. A value of LER greater than 1 indicates an overall biological advantage of intercropping. When LER is less then 1, more land is needed to produce a given yield by each component of intercrop. When LER is equal to 1, there is no advantage to intercropping over sole cropping. The LER has also been used to calculate monetary advantage, the relative land area required as sole crop to produce the value of yields achieved in intercropping.

Monetary advantage = value of combined intercrop yield $\times \frac{LER \quad 1}{LER}$

(ii) **Relative crowding co-efficient (RCC):** This is one of the competition functions proposed by de Wit (1960) and examined in detail by Hall (1974). On the basis of yield per unit area, RCC can be written as:

$$Kab = \frac{Yab}{Yaa \quad Yab} x \frac{Zba}{Zab}.$$

Where, Kab = RCC of crop a intercropped with crop b

Yab = Yield per unit area of crop a intercropped with crop b (expressed over the area occupied by both crops)

Yaa = Yield per unit area of sole crop a

Zab = Proportion of intercropped area initially allocated to crop a

Zba = Proportion of intercropped area initially allocated to crop b

and Zab+Zba = 1

For a given crop, the index will be greater than equal to or less than unity, which indicates that there is a yield advantage, no effect or yield disadvantage of intercropping, respectively. With the index more than a unity will mean that the given crop is more competitive than associated crop or component and vice-versa. This approach of RCC now extended recently to agroforestry also.

(iii) **Aggressivity (A):** A function which has attempted to measure intercrop competition by relating the yield changes of both component crops is the aggressivity proposed by Mcgilchrist (1965).

$$Aab = \frac{Yab}{YaaxZab} - \frac{Yba}{YbbxZba}$$

Where, Aab = Agressivity of crop a with b

Yab, Yaa and Zba are the same as in case of RCC

Yba = Yield per unit area of crop b intercropped with crop a (expressed over the area occupied by both crops) and

Ybb = Yield per unit area of sole crop b

More comprehensively,

$$A_{ab} = \frac{\text{Actual yeild of } a \text{ when intercroped}}{\text{Expected yield of } a \text{ when intercropped}} \quad \frac{\text{Actual yield of } b \text{ when intercropped}}{\text{Expected yield of } b \text{ when intercropped}}$$

An aggressivity value of zero indicates that the components species are equally competitive. For any other situation both species will have the same numerical value but the sign of the 'dominant' species will be positive and that of 'dominated' negative; the greater the numerical value the bigger the difference in competitive abilities and bigger the difference between actual and expected yield.

(iv) **Competitive Ratio (CR):** Considering the improvement over aggressivity Willey and Rao (1980) suggested that instead of taking the difference of two terms in aggressivity, take the ratio of these terms which they designated as Competitive Ratio (CR). Accordingly

$$CR = \frac{Yab}{YaaxZab} \quad \frac{Yba}{YbbxZba}$$

This can be written as,

$$CR = \left(\frac{Yab}{Yaa}\right) \div \left(\frac{Yba}{Ybb}\right) \times \left(\frac{Zba}{Zab}\right) or \left(\frac{La}{Lb}\right) \times \frac{Zba}{Zab}$$

All the abbreviation are the same as in case of aggressivity. However, La and Lb are the LER for crops a and b, respectively.

(v) **Tree-crop interaction (TCI) equation:** The approach is to quantify effects of each factor in relation to performance of sole crop control in the absence of any interference by trees assuming that main factors are not confounded experimentally. Overall tree-crop interaction (I) on crop yield can be expressed as:

$$I = F - C + M + P + L$$

Where, F is the benefit of tree pruning on soil fertility and microclimate of the soil,

C is the reduction in crop yield due to interspecific competition between the tree and crop,

M is the consequence of above-ground changes in temperature, light, humidity and wind speed,

P is the consequence of change in soil properties, and

L is reduction of losses of nutrient or water.

The effect of each factor is expressed as the percentage of the control crop.

3.5. MANAGEMENT OF AGROFORESTRY SYSTEM

3.5.1. Characteristics of Tree Species for Agroforestry:

For choosing appropriate tree species for agroforestry systems, the effect of trees on crops must be considered. Trees to be grown in conjunction with agricultural crops should have inter alia the following characteristics (King, 1979):

- amenable to easy wide spacement.
- tolerate relatively high incidence of pruning, i.e., their photosynthetic efficiency should not decrease with heavy pruning.
- a low crown diameter to bole (trunk of a tree) diameter ratio, i.e., the width of the crown should be small relative to bole diameter.
- permit the penetration of light on the ground.
- toterant of side shade
- phenology (with reference to leaf flushing and leaf fall) advantageous to the growth of the annual crop.
- the rate of litter fall and its decomposition should have positive effect on soil.

- the above-ground changes over time in structure and morphology be such that they refine and improve characteristics reducing competition for solar energy, nutrients and water.
- their root system and root growth characteristics should be such that they can explore the soil layers that are different to those being trapped by agricultural crops.
- should be efficient nutrient pumps.

3.5.2 Suitable Areas for Agroforestry:

The following types of lands can be taken for agroforestry: (i) Cultivable land (ii) Field boundaries (iii) Along with farm roads, rivers, canals and *nallah* sides affected by erosion (iv) Pockets within cultivated holding where cultivation is not possible (v) Old fallows (vi) Cultivable waste, and (vii) other areas like community or panchayat land in which agroforestry can be incorporated.

3.5.3. Management Practices in Agroforestry

The management practices for agroforestry are more complex because more than one species having varied phenological, physiological and agronomical requirements are involved in agroforestry than in case of monocropping. Management aspects of agroforestry system include sharing of growth inputs, such as space, water, nutrients and light by the component species and tree management aspects such as density, geometry and pruning/lopping etc.,

3.5.3.1 Management of Component Species

(a) **Space:** The component species in agroforestry share the same space for their growth and development. The competition with crops increases linearly with increasing tree population. The reduction in crop yield is higher under closely spaced trees. This competition in agroforestry system can be reduced by modifying the spatial arrangement of trees. Depending upon the growth behaviour of the mixed species competition starts from the very beginning. The performance of *Prosopis juliflora* with and without interplanted *Leptochloa fusca* grass studied by Singh *et al.* (1988) revealed that the tree growth was adversely affected during the establishment stage because of luxuriant growth of grass that smothered the tree saplings which were much smaller in height than the grass. Such adverse effects of fast growing intercrops can be minimised by delaying the growing of such crops by one year. Use of taller saplings may also help in reducing such competition during establishment stage.

(b) **Water:** In agroforestry, the water supplies must meet the requiement of component species. In case of water scareity, it is the understory annual crop which suffers the most. The upper storey woody perennials after few years may extend their roots deep and thus, meet their water needs from the ground water. Comprehensive information on rooting pattern of the component species is of paramount importance. Some of the management practices for governing moisture distribution in agroforestry systems include: trenching, root pruning, deep cultivation and inserting polythene sheet barriers.

(c) **Light:** When water is not limiting, the dry matter (W in g per sq m) produced by vegetation is linearly related to total intercepted radiation. Squire *et al.* (1987) represented this relationship by W= Sfe dt, where S= is the total radiation (man of daily totals, MJ / m^2); f = the fraction of man daily insotation intercepted by the canopy, e = the amount of dry matter formed per unit radiation intercepted (conversion coefficient, g / MJ); and t = the duration of crop growth in days. The Value of S varies from 12 to 30 MJ / m^2 in the tropics.

The competition for light between upper storey and lower storey crop combinations start after few years of growth. Trees having spreading canopy like *Acacia, Dalbergia. Albizia, Leucaena, Prosopis,* etc. curtail more light than the erect type species such as *Eucalyptus, Populus, Hardwickia,* etc. However light competition can be managed through:

(i) Selection of intercrops as per their requirement for light. More light demanding crops, such as wheat, rice, maize, etc. can be taken during initial 2 to 3 years without much reduction in their yield. Later on, these crops may be replaced by shade loving commercial crops, such as turmeric, colocasia, potato, etc.

(ii) Reduction of tree density through selective felling after 4-5 years of planting.

(iii) Preodic pruning / lopping of trees.

(d) **Nutriants:** Growing of leguminous trees in association with agricultural crops result in build-up of nutrients in the soil. The degree of nutrient accretion to the soil depends upon the nature of tree species, their rooting capacity and litter production behaviour both in terms of quantity and quality. Yadav *et al.* (1972) found that application of 700 g of single super phosphate in one dose and 400 g of ammonium sulphate in two equal splits, one in March and the other in May, boosted growth of *Acacia arabica, Eucalyptus* hybrid, *Prosopis juliflora* and *Terminalia arjuna* except *Albizia lebbeck*. In a poplar based agroforestry system, sowing of wheat in first week of November, additional application of N and higher seed rate of wheat resulted in significant higher yield of wheat. In case of *Eucalyptus* based agroforestry, boundary plantation of *Eucalyptus* in North-South row orientation is better as its effect on associated crops is minimum. Fodder crops should be preferred over grain crops near boundary plantation of *Eucalyptus* trees.

3.5.3.2 Nursery management

(i) **Collection of seeds:** In order to produce quality seedlings good and viable seeds should be collected from best shaped, middle stage mature tree. Young and over matured trees should be avoided. It is also important to find out the germination percentage. The collected seeds should be dried and stored in air-tight containers to maintain viability. The seeds can be dried at low miosture (5%) and stored at low or sub-normal temprature for a longer period. The viability of seeds varies with the species from 2 weeks to 2 years and more. Seeds of *Acacia, Albizia, Cassia, Leucaena* and *Prosopis* have longer viability and thus, can be stored for more than 20 years as compared to *Azadirachta* and *Shorea* stored for only two weeks.

(ii) **Seed treatment:** Seeds show some degree of dormancy. The seeds with hard seed coat should be subjected to scarification. Scarification is done with hot or cold water as well as chemicals (e.g. *Acacia spp., Albizia spp.*). *Terminalia arjuna,* and *Cassia spp.* require mechanical scarification. Seeds of oaks and chestnut respond well to immersion in boiling water for 30 seconds.

Rhizobium commonly used for supplying N is used for legumes. Franckia is used for supplying N for the seedlings of

Casuarina and *Alnus*. *Azospirillus* is used as seed treatment and seedling dip for non-legume tree species, e.g. neem. Ectomycorrhizal fungi (VAM) are used for temprate forest species, e.g. Eucalyptus, Pines, etc.

(iii) **Sowing:** Generally larg-sized seeds are sown in the containers and small sized seeds and those with poor germination are sown in the nursery beds, from where they are taken to the main field for planting. The nursery bed is prepared by mixing soil, sand and farm yard manure (FYM) in the ratio of 3:1:1. Seeds are broadcasted on the surface and lightly covered with a thin layer of fine earth. Light watering is done using rose can. The seedling should be protected from hot sunshine by providing shade. Proper manuring with organic manures (viz., compost, sheep and goat manure, oil cakes) and inorganic fertilizers (viz., urea, ammonium sulphate and ammonium nitrate) is important for healthy growth of seedlings.

(iv) **Nursery Care:** It includes watering, weeding, culling out and pest and disease control. At the initial stage, watering should be done twice a day for 15 days and after that once a day for a month. The seedlings should not be allowed to suffere in want of water. Culling should be done to remove weak and unheathy seedling suffering from pests and diseases. Seedlings should be kept free from root competition, for this container plants are required to be shifted periodically to avoid roots striking the ground. Appropriate timely measures should be taken to protect the seedlings from insects and diseases.

3.5.3.3 Planting and after care

The most appropriate time for planting tree seedlings is on onset of monsoon. Soil around the plant should be packed well after planting and a convex mound is formed the plant with a circular depression around for rain water to collect. The stump planting has been found to give better results in *Tictona grandis*. Plants with a collar diameter of 1.5-2.0 cm and taproot with a length of not less than 30 cm are selected. Shoots are cut off at 3 cm and all the lateral roots are pruned.

After planting, weeding should be done when the soil is dry. Loosening the soil around the plant is important for better aeration. Mulching around the tree seedlings helps in cheeking evaporation losses, maintaining and increasing soil organic matter, supressing weed growth, and protecting the soil surface from the erosive impact of high rainfall. In case of mortality re-filling should be done, since survival percentage normally may not exceed about 40 to 50 per cent.

Under agroforestry system, trees are protected as long as the crop is not harvested. After the harvest of crop, trees should be protected against cattle. For protection of plants, biofence can be raised through seeds and seedlings of *Acacia farnesiana, A. melligera, Albizia amara* and *Prosopis juliflora.* Stem cutting for vegetative propogation can also be used for biofence, for which the suitable species are *Agave americana, A. sisalana* and *Aloe barbadense.*

3.5.3.4. Management of Trees

(a) **Tree density / spacing:** The nature and growth habit of the species and objective of plantation largely determine plant to plant and row to row distance in planting of trees. Closer planting covers the risk of mortality during establishment. If the purpose of a plantation is to produce firewood (energy plantation), than the planting distance may be kept ats 1-2 m between rows, and 1 m between plants. It the objective is both firewood and timber production, then high density plantation may be preferred. After about 3-4 years, 50 per cent plants may be harvested for firewood and remaining may be kept for timber production.

In agroforestry, minimum suggested spacings between rows and plants are 4 m and 2 m, respectively, Light is more critical factor limiting agricultural production in agroforestry, in such case shade tolerant crops would allow higher tree stocking rates than less tolerant crops. The other consideration are management efficiency, e.g., machinary and tools movement problems which may be partly overcome by a carefully planned planting arrangement. A low density rate is the best solution. The personal preference of a farmer is an important factor in selecting the optimal tree density. Many farmers prefer a 'parkland' environment by having fewer trees for more agricultural production, whereas others may prefer the forest appearance of higher density of trees.

(b) **Fertilization:** Application of fertilizer in agroforestry system is an important aspect in order to obtain higher production of agricultural crops. The results of experiment (Sahu *et al.*, 1996) revealed that N application had beneficial effect on yield of wheat as well as tree species. However, in case of N-use efficiency, i.e., recovery factor, physiological efficiency and agronomic efficiency decreased with increasing level of N application. Karim *et al.* (1991) recorded the higher yield of maize, cowpea and sweat potato grown under *Leucaena leucocephala* with the application of N fertilizer. Shukla *et al.* (1998) attributed higher fodder yield of berseem under *Sesbania* alleys to the application of higher level of P and S fertilizers and maximum yields were obtained at 120 kg P_2O_5 per ha and 60 kg S per ha.

(c) **Canopy management:** Tree canopy or tree cover has enormous effect on underneath natural vegetation and/or intercrop grown in association with trees. The tree having dense canopy permits very little solar radiation to fall on the underneath crop to meet its requirement. There are several canopy management options as given below:

Pruning (Cutting back of plant growth including side branches or roots): This practice allows the grower to manipulate the growth and development of trees to improve the quality of tree production and to increase agricultural return. General pruning is performed once or twice during the year in which small branches and twigs are removed. If it is not performed, the quality of timber is greatly reduced by the presence of large knots which develop at the base of each branch due to its unrestricted growth of branches. There are many points to be examined on pruning which involve how much to remove and when and how often to prune. These points may be decided by the products required, i.e., fodder, fuel, timber or green mulch, etc. This practice is also adopted for maintaining trees / shrubs in live-fence and alley cropping. The biomass and foliage obtained due to this operation is used as fodder.

Pruning practices can be devided into:

(i) Bending and coppicing (ii) Pollarding and bushing (iii) Branch pruning and (iv) Root pruning (Carnell, 1983).

(i) **Bending and Coppicing:** These practices involve regular or repeated harvesting of vertical shoots of trees which are rabidly replaced after cutting. Coppincing means the ability to regenerate by shoot or root suckers of an established forest tree. Regular coppicing promotes the growth of leafy shoots which are acessible for harvesting or browsing. *Salix* spp. in kashmir valley is coppiced annually, tea plants are coppiced every 4 to 6 years.

(ii) **Lopping (Cutting one or more branches) / Pollarding (Cutting back the crown of a tree):** In this practice, growing trees or major branches may be lopped for either fodder or green leaf manure and twigs may be used as fuel. It controls crown spread in the agroforestry system. Bushy and spreading trees like mesquite and *Acacia* need regular lopping, whereas erect trees, e.g., *Eucalypus, Populus, Casuarina,* etc., need little or no lopping. In pollarding, branches are removed usually at a height about 1-3 m at periodical intervals. High pollarding will produce tree with a greater potential for wood production while low level pollarding and lopping will produce spreading crown with a greater potential for fodder production.

(iii) **Branch pruning:** This practice involves simply the removal of lower branches thereby lessening the stem taper and decreasing the number of nodes in the bole wood without greatly decreasing the bole volume increment. This practice is extensively followed in case of poplars. In case of fruit trees branch pruning is highly selective, i.e., restricting height growth encouraging development of spreading, well illumnated canopies and enhancing fruit bud production.

(iv) **Root pruning:** Root pruning checks root growth and have following beneficial effects:

- Enhanced flower bud production at the result of checked shoot growth.
- The regrowth of more fibrous root system, with a greater access to immobile soil nutrients.

Coppicing (Cutting tree species to ground level to produce new shoots from the stem): Coppicing means the ability to regenerate by shoot or root suckers of an established forest tree. Coppice sprouts usually grow vigorously because they are served by roots big enough to feed the former tree.

Thinning: It involves the removal of selected trees from the site. This leaves a greater share of the limited resources of the site-light, moisture, nutrients and space for the remaining trees and the agricultural crop. Some thinning may also be necessary to remove diseased or damaged tree. In order to reduce the effect of thinning on production of agricultural crops, thinning may be done immediately after harvesting a crop.

3.6 DESIGN AND ANALYSIS OF AGROFORESTRY EXPERIMENTS

3.6.1 Criteria for Good Agroforestry Design

A good agroforestry design should fulfil the following criteria:

(i) **Productivity:** There are many different ways to improve productivity with agroforestry, for example, increased output of tree products, improved yields of associated crops, reduction of cropping system inputs, increased labour efficiency, diversification of production, satisfaction of basic needs and other measures of economic efficiency or achievement of biological potential.

(ii) **Sustainability:** By seeking improvements in the 'sustainability' of production systems, agroforestry can achieve its conservation goals while appealing directly to the motivation of low-income farmers, who may not always be interested in conservation for its own sake.

(iii) **Adoptability:** No matter how technically elegant or environmentally sound an agroforestry design may be, nothing practical is achieved unless it is adopted by its intended users. This means that technology has to fit the social as well as environmental characteristics of the land-use system for which it is designed.

3.6.2. Important Parameters of Agroforestry Research:

(i) **Tree density:** Number of trees per ha.

(ii) **Tree rectangularity:** In agroforestry experiments, row to row distance should be in east-west direction and tree to tree distance in north-south direction. This will ensure proper utilization of light in agroforestry.

(iii) **Tree arrangement:** Whether in a given rectangularity, trees of adjuscent tree rows are opposite one onother i.e. a square arrangement, or are offset in someway.

3.6.3 Features of Agroforestry System and Measurment

There are several features in agroforestry systems that necessitate modification of mearurment methods in agriculture.

(i) **Presence of interfaces:** Due to tree-crop interaction intercrop performance is not expected to be uniform across the crop area. Crop next to a tree or hedge will differ from those not subjected to the influence of tree.

(ii) **Time:** All agroforestry systems have to be monitored over a longer time because the effects of tree components to change over time and take several years to become apparent.

(iii) **Large plot size:** Large area is needed to establish agroforestry system and also large variability in crop performance is expected. The spatial variability in crop yields is expected to be larger than in sole crop on the same site because of competition between the components and complexity system making uniform management practices difficult.

(iv) **Border effects:** If an edge effect is caused by the treatment applied to the same plot or the presence of uncropped path between the plots, it is referred to border effect. A tree can extend its influence through radiation over a much greater distance.

(v) **Neighbour effects:** If the edge effect is caused by the treatment applied to the neighbouring plot, it is referred to as neighbouring effect. Neighbour effects involve the interaction of two plots which are more complex and difficult to estimate. Neighbouring effect can be particularly more important in nutrient and irrigation experiments because of tree height which influences the neighbouring plots through variable quantity and quality of light.

3.6.4 Design Requirements

Design feature changes according to the aim of the experiment and the enviroment in which it is done. Variation caused by enviromental features and influence of farm management will cause major differences in design and analysis requirements.

(i) **Clear objectives:** The kind of layout used in agroforestry experiments will depend on the precise research objectives and resources available including extent and nature of site. Objectives need to be specified very clearly so that they may be echieved consistently and with desired precision.

(ii) **Control of variability:** Design of any long-term trial should aim to maintain control of both spatial and temporal variabilities and to minimise input resources. When genetic, envitonmental and management factors are considered separately, the pattertn in variability caused by trees, plots, experimental blocks, farms and regional loeations are found different in relation to each other.

(iii) **Sampling strategies:** To ensure unbiased data, a sampling satrategy need to be determined both in space and time.

(iv) **Good trial management:** Care in design, sampling, management and uniform data recording is needed together with commitment to ensure good results and proper interpretation.

(v) **Choice of indicator:** Indicator type clearly depends on the research objectives and on the type of research but it must also be meaningful to farmers.

(vi) **Scale and frequency:** In agroforestry trials, the data of tree component should be measured once before sowing and onother after intercrop harvest depending upon facilitces and chosen variables.

3.6.5 Design for Agroforestry

The basic experimental designs are just as appropriate for agroforestry experiments as for other field experiments and the same analysis are, in general, equally valid whether the variable being measured is yield or is an environmental variable. These designs, which seems to have the most potential for use in agroforestry, are briefly discussed below:

(a) **Simple design for investigating the tree-crop interaction:** Generally two designs are used for tree-crop interaction:

(i) Parallel rows of tree and crops

(ii) Concentric rows of crops around a single tree

The design should be replicated unless they are purely observational and the orientation should be radomised. In both cases analysis involves fitting the response curve describing the change in variable measured with the increasing distance from the interaction.

(b) **Systematic designs:** In these designs, spacing varies in regular fashion, thus removing much of the need for growth areas.

Suitably replicated, the effect of spacing can be investigated by examining the change in yield, or any other variable, along the length of the plot by fitting an appropriate response curve.

A systematic design for monocropping situation is a fan design. The shape of the area available to each plant is constant but its size increases radically. Onother systematic design varies the space but keeps the area constant. These design suffer from the disadvantage that the crops are not rectangularly arranged but lie on the area of circles. However, paralleled row design avoids this allhough at the cost of confounding the effects of area and shape. However, two superimposed parallel row designs in which vertical lines indicate rows of trees and horizontal lines indicating rows of crops, allows the effect of independent variation of the densities of two components to be examined. It consists of two parallel row design arranged at right angles; the scale of two designs can, of course, be different to allow for differences in size and desired density between the two crops. Analysis is again by examining response curves but since both densities are varied, the response surface in two diamensions is now of intrest.

(c) **General designs:** Agroforestry experiments can use the same designs as used for agronomic intercropping experiments. However, one or two more unusual designs which are particularly useful for agroforestry are:

(i) **A nearest-neighbour design:** In this design based on Latin square, each treatment occurs next to onother treatment (i.e., to the right, left, above or below) on four occasions. Thus, any deletarious or beneficial effects of one tretment on its neighbours should balance out. This may permit some reduction in the area needed for guard rows although it is still preferable to have an adequate guard rows inorder to minimise interactions. Analysis of this design can be by the usual methods for Latin squares. Further information about this design may be found in paper published by Ramanatha Chetty, (1989).

(ii) **A beehive design:** In most experiments plants are arranged in rows but if the interactions between individual plants are of intrest, a hexagonal grid can be useful particularly in competition experiments. In this design, different plants have different numbers of neighbours of the other species, thus allowing the measurment of competitive effects (Veevers and Boffey, 1975; Martin, 1973)

(iii) **Y design:** The design is more important if a regular linear plant arrangement is used to consider whether the effects of row orientation are to be investigated.

3.7 STRATEGY FOR AGROFORESTRY DEVELOPMENTS

There is little information on the quantitative amelioration of various agroforestry systems. Similarly, no adequate and sound information on the ideotype of trees/shrubs including fruit trees, palm, bamboo, etc., and crop plants needed for a specific agroforestry system. The common set of objectives of all efforts for agroforestry development are: (a) enhancing output of specific products, such as fodder, fuelwood, small timber and crop yield; (b) improving productivity of small farms with minimum external inputs in different agro-ecological regions; (c) divising management techniques that can facilitate profitable use of degraded and unproductive land; and (d) providing environmental amelioration through watershed protection, soil conservation, etc. The additional needs of fuel, timber and forage can be met through three pronged attack on the problem:

(i) Improving productivity of lands already under agriculture and forests and reclaiming deteriorated lands so as to reduce the needs for deforestation.

(ii) Growing timber and fuel wood on wasteland and under-utilized marginal lands through improved technology of production.

(iii) Including trees and grasses in agricultural system by adopting agroforestry system aimed at increasing productivity and sustainability of both, crops and trees.

3.7.1 Future Thrust

- To develop agroforestry systems for different agro-climatic regions of the country for higher productivity, sustainability and better economic returns to the farmers.
- To develop statistical and mathematical models after evaluafion of successful agroforestry systems/practices and to upgrade and refine already identified agroforestry

models/practices for value addition, quality improvement and quick returns coupled with higher productivity.

- To explore unexploited and under-exploited plants of high economic value, such as aromatic, oil yielding plants.
- To study the role of agroforestry on physical soil properties in mitigating reversion of environmental degradation processes, such as control of wind and water erosion, sand dune establization, reclaimation of salt affected and acidic soils, rehabilitation of revines and gully lands.
- Agroforestry systems as a source of sink for atmospheric carbon sequestration to balance the predicted climatic changes should be one of the key areas of research in the near future.
- To identify sutable forest and fruit species for balancing ground water level without jeopardising productivity and economic returns to the farmers, since hydrologists are predicting major fluctuations in ground water hydrology of major food production zones.
- To study the role of trees in improving soil quality for sustainable agricultural production aiming to reduce depensence on artificial inputs, such as fertilizers.
- Development and standardization of agronomic/management practices, such as irrigation, nutrient management, spacing, pruning, and plant protection, etc. for the already identified agroforestry systems for improving productivity.
- To improve already identified promising tree species, such as Sishoo (Shisham), Teak, Poplar, Babul, Eucalyptus, Tamarindus, Neem, Khejri, Maharukh, Alnus, Amla, and Kadam, etc. using latest plant breeding and genetic tools for improved productivity, value addition and quick returns.
- To strengthen basic research on agroforestry covering following aspects:
  - (i) Isolation of tree-intercrop-soil-root relationships in terms of sharing space, moisture, nutrients, light and other growth factors amongst the agroforestry components.
  - (ii) Nitrogen fixation, role of VAM, physical, chemical and biological transformations in soil as governed by the species mixtures.

(iii) Developing agroforestry models using basic information generated at (i) and (ii) above for wider application.

(iv) Research on above and below ground interactions between tree and crop including interactions between soil micro-organisms in the rhizosphere.

(v) Developing symbiotic microbial associations for improving sustainability.

- To identify high value medicinal, aromatic, industrial and aesthetic trees, bushes, grasses and crops and to develop technology for their cultivation in a unified agroforestry system wherever possible for export purpose.
- To develop mechanism for transferring agroforestry technologies from the research centres to the user agencies, i.e. Farmers.
- To study on rationalistion of the use of chemical imputs for sustained yield and adequate pest and disease control leading to increased use of integrated pests and disease management using biological control wherever possible.
- Use of tissue culture and biotechnology to overcome barriers to plant improvement.
- Development of improved soil and water management systems including *in situ* water harvesting and recycling.
- Development of efficient models for homestead agroforestry system.

3.7.2. Research Needs and Priority areas

There is an urgent need of research on various aspects of agroforestry systems and the future line of research under major agroforestry systems (Deb Roy and Pathak, 1983; Shankar, 1988; and Deb Roy, 1989).

(a) **Silivipastoral system:** Major portion of our vast wasteland could be profitably developed through silvipastoral systems. Major thrust should be given for refining and improving the existing system wherever available and studying new models in each of the agro-climatic region. Suggested lines of research are:

(i) Selection, introduction and evaluation of suitable multipurpose tree species, preferahly nitrogen fixing trees / shrubs with grass-legume mixture for each of the agro-climatic region.

(ii) Site-specific silvipastoral studies on problematic soils for increased biomass and secondary production.

(iii) Evaluation of silvipastoral system with mixed block of cattle, sheep and goat on year long basis by lopping / pruning of the fodder trees / shrubs.

(iv) Breeding and selection of desirable plant types of trees / shrubs along with compatible grasses and legumes.

(v) Standardising optimum tree density by gradual manipulation of thining for sustained production of pasture, wood including fuel wood, and leaf fodder.

(vi) Depelopment of prediction equation for indirect estimation of leaf fodder and fuel wood through lopping and wood and leaf fodder at felling.

(vii) Impact of silvipostoral system on soil structure and dynamics of nutrient, moisture and on macro and micro environment especially in arid and semi-arid regions.

(viii) Long term performance of silvipastoral system *vs.* dryland agriculture especially under arid conditions.

(b) **Agri-silviculture:** Ideal characteristics of tree / shrubs for growing agricultural crops have been given by King (1979) which will provide guidelines for choosing the right tree.

(i) Selection of fast growing, nitrogen fixing trees, their vegetative propogation and improvement through breeding.

(ii) Development of improved nursery establishment technology.

(iii) Determination of most farvourable combination of tree species / varieties, their density and agricultural crop with specific varieties for different agro-climatic zones.

(iv) Determination of shade tolerent species / varieties of crops and their optimum light, moisture and nutrient requirement so as to provide canopy manipulation of the overhead trees for optimum production for each agro-climatic zone.

(v) Standardisation of alley cropping practices in relation to different crops / varieties with fast growing trees / shrubs for each agro-climatic zone and minimisation of input from outside the system.

(vi) Development of desirable agricultural implements suiting the specific agri-silvicultural system.

(vii) Sandardisation and effect of micro and macro shelter-belt and wind-breaks on crop production and micro-climatic improvement.

(viii) Plant protection studies in relation to specific agro-forestry system

(ix) Improvement in practice of shifting cultivation.

(x) Studies on nitrogen fixation by trees, shrubs and other woody perennials by mycorrhyza and symbiotic bacteria.

(c) **Horti-agri-silviculture:** As the fruit trees are widely spaced, there is ample scope for introducing quick growing nitrogen fixing multipurpose trees in between the two fruit trees besides crops in between rows of trees. Here, trees may provide shelter, shade and soil improvement like those of plantation crops. For the success of this system, following suggestion are made:

(i) Selection of horticultural tree / shrub variety and other desirable trees and the ideal crop combinations for temprate, troical and subtropical regions.

(ii) Standardisation of ideal horti-agri-silvicultural system for each of the important horticultural and plantation crops for each zone.

(iii) Development of technology for agri-horticulture and horti-silvi-pasture in problematic soil and arid areas.

(d) **Coastal Zone agroforestry:** Pathak *et al* (2006), suggested following steps to improve status of agroforestry research with emphasis on coastal zone agriculture and agroforestry which can impart sustainability to productivity and save fragile coastal ecosystems from degradation:

- A time frame programme must be chalked-out for rehabilitation of mangroves along coasts with available techniques without further waste of time.
- Aqua-culture research keeping managroves intact.

- Genetic improvement of identified potential multipurpose trees.
- Studies on organic farming, mycorrhizae in relation to agroforestry and integrated pest-management in agroforestry.
- Research to explore marine algae or seaweeds for food, medicine and green manure.
- Developing viable techniques for rehabilitation of degraded salt affected areas involving good quality forages, MPTs, plantation crops and plants of industrial application. Research on quality assessment of the products obtained in saline habitats.
- Research on developing halophytic crops involving fish, shrimp culture and poultry with agroforestry systems, genetic improvement of salt tolerant plants, raising nursery with saline water and multiplication and conservatiopn of useful gentic material.
- Developing more agro-based industries in coastal area to create more employment through value addition to the product based on coconut, oil palm, honey, cashewnut, rubber, fruit, fish and shrimp, milk, beverage, poultry, sea food and mangrove products.

REFERENCES

Balasubramaniyan, P. and Palaniappan, S.P. 2001. Principles and Practices of agronomy. Agrobios, India, pp: 440-441.

Bene, J.G., Beall, H.W. and Cote, A. 1977. Tree, food and people: Land management in the tropics. International Development Research Centre, No. IRDC-084:52.

Blanford, H.R. 1958. Highlights of one hundreds years of forestry in Burma. Empire Forestry Review, 47(1): 33:42.

Brookfield, H and Padoch, C. 1994. Agrodiversity, Environment 36(5): 7-11

Brownrigg, L. 1985. Home gardening in international development: What the literature Shows? League for International Food Education, Washington DC.

Canell, M.G.R. 1983. Plant management in agroforestry: Manipulation of trees, population densities and mixtures of trees and herbaceous crops. Plant Research and Agroforestry (*ed* Huxley, P.A.). International Council for Research in Agroforestry. Nairobi, Kenya, pp: 455-486.

Chaturvedi, A.N. 1982-83. Poplar farming in Uttar Pradesh (India). U.P. Forest Bulletin No.45, Lucknow (UP), pp:42.

Chinnamani, S. 1969. Natural grasslands (Kanchas) of red soils of Hyderabad. Annal of Arid Zone, 8(1): 52.56.

Chinnamani, S. 1989. Agroforestry Research in India- Reseach Highlights. *In* Agroforeasty Systems in India –Research and Development. India Society of Agronomy, IARI, New Delhi (India): 93-108.

Chinnamani, S. 1990. Kanchas: The silvipasture of Deccan-a repository of agroforestry. Range Management and Agroforestry, 11(1):67-70.

Conklin, H.C. 1957. Hununoo Agriculture: A report on an integral systems of shifting cultivation in the Phillipines. FAO, United Nations, Rome.

Dadhwal, K.S. and Katiyar, V.S. 1996. Land degradation and environmental pollution by mining and land-slides. Soli Management in Relation to Land – Degradation and Environment (*Eds.* Biswas, T.D. and Naryanswamy, G.) Indian Society of Soil Science, New Delhi, India, pp:114-126.

Deb Roy, R and Pathak, P.S. 1983. Silvipastoral research and development in India. Indian Review. Life Science, 3:247-264.

Dev Roy, R.1989. Basic research need in agroforestry: Indian context. In Agroforstry Systems in India-Research and Development Indian society of Agronomy, IARI, New Delhi, India, pp:104-108.

Dommergues, V.R. 1987. The role of biological nitrogen fixation in agroforestory *In* Agroforestry: A decade of Development. ICRAF Nairobi, Kenya, pp: 245-272.

Dhyani, S.K., Singh, B.P., Chauhan, D.S. and Prasad, R.N. 1994. Evaluation, D.S. and Prasad, R.N. 1994. Evaluation of MPTs for agroforestry systems to amaleorate fertility of degraded acid alfisols on sloppy lands. *In* Agroforestry Systems for degraded lands (*eds.* Singh, P., Pathak P.S. and Roy, M.M.). Oxford & IBH Publishing Co. Pvt. Ltd. New Delhi, India, 1:241-247.

Dwivedi, A.P. 1992. Agroforestry principles and practices. Oxford and IBH. Publication Pvt. Ltd. New Delhi, India.

Forde, D.C. 1937. Land and labour in a cross river village. Geographical Journal, Vol. XC (1): 25-29.

Gill, A.S. and Bisaria, A.K. 1997. Agroforestry for sustenance of agriculture and global environment. *In* Agros 'Annual Review of crop ecology', (*ed.* Dahama, A.K.), 1:55-78.

Gill, A.S. and Bisaria, A.K. 1999. Agrisilviculture research under different agro-climatic regions of India. *In* Sustainable Alternate Land Use Systems for Drylands (*Eds.* Singh R.P. and Osman, M.). Oriental Enterprises, Dehra Dun, pp: 84-101.

Gliessman, S.R. 1981. Los sistemas agroforestales como sistemas agroforestales en tropico humedo en Maxico. *In* Investigation de technicas agroforestales tradicionales (*eds* Beer, J.W. and Somarriba, E). Technical Bulletin No.12. Centro Agronomico Tropical de Investigation Ensenanza. Turralba, Costa Rica.

Heliley Lord, 1957. An African Survey, Oxford University Press, Oxford.

Huxley, P.A. 1983. The tree-crop interface of simplifying the biological / environmental study of mixed cropping agroforestry systems (Working Paper). International Council for Research in Agroforestry, Kenya, No. 13, pp: 3-20.

ICRAF. 1985-86. Diagnostic of land use system and designing of agroforestry interventions and research proposals of ICAR / ICRAF team, published by International Council for Research in Agroforestry, Nairobi, Kenya.

Karim, A.B., Savill, P.S. and Rhodes, E.R. 1991. The effect of yound *Leucaena leucocephala* hedges on growth and yield of maize, sweet-potato and coapea in Sierra Leone. Agroforestry Systems, 16:203-211

King, K.F.S. 1979. Key-note address-some principles of agroforestry. Proceedings of Agroforestry Seminar held at New Delhi, India. ICAR, New Delhi, India, pp: 17-26.

King, K.F.S. and Chandler, H.J. 1978. Wasted Land, ICRAF, Nairobi, Kenya, pp: 36.

Lundgren, B.O. 1982. Agroforestry approachas to land use in the tropics. ICRAF, Nairobi, Kenya, pp:14

Lundgren, B.O. and Raintree, J.B. 1982. Sustained agroforestry. *In*. Agroforestry Research for Development – Potential and Challenges in Asia (*Ed* Nestel, B), IS NAIR, pp:37-41.

Mac Dicken, K.G. and Vegara, N.T. 1990 Agroforestry: classification and management. John Wiley & Sons, New York, USA, pp:382.

Martin, F.B. 1973. Behive design for observing variety competition. Biometrics, 29: 397-402.

Megilchrist, C.A. 1965. Analysis of competition experiments. Biometrics, 21: 975-985.

Nair, P.K.R. 1979. Agroforestry research: a retrospective and prospective appraisal. Proceedings of International Conference on Agroforestry. ICRAF, Nairobi, Kenya, pp: 275-296.

Nair, P.K.R. 1984. Soil Productivity Aspects of Agroforestry, ICRAF, Nairobi, Kenya.

Nair, P.K.R. 1985. Classification of agroforestry systems. Agroforestry systems, 3 (2): 97-128

Nair, P.K.R. 1989. Agroforestry defined. Agroforestry systems in the Tropics. Published in Cooperation with ICRAF. Kluwer Academic Publishers, Derodrecht, Netherlands.

Nair, P.K.R. 1993. An interoduction to agroforestry. Kluwer Academic Publishers, Dordrecht, Netherlands, pp: 499.

NCA. 1976. Report of the National Commission on Agriculture, Part IX, Forestry, Govt of India, Ministry of Agriculture and Irrigation, New Delhi

Osman, M; Venkateswarlu, B and Singh R.P. 1989. Agroforestry systems in semi-arid tropics – a review. *In* Agroforestry Systems in India –Research and Development. Indian Society of Agronomy, IARI, New Delhi, India, pp: 18-35.

Pathak, P.S. and Hegde, N.G. 1996. Agroforestry and biodiversity work intended to benefit India. Diversity 12 (3): 47-48.

Pathak, P.S., Solanki, K.R., Rai, P., Handa, A.K. and Peteria H.M. 2006. Hand book of Agriculture (5 th edition). ICAR, New Delhi, India, pp: 1076-1127.

Prasad, R.N. Singh, A. and Verma Arun. 1986. Problems of Hill Lands and their Management in North-East India, Indian Journal of Soil Conservation, 14 (3): 66-72.

Ramanatha Chetty, C.K. 1989. Design for agroforestry experiments. *In* Agroforestry Systems In India – Research and Development. Indian Society of Agronomy, New Delhi, India, pp: 197-201.

Raghavan, M.S. 1960. Genesis and history of the Jumri system of cultivation. *In*: Proceedings of the ninth Silviculture conference, held at Dehra Dun. Forest Research Institute, Dehra Dun, India.

Rao, D.L.N., Gill, H.S. and Abrol, I.P. 1990. Regional experience with perennial *sesbania,* ICRAF, Nairobi, pp:189-198,

Sahu, K.K., Prakash, O and Joshi, C.S. 1996. Productivity and nitrogen use efficiency of Wheat in agro-silvicultural system with some multipurpose tree soecies. *In* Proceedings of IUFRODNAES International Meet on Resource Inventory Techniques in support of Agroforestry Environment Activities (*Eds*. Kohli, R.R. and Arya, K.S.), Chanidigarh India, pp: 227-273.

Shankar, V. 1988. Slvipastoral research in India. Pasture and Forage Crops. *In*. The State of Knowledge Report (ed. Punjab Singh). Range Management Society of India, IGFRI, Jhansi, India.

Shukla, N.P., Rakib, A and Burman, D-1998. Production of *Sesbania* alley intercropped with Berseem and maize at defferent moisture regimes and fertilizer levels. *In* Extended Summaries of First International Agronomy Congress on Agronomy, Environment and Food Security in 21st Century, pp:333-334, 23-27 November 1998, New Delhi.

Singh, G., Abrot, I.P. and Cheema, S.S. 1988. Agroforestry on alkali soil: Effect of planting Methods and amendments on initial growth, biomass accumulation and chemical composition of mesquite (*Prosopis juliflora*) with interspace planted with and without karnal grass (*Diplachna fusca Linn. P. beauv.*). Agroforestry System, 7: 135-160.

Singh, G., Yadav, J.S.P. and Singh, G.B. 2000. Multipurpose tree species management. *In*: Natural Resoure Mangement for Agricultural Priduction in India. (*eds.* Yadav, J.S.P. and Singh, G.B.). International Confernce on Managing Natural Resources for Sustainable Agricultural Production in the 21 st century, held on Feb. 14-18,2000 in New Delhi, India pp: 221-254.

Singh, P.1988. Forage production: present status and future strategies. *In*: Forage Production in India (ed. Punjab Singh), RMSI, Jhansi India, pp:1-10

Singh, P 1994. Land degrtadation a global menance and its improvement through agroforestry. Agroforestry Systems for Surtainable Land Use (*eds* Singh, P., Pathak, P.S. and Roy, M.M.) Oxford and IBH Publishing Co. Pvt. Ltd., New Delhi, India, pp:4-10.

Singh, P. and Roy, M.M. 1993. Silvipastoral systems for ameliorating productivity of degraded lands in India. Annals of Forestry, 1: 61-73.

Singh, P. and Roy, M.M. 1998. Agroforestry and rangeland development *In:* Fifty years of Agronomic Research in India (*eds.* Yadav, R.L. Singh, P. Prasad R., and. Ahlawat I.P.S). Indian Society of Agronomy, IARI, New Delhi, India, pp: 221-254.

Squire, G.R., Ong, C.K. and Monteith, J.L. 1987. Crop growth in semi-arid environment. *In* Proceedings of the International Pearlmillet workshop, 7-11 April, 1986 at Hyderabad, India ICRISAT, pp:219-231.

Steppler, H.A. and Lundgren, B.O. 1988 Agroforestry; Now and in the future. Outlook on Agricultue,17(4):146-151.

Swaminathan, M.S. 1987. A promise of agroforestry for ecological and nutritional security. *In*: Agroforestry – a decade of Development (eds. stepples, H.A. and Nair, P.K.R.) International Council of Research in Agroforestry, Nairobi, Kenya.

Tewari, D.N. 1995. Agroforestry. International Book distributors, Dehra Dun, India.

Veevers, A. and Boffey, T.B. 1975. On the existance of levelled behive designs. Biometrics,31:963-976.

Virginia, R.A., Jenkins, M.B. and Jerrell, W.M. 1986. Development of root symbiont occurance in soil. Biology and Fertility of soil, 2:127-130.

Willey, R.W. 1979. Intercropping, its importance and research need. (I) competition and yield advantage. Filed crop Abstracts, 31 (1): 1-10.

Willey, R.W. and Rao, M.R. 1980. A competitive ratio for quantifying competition between intercrops. Experimental Agriculture, 16:117-152.

Yadav, J.P.S., Bhumbla, D.R., and sharma, O.P. 1972. Performance of certain forest species on saline sodic soils. Proceeding of symposium on New Development in the field of Salt Affected Soils. Society of Soil Science, Cariro, pp: 683-690.

CHAPTER IV

ORGANIC FARMING

4.1 INTRODUCTION

Since its inception agriculture has tested the resiliency of nature. In olden times, agriculture was practiced in a very simple way on natural resources using less energy. Existence of organic farming systems in India is traced back from ancient time as evident in its ancient literature of *Rigveda*. Before advent of synthetic chemicals most food in the world was organically produced, which primarily aimed at cultivating the land and raising crops in such a way as to keep the soil alive and in good health by use of organic wastes of crops and animals, farm wastes, aquatic wastes and other biological materials along with beneficial microbes to release nutrients to crops for increased sustainable production and an eco-friendly pollution free environment. This system was sustainable for centuries in terms of their ability to maintain a continuing stable but low level of production and simultaneously maintaining the ecological balance to a great extent by controlling pests and diseases and maintaining soil fertility without the use of chemical pesticides and fertilizers. The soil health and pest control were achieved by growing mixture of crops, introduction of legumes in crop rotation and using animal manures. These practices also helped in avoiding total crop failure by fluctuation of weather and pest epidemics.

The needs and growing aspirations of increasing population have forced changes in production practices that have imposed excessive demands on natural and renewable resources leading to exploitive agriculture. This has created energy crisis and has adversely affected the environment and soil health. Indeed the much vaunted green revolution in India relying on the high yielding but highly vulnerable varieties, high dose of fertilizer and an abundant supply of water has hardly failed to ward off the threat of droughts. The havoc unleased by this hitech framing based on costly and environmentally hazardous inputs like fertilizers, insecticides and herbicides have led to a worldwide insurgence in natural or organic farming that seeks to derive yield from the earth without despoiling the ecosystem at large.

In USA, the growing popularity of organic farming close to natural farming is reflected in a document prepared by the National Academy of Science, a pre-eminent body of researchers comes to the conclusion that the application of pesticide /synthetic pesticides does not necessarily lead to better farming than the natural, traditional farming techniques perfected over century. A study initiated in 1984 comes out with the instances of farmers around USA who have developed effective natural methods of agriculture sans synthetic chemicals.

'Biodynamic agriculture', an innovative natural farming method, relies on non-chemical preparations made out of cow dung, sand, clay and other organic substances. The term 'biodynamic' has been derived from Greek words, *viz.* 'bios' (life) and 'dynamics' (energy). Hence, the name 'biodynamics' indicates 'working with energies which create and maintain life'. The concept of biodynamics is not new. Dr. Rudolf Steiner, an Austrian philosopher delivered lectures on biodynamics in 1924 at Koberwitz in Germany. In later years, he and his colleagues started laboratory research in Donarch, Switzerland. 'Biodynamic agriculture' was also experimented in Philippines' city of Luzon. According to Luciano olden, who conducted the biodynamic demonstration, vegetables grown tested little different and he felt more energetic as a result of regular eating biodynamically produced vegetables.

The credit for putting the natural farming on practical footing on a global scale goes to Japanese scientist Masanobu Fukoka (1985) who secured a FAO award for the single service rendered to the cause of natural farming. His method appreciates minimum interference with naturae and believes in nature's inherent potential to regenerate itself since soils nourishing qualities stem from organic matter that returns to soil periodically.

Significantly, natural/organic farming assumes importance in view of the fact that about 10,000 human lives are lost, preponderant majority of them in developing world due to misuse of agro-chemicals (Rao, 1990). Indiscriminate use of pesticides adversely affects the ecological balance resulting in pest resurgence, pesticide resistance, pollution of environment and accumulation of pesticide residues at toxic levels in the food chain, whereas excessive and imbalance use of fertilizer cause soil degradation, degeneration of soil fertility and pollution's of soil, water and air. Hence, it is the need of the day to save our environment from pollution and sustain crop production. Organic farming, which refers to growing of crops without using chemical pesticides (insecticides, fungicides, herbicides etc.) and chemical fertilizers, is the best possible way to overcome these problems.

4.1.1 Use of Chemical Inputs and their Adverse Effects

Agricultural chemicals (insecticides, fungicides, herbicides and chemical fertilizers) including hormones and antibiotics leave residues in food that may be hazardous to human and animal health and may cause genetic disorder. High yield of crops are heavly dependent on use of chemical fertilizers, but in the long run many problems, such as occurence of Zn and S deficiency in rice growing areas and adverse effect on soil biotic life occur.

4.1.1.1 Pesticides

The most important pesticides used in agriculture are insecticides, herbicides and fungicides. According to Nigam and Murthy (2000), in terms of consumption on worldwide basis herbicides rank first (45%) followed by insecticides (36%), fungicide (14%) and others (2%). However in India insecticides (71%) rank first, followed by herbicides (18%) fungicides (6%) and others (5%). Singhal (1999) reported that during 1997-1998 India used 45266, 28782 and 12282 tonnes of insecticides, fungicides and herbicides, respectively. In India, maximum use of pesticide is in cotton followed by rice, and vegetable crops. India used about 300 g per ha of pesticides while in Japan it is as high as 12000 g per ha (Handa and Walia, 1996).

The adverse effects of pesticides and fertilizer on environment, animal and human health are given below:

4.1.1.2 Insecticides and Fungicides

- The pesticides applied to control target insect-pests and diseases also many a time kill non-target organisms ranging from the soil micro-organisms to the farmer friendly insects.
- Organochlorine insecticides (e.g., DDT, HCH, aldrin, lindane, etc) persist longest in soil, followed by organophosphate, carbamates and synthetic pyrethroid insecticides (Singh and Dhaliwal, 1992).

- Residues of organochlorine insecticides like DDT, HCL aldrine, endosulfan, heptachlor, and organophosphate insecticides, like malathion and monocrotophos were found in groundwater as well as in surface and sub-surface soils (Handa and Walia, 1996).
- Ecological disaster results due to indiscriminate use of pesticides. Many target and non-target pests become genetically resistant. Contamination of soil and water by long-lived pesticides affected wildlife population with some species nearing extinction (Funderburk *et al.*, 1993)
- According to Tewari and Sharma (1996) about 504 insects and mite species and nearly 150 plant pathogen species are known to have developed resistance to pesticides all over the world.
- Insecticides applied to some crops like vegetables just a few days before their harvest may enter into human stomach through foodchain. The systematic insecticides cannot be removed by washing or peeling of vegetables and fruits. Not only agricultural produce but even milk and milk products, and human milk are also known to contain some of the pesticides in quite high amount (Singh and Dhaliwal, 1992). Pesticide residues in food in India are quite high as compared to other countries.
- Only 12 out the numerous persistent organic pollutants (POPs), that are prevalent in our environment, are the most persistent bio-accumulative chemicals, and have been identified for banning worldwide. These 12 chemicals include nine pesticides: Aldrin, Chlorodane, DDT, Dialdrine, Endrin, Heptachlor, Mirex, Taxafene, PCBs, HCBs, Dioxins, Furan.

Degradation of pesticide residue: Pesticide persistence is usually specified in residual half-life (RL_{50}). The RL_{50} is a measure of the time in days required to reduce pesticide residues to half of its initial deposits (pesticide present on any surface, e.g. leaves, skin etc., after application). The amount of pesticide lost per unit of time is proportional to the total amount present in a particular time. Degradation of this type is called "first-order degradation" and can be represented mathematically as: $dN/dt = -\lambda N$, where λ = decay constant; N = mass present at time t.

4.1.1.3 Herbicides

- Nitrogen-fixing rhizobia, nodulation and biological nitrogen fixation in grain legumes are adversely affected by herbicides (Singh and Wright 1999).
- Certain herbicides having long persistence in soil, when applied to crop, may adversely affect the growth and development of the succeeding sensitive crops in the rotation (Moyer, 1995).
- Groundwater may be contaminated by leaching down of herbicides, like atrazine, cyanazine and simazine (Pinoka *et al.*, 1988)
- Crops like tomato, cotton, etc. are very sensitive to some herbicides, e.g., 2, 4-D. The drifting of these herbicides in the adjoining field may result in loss of the sensitive crops (Singh & Sekhon, 2002).
- According to Moss (1993), by 1991 herbicide resistance has evolved in 84 weed species in 31 countries. Example of herbicide resistance in weeds are: resistance in *Phalaris minor* (little canary grass) to isoproturon in India (Malik and Singh, 1995), *Lolium rigidum* (ryegrass) to many herbicides in Australia (Powles, 1993) and *Alepocursus myosuroides* (black grass) to chlorotoluron and isoproturon in the United Kingdom (Moss, 1993).

4.1.1.4 Fertilizers

- The contamination of underground water by nitrate takes place by leaching down deep into the soil of nitrogen applied to the crops. According to World Health Organisation (WHO) the safe limit of nitrate content in the drinking water is only 45 mg per liter. In India, the nitrate content in well water is very high particularly in the areas were consumption of nitrogenous fertilizer is high (Majumdar & Gupta, 2000). It is also found in food items and their consumption may cause problems to human health.
- The rapid hydrolysis of urea, a nitrogenous fertilizer, in soil can result in high soil pH and high ammonium ion concentrations which is lost as gas and also causes toxicity to germinating seedlings (Gould *et al.*, 1986).
- The continued application of fertilizer phosphate to agricultural soils may result in the build up of natural trace

contaminants, such as cadmium, uranium and radium contained in fertilizer (Sharplay and Menzel, 1987).

4.1.2 Need for Organic Farming

The organic production system is adopted to:

- reduce soil degradation and increase the organic matter content of soils and thus, reducing the bulk density and decreasing compaction.
- maintain long term soil fertility.
- promote soil and water conservation.
- enhance biological diversity within whole system.
- increase biological activity.
- recycle wastes of plant and animal origin in order to return nutrients to the land and thus, minimizing the use of non-renewable resources.
- rely on renewable sources in locally organized agricultural systems.
- minimize all forms of pollution by promoting the healthy use of soil, water and air.
- handle agricultural products with emphasis on careful processing method in order to maintain the organic integrity and vital qualities of the products at all stages.
- promote sustainability in agriculture.

4.2 CONCEPT DEFINITIONS, CHARACTERISTICS AND ADVANTAGES

4.2.1 Basic Concept

The concept of organic farming implies the use of organic manures and natural methods of plant protection instead of using chemical fertilizers and pesticides. Many people regard organic farming involving the integrated use of fertilizers and organic manures as well as integrated chemical and natural pesticides and herbicides.

In fact, organic farming is the agricultural system which aims at cultivation of the land in a way so that the soil is kept dynamic with living activities and in good health, at the same time keeping the environment clean, maintaining the ecological balance and providing stability to the production level without polluting soil, water and air. In this system, the use of chemicals is kept at its minimum, i.e. to the level of bare necessity. This system is self-sufficient and self-dependent as compared to modern chemical farming relying more on chemical inputs. In this system, soil is not only considered to be physical medium for standing crop but also as a shelter of innumerable living entities. It is the only means to achieve the stable/sustainable target of production. The key strongly integrated principles of organic agriculture systems are mixed farming, crop rotation and organic cycle optimization.

According to Bhattacharya and Chakraborty (2005) the success of organic production systems largely depends on conservation of natural resources, *viz.* use of conservation tillage, organic manures, green mauring, biofertilizer, crop rotation, intercropping with legume components, integrated pest management (cultural, mechanical and biological), judicious use of irrigation water (free from nitrates and heavy metals) and its certification, post harvest handling and marketing network.

4.2.2 Definitions

Organic farming has been defined differently but the definition proposed by United States Department of Agriculture appears to be most comprehensive.

- According to the USDA (1980), organic farming is the production system which avoids or largely excludes the use of synthetically compounded fertilizers, pesticides, growth regulators, and livestock feed additives. To the maximum extent feasible, organic farming systems rely upon crop rotations, crop residues, animal manures, legumes, green manures, off-farm organic wastes, mechanical cultivation, mineral-bearing rocks, and aspects of biological pest control to maintain soil productivity and tilth, to supply plant nutrients, and to control insects, weeds and other pests.
- Fantilanan (1990) describes "organic farming is a matter of giving back to nature what we take from it. It is safe, inexpensive, profitable, and sensible. Organic farming is not mere non-chemicalism in agriculture. It is a system of farming based on integral relationships among soil, water, plants and microflora and the overall relationship between plants and animal kingdom, of which, man is the apex animal. It is the totality of these relationship which is the backbone of organic farming.

- Organic farming is basically a simple idea, beginning with soil, compost, natural cycles that need to return garbage, sludge and wastes back to the land; the hazards that pesticides and artificial fertilizers cause to the environment, and personal health benefits that result from eating quality nutritious food (Oldeman *et al.*, 1990).

4.2.3 Organic Versus Natural Farming

(i) The principle elements considered in organic farming are maintaining a living soil, making available all essential nutrients and organic mulching, whereas the essential principles of natural farming are no cultivation and use of no chemical fertilizers, weeding and plant protection.

(ii) Organic farming does not totally exclude elements of modern farming and involves limited and essential use of tillage implements for ploughing, weeding and use of chemicals, whereas natural farming totally eliminate all components of modern farming.

(iii) In organic farming, application of organic source of plant protection and use of organic manure are allowed as against 'do nothing' approach of natural farming.

(iv) Organic farming does not have the philosophical overtone of natural farming which involves continuous search to know the true spirit and form of nature.

4.2.4 Characteristics

According to Palaniappan and Annadurai (2006) the most important characteristics of organic farming are as follows:

- Maximum but sustainable use of local resources.
- Minimum use of purchased inputs, only as complementary to local resources.
- Ensuring the basic biological functions of soil-water-nutrients-human continuity.
- Maintaining a diversity of plant and animal species as a basis for ecological balance and economic stability.
- Creating an attractive overall landscape which gives satisfaction to the local people.

- Increasing crop and animal diversity in the form of polycultures, agroforestry systems, integrated crop/livestock systems, etc. to minimise risk.

4.2.5 Options

Following options are available in organic farming:

(i) **Pure organic farming**: It advocates the use of organic manures and biological pest control methods. Use of both inorganic fertilizers and chemical pesticides is excluded.

(ii) **Integrated green revolution farming:** In this farming, the basic trends of green revolution, such as intensive use of external inputs, increased irrigation, use of high yielding varieties and hybrids and mechanisation of labour are retained, but to limit the damage to the environment and human health both organic and inorganic inputs are integrated to achieve much greater efficiency and thus, reducing the needs for chemical fertilizers and pesticides.

(iii) **Integrated farming system**: It is the low input organic farming depending o local resources, such as recycling agricultural wastes and crop residues.

4.2.6 Objectives

Following are the objectives of organic farming:

A. Organisation

1. Ecological orientation, economy, efficient labour input.
2. Diversification and balanced combination of enterprises resulting stability.

B. Production

1. Cycle of nutrients predominantly through farm produced materials.
2. Weed control by crop rotation and cultural practices.
3. Pest control based on non-polluting substances.
4. Farming combined with livestock for production and health.
5. Production of quality food and other produces free from adverse effect of inorganic.

C. Influence on life process

1. Production is integrated into healthy environment.
2. Harmonic balanced conditions for crop production and animal husbandry.

D. Social values

1. Optimum input: output ratio.
2. No or low environmental pollution and soil degradation.
3. Maximum conservation of soil and water, and soil fertility.
4. Supports biodiversity in production system.
5. Not only economic motivation but holistic approach.

4.2.7 Components

The components of organic farming are:

(i) **Nutrient management:** Organic manures, such as green manures, legumes in cropping systems (biological N fixation) farmyard manure, biogas slurry, compost, crop residues, biofertilizers, organic fertilizers, etc.

(ii) **Non-chemical weed control:** Such as control of weeds by mechanical means. No herbicides are applied.

(iii) **Non-chemical pest management:** Pesticides of organic origin and biological pest management are encouraged.

4.2.8 Organic Vs Conventional Farming

Organic Farming	Conventional Farming
A. Yield	
The nutrient release from organic source is slower as compared to chemical fertilizers and hence, the yield is expected to be lower.	The high yielding varieties fertilized with chemical fertilizer give more yield as compared to organic source of nutrients.
Organic sources of nutrients release nutrients slowly and there is little scope of nutrient loss.	Nutrient from chemical fertilizers are released fast resulting escape of nutrients to atmosphere and lower zone of the soil and thus, become partly available to the crop.
Weeds are not effectively controlled due to no use of herbicides resulting in lower crop yields	Chemical weed control is considered more effective and economical to that with cultural, mechanical or biological methods used in organic farming.

Organic Farming	Conventional Farming
The severity of diseases may be higher under organic manures coupled with high moisture in soil.	Chemical methods of insect and disease control is more effective than non-chemical methods.
B. Quality	
Organically produced crops including vegetables and fruits contain lower or no concentration of nitrate and pesticide residues.	Chemically produced food, sometimes, contain undesirable ingredients, like nitrate and pesticide residues affecting quality of the produce.
C. Energy	
Substituting manure for chemical fertilizers, rotary hoe for herbicides, and biological pest control for pesticides can save huge energy (Steinhart and Steinhart, 1974).	Energy requirement is more. Twenty-five percent more energy is needed with a conventional cropping system in UK (Jones, 1986).
D. Economics	
Costly inputs are not used. Moreover, organically grown produce fetches a higher price in the market.	The cost of input is expensive due to energy intensive chemicals, like fertilizers and pesticides.
More labour is needed for operations, like manual weeding and application of organic manures (e.g., Farmyard manure).	Less labour is needed.

4.2.9 Advantages

Advantages of organic farming are given below:

- Produce optimal conditions in the soil for stable yields and quality crops.
- Supply all macro and micro-nutrients required by crops.
- Improve growth and development of plants.
- Improve (i) soil physical properties (such as granulation, good tilth, aeration and water holding capacity, etc.), (ii) chemical properties (such as supply and retension of soil nutrients in available form), and (iii) biological properties (such as improvement in environment of rhizosphere for growth of beneficial microorganisms).
- Reduce the need for chemical inputs and thus, minimizing pollution.
- Moves the pH of soil towards neutrality and thus, restore appropriate soil condition for crops spoiled either due to acidity or salinity and alkalinity.

- Provides more resistant to diseases and insects and hence, a few chemical sprays or other protective treatments are required.
- Provides food free from toxic chemical residues.
- Avoids chain reaction in the environment from chemical sprays and helps to prevent environmental degradation.
- Provides more income as consumers pay more for organic foods.

4.2.10 Limitations

Advantages of organic farming are given below:

- Scope for organic farming is still limited due to general low yield per unit area.
- Crop loss generally occurs in changing over to organic farming, since the soil takes 3 to 4 years to build up from weaken or destroyed condition created by continuous use of chemical inputs.
- Organic food is no healthier and provides no significant benefits compared with conventionally produced food as concluded recently by Dr. Alan Dangour (who led the review by the London School of Hygiene and Tropical Medicine) from his studies based on 50 years of published works. It was funded by U. K.'s Food standard Agency. The findings partly published on July 29, 2009 in the American Journal of Clinical Nutrition, which revealed that there is no good evidence that consumption of organic food is beneficial to health based on the nutrient content. However, the European Union study co-ordinated by Dr. Leifert which ended in May 2009 involved 31 Research and University institutes. It was found that levels of nutritionally desirable compounds, such as antioxidants and vitamins were higher in organic crops (Reported by Karen Mc Veigh, Guardian News Service, 2009).
- Carbon-dioxide, methane and nitrous oxide - the green house gases are reported to accentuate the potential for global and climate change. Carbon-dioxide emissions are due to the decomposition of soil organic matter and crop residues, besides from burning fossil fuels. Methane originates mainly from cattle manure. Nitrates formed in excess of the crop demand may further denitrify to yield nitrous oxide and its emission is further increased as denitrification is positively

correlated with microbial biomess and organic matter which tend to increase on manuring. Proper management that prevents large pools of available soil NO_3 will minimize the negative effect (Tiwari *et al.*, 2005).

- Heavy metals concentrations in animal wastes (manures) and sewage sludges can be at times very high and often exceeds concentration normally found in chemical fertilizers. Applications of such organic manures in very high quantities to meet the requirement of major plant nutrients may lead to heavy metal accumulation in soil polluting arable land.
- Faulty storage of compost / FYM results in 40 to 60 per cent losses in nutrients specially nitrogen due to exposure to sun waterlogging. Leaching of nitrate polluting ground and surface water resources id observed from cattle dung pits.

4.3 PRINCIPLES AND PRACTICES OF ORGANIC FARMING

4.3.1 Principles of Organic Farming

Basic principles of organic farming (Vogtmann, 1984) are given below:

(i) Organise the production of crops and livestock and the management of farm resources, so that they harmonise rather than conflict with natural systems.

(ii) Use and development of appropriate techniques based upon an understand of biological systems.

(iii) Achieve and maintain soil fertility for optimum production by relying primarily on renewable sources.

(iv) Use of diversification to pursue optimum production.

(v) Aims for optimal nutritional value of staple food.

(vi) Use decentralised structures for processing, distributing and marketing of products.

(vii) Strive for equitable relationship between those who work and live on the land.

(viii) Create a system which is aesthetically pleasing for those working in this system and for those viewing it from the outside, e.g., it should enhance rather than scare the landscape of which it forms a part.

(ix) Maintain and preserve wildlife and their habitats.

4.3.2 Practices of Organic Farming

Various practices which are used in organic agriculture include (Vogtmann, 1984):

(i) Balance crop rotation with legumes and use of catch crop.

(ii) Green cover of the soil is maintained as much as possible with green manures and under sowing.

(iii) Animals are preferably included in the farming operation.

(iv) Shallow ploughing (approximately 12 to 15 cm) is practiced with deep loosening of the soil to a depth of 30 cm.

(v) The purpose of fertilizing is to activate soil life and improve soil structure (no use of synthetic mineral - N). Nitrogen is supplied through different organic sources and other main elements are supplied through basic slag, rock phosphate, potash of magnesia, and for the trace elements mainly through various types of rock dusts, e.g., basalt, etc.

(vi) Weed control is conducted through the cropping system, tillage and mechanical as well as thermal methods e.g., flame gun, etc. (no use of herbicides).

(vii) Pest control is conducted through careful selection of crop varieties that are best adopted to local conditions, cropping systems, and tillage. Various plant extracts as preventives, biological and biotechnological methods are used. Direct application of certain compounds (e.g., Rotenone, Cu, S, etc.) is allowed (no use of synthetic pesticides).

4.4 NUTRIENT MANAGEMENT IN ORGANIC FARMING

Nutrient management in organic farming systems depends solely on organic inputs for nutrient availability through mineralization or conversion of organic into inorganic forms by microbes. The source of organic form of nutrients are green manures, crop residues. Compost, bio-fertilizers and bio-solids from agro-industries and food processing wastes.

Apart from providing nitrogen either through nitrogen fixation or crop residue, inclusion of legumes in the rotation plays other very important roles including changes in soil properties, reduction of diseases, reduction of toxic substances in crop residues and / or release of growth promoters from residues (Heichel and Barnes, 1984).

Various practices can play an important role in nutrient management in organic farming system as discussed briefly below:

4.4.1 Green Manuring (GM)

4.4.1.1 Background and scope

Green manuring is an age-old practice. The first test on green manuring in India was made in 1882 at Kanpur followed at Nagpur and Damraon in 1882 and 1987, respectively. The European planters were the pioneers who gave a systematic practice of green manuring as far back as 1890 in the coffee estates of southern India. It started losing its importance with the availability of chemical fertilizers and their fast and immediate effect on crop production from late 60s. In the present era of intensive cropping system, heavy input technology, environmental degradation and other related problems, its re-inclusion in nutrient supply system is highly valued.

4.4.1.2 Definition

Green manuring can be define as a practice of ploughing or turning into the soil of undecomposed green plant material for improving the physical condition of the soil or for adding N when the green manure crop is a legume.

4.4.1.3 Types of GM

For green plant material, the crop is either grown *in situ* or brought from outside and incorporated.

(i) **Green manuring in situ:** In this practice, green manure crops are grown and buried in the same field, which is to be green manured, either as a pure crop or an intercrop with the main crop. The most common green manure crops grown are sunnhemp *(Crotalaria juncea), dhaincha (Sesbania aculeata,* pillipesara (*trilobus*), clusterbean or gaur (*Cyamopsis tetragonoloba*), wild indigo *(Tephrosia purpurea)*, cowpea *(Vigna unguiculata)*, green gram *(Vigna mungo)* and berseem *(Trifolium alexandrinum*).

(ii) **Green leaf manuring:** This refers to turning under into the soil green leaves and tender green twigs collected from shrubs and trees grown on bunds, waste lands and nearby forest areas.

The common shrubs and tree species used are *Glyricidia maculata, Sesbania speciosa, karanj (Pongamia pinnata),* etc. Tree species like *Leucaena leucocephala* also have tremendous potential. Vegetative parts of other trees like *Casuarina, Calotropis, Tephroisa, Indgofera* and *Cassia* can also be used as green manures.

4.4.1.4 *Subsidiary objectives of GM*

These are given below:

(i) **Shade crops:** With the object of shading the soil surface and preventing the rise of temperature, green manure crops are sown in young orchards and later on incorporated as green manure.

(ii) **Cover crops:** With the object of covering the surface with a vegetation in plantation crops like tea and coffee, green manure crops (*e.g., Gliricidia*) are grown in hill slopes during the rainy season to prevent runoff and soil erosion. Later on it is used as green manure.

(iii) **Catch Crop:** With the objective of utilizing left-over nitrate and moisture in soil, legumes are inter-sown in main standing crop either before or after harvest. The catch crops are ploughed in as green manures.

(iv) **Forage crops:** Some legumes, like *Philippesara* are grown for taking a few cuttings of green fodder for cattle in early stages. The seeds of legume are broadcasted in the standing rice crop, 3 to 5 days before harvest in coastal Andhra Pradesh.

4.4.1.5 *Advantages of GM*

The benefits generally attributed to green manuring are:

- Green manuring adds organic matter to the soil which improves physical conditions of soil by
  - (a) building up soil structure and improving tilth,
  - (b) promoting formation of crumbs in heavy soils leading to better aeration and drainage,
  - (c) increasing the water holding capacity of light soils, and
  - (d) providing canopy cover over the soil and thus, reducing soil temperature and protecting the soil from erosion.
- Green manure crops improve fertility of the soil by

(a) preventing leaching of nutrients to lower layer of soils,

(b) absorbing nutrients from lower the layer of soils and leaving them in root-zone of crops when ploughed in, for use by the succeeding crops,

(c) increasing the availability of certain plant nutrient like phosphorus, calcium, potassium, magnesium and iron through activities of soil microorganisms and by producing organic acids during decomposition, and

(d) by fixing nitrogen (on an average 60 to 100 Kg per ha in single season under favourbale conditions) through nitrogen fixing bacteria, *rhizobia*, harboured by leguminous green manure plants.

- Green manures may also play a key role in the reclamation of problem soil. *Dhaincha (Sesbania aculeata)* applied to sodic soils continuously for four to five seasons improves the permeability and help to leach out the harmful ions of sodic soils. Decomposition products of *Sesbania* help in removal of exchangeable Na from salt affected soils. *Argemone mexicana* and *Tamarindus indica* have a buffering effect when applied to sodic soils.

- Green manuring have following positive effect:

  (a) In increases the yield of crop to an extent of 20 per cent compared to no green manuring. According to Tiwari *et al.* (1980) 40 kg N per ha along with green manuring compared well with the application of 120 Kg N per ha without green manuring with regards to availability of nutrient in the soil, their uptake and rice yield. An increase of 54 per cent in grain yield of wheat succeeding rice with green manure, was also observed, indicating notable residual effect of green manure. Rye, maize and wheat produced higher yield after green manured crops than after summer fallow (Reddy *et al.*, 1986)

  (b) It increases vitamin and protein content of rice.

- Certain green manure like Pongamia and Neem leaves are reported to have insect control effect.

4.4.1.6 Disadvantages of GM

When the proper technique of green manuring is not followed or weather conductions become unfavourable, the following disadvantages are likely to become evident:

- Under rainfed conditions, if sufficient rainfall is not received, proper decomposition of green manure crop and satisfactory germination of succeeding crop may not take place.
- In some cases, green manuring means loss of *kharif* (rainy season) crop, The practice of green manuring may not be always economical.
- An increase of diseases, insects and nematodes is possible.
- The cost of growing green manure crops may be more than the cost of commercial N fertilizers.
- A risk is involved in obtaining a satisfactory stand and growth of green manure crops in rainfed areas.

4.4.1.7 *Classification of green manures*

Green manures can be classified into two groups *viz.*, legumes and non-legumes and further sub-divided into two groups in each *viz.*, green manure and green leaf manure.

Green manures	**Legumes**	1) Green manure, e.g., Sunnhemp *(Crotalaria juncea)* 2) Green leaf manure, e.g., glyricidia *(Glyricidia macutata)* Cassia auriculata
	Non-Legumes	1) Green manure *e.g.,* Sunflower (*Helianthus annuus*) Jowar (*Sorghum vulgare*) 2) Green leaf manure *e.g., Calotropis gigantea, Thespesia populnea*

Legumes fix free nitrogen from the atmosphere, whereas in case of non-legumes free nitrogen is not fixed except in specific plants which have root nodules, produced by bacteria or fungi, e.g., *Casurina*, Cycas. Non-legumes are not etc. require more soil moisture and time for decomposition.

4.4.1.8 *Plant suitable for GM*

- Preferably be a legume with good nodulation capacity for rapid nitrogen fixation.
- Should have low water requirement for its own growth.
- Should have deep root system to tap lower region for plant nutrients.
- Should be of leafy habit.
- Should contain large quantities of non-fibrous tissues of rapid decomposibility.

4.4.1.9 The methods of growing GM crops in situ

Both legumes and non-legumes are used as green manuring crops. Legumes due to their additional advantage of adding atmospheric nitrogen to the soil are most commonly used as green manure crops, while non-legume crops are used to a limited extent.

The green manure crops are grown in different ways in India to suit local rotations and climatic conditions. The common methods are:

(i) **Summer-sown catch crop:** The quick growing green manure crops like Sunnhemp *dhaincha*, pillipesara, etc. are grown in May to June and buried in the field in the month of July to August before the planting of main crop. In tropical rice (Oryza sativa) lowlands, nitrate assimilation by plants during the transition from the dry to the wet season can preelude nitrate loss upon soil flooding and permit recycling of this N (George *et al.*, 1994). They concluded that though weeds were also effective in conserving soil N during the dry to wet transition, but legumes (*Vigna radiata* and *Sesbania rostrata*) were most suitable nitrate catch crops because they allow the harvest of an economic product or recycle of more N to a subsequent flooded rice crop. This method is most common with rice in Andhra Pradesh, Mysore, Tamilandu, Kerala, Orissa, Bihar, Jharkhand and West Bengal.

(ii) **Undersown crop:** The quick-growing green manure crops are sown along with the main crop between the lines of the crop, e.g., *Dhaincha* intercropped with rice; sunhemp, cowpea or horsegram sown in between the rows of irrigated cotton and sugarcane. This method is followed in Punjab, Haryana, Uttar Pradesh, Karnatka, Gujarat, Chattishgarh and part of Madhya Pradesh.

(iii) **Crops taken on bare fallow:** In wheat tracts, or with vegetables, green manure crops like sunhemp, *guar*, cowpea and *dhaincha* are sown and buried during the *Kharif* season, in place leaving field fallow or alternative *Kharif* crop being raised. With this method, the cultivator lose one *Kharif* season crop. This method is followed in Punjab, Uttar Pradesh, Rajasthan, Bihar and Madhya Pradesh.

(iv) **Green manure as a main crop:** This is only practiced on very poor sandy soils as a preliminary to more profitable rotations, or a green manure crop, mainly dhaincha, is raised as a main crop in the programme of reclaiming saline and alkali soils.

4.4.1.10 *Selection of GM crops and their agronomy for growing in situ*

Selected crops should be best adapted to the climate, soil and crop rotation. Clement et al., (1998) have clearly shown the suitability of some green manure species as an alternative to mineral fertilization as most of the green manure species performed as well as urea supergranules. They have shown that higher nitrogen uptake rates by rice with green manure than with urea from 52 to 69 days after transplanting indicate a better synchrony between green manure -nitrogen availability and nitrogen uptake by rice.

(i) **Sunnhemp (*Crotalaria juncea*):** As a quick growing and most outstanding green manure-cum-fiber crop is well suited to almost all parts of the country, provided the field is not swampy or waterlogged and the area receives sufficient rainfall or has an assured irrigation. The crop can be cut even when it is 45 days old. The seed rate is 25 to 40 Kg per ha and the yield of green matter may very ranging from 12000 to 25000 Kg per ha. It is used with sugarcane, potatoes, second crop of rice in south India and irrigated wheat in north India.

(ii) **Dhaincha (*Sesbania aculeata*):** It is quick growing succulent green manure crop occupying the second place next to sunnhemp. It comes up well in loamy and heavy soils. It has the advantage of growing under adverse conditions of drought, water logging, salinity and acidity. It corrects alkalinity if grown repeatedly for four to five years. The recommended seed rate is 25 to 30 Kg per ha. It makes good growth in two to four months and produces green matter ranging from 10 to 20 tonnes per ha. The plants ploughed in for single crop rice. It is also in wide use in Assam, West Bengal, Bihar and Tamilnadu with sugarcane, potatoes and rice.

(iii) **Guar or clusterbean (*Cyamopsis tetragonoloba*):** It flourishes well in areas of low rainfall and poor fertility. It is the most common green manure crop in Punjab and Rajasthan.

(iv) **Sesbania or Wild Dhaineha** (*Sesbania speciosa*): It can be grown in different soil conditions including alkaline soils. The growth is very slow in the first 30 to 40 days, it picks up subsequently making rapid growth. Sesbania seeds are broadcasted at 50 kg per ha. three to five days prior to the harvest of rice crop. With the available soil moisture, the sesbania seeds germinate. This method involves no preparatory cultivation. The seeds get thrust into the soil while labourers move during harvest of rice crop. In another method, the field is ploughed and sesbania seeds are broadcasted at 35 to 50 kg per ha. A good stand can be obtained by irrigation. Three weeks old seedlings of sesbania can be grown along the borders of the field during the first crop season of rice and utilized as green manure for the second crop. Such border planting of sesbania at a spacing of 5 to 10 cm in one ha will give 5000 to 8000 kg of green matter for the second crop.

(v) **Pillipesara (*Phaseolus trilobus*):** It is a herbaceous creeper growing into a short dense cover crop. Initially, adequate soil moisture is essential for its early growth. Though it does not produce a bulky yield, it is capable of being cut twice or thrice before being ploughed into the field. The seed rate is 20 to 25 kg per ha and yield of green matter is 10,000 to 12,000 kg per ha.

(vi) **Wild Indigo (*Tephrosia purpurea*):** It is a slow growing green manure crop suitable for light soils. It does not withstand water stagnation. The seeds are broadcasted as catch crop in the standing crop of rice just a week before harvest. Wild Indigo has hard seed coat and hence to hasten germination, the seeds are to be pounded with sand or steeped in hot water at 55°c for two to three minutes. The seed rate is 25 to 40 kg per ha. The yield of green matter varies from 3500 to 6000 kg per ha.

The leguminous green manure crops, like dhaincha (*Sesbania*), sunnhemp (*Crotolaria juncea*), cowpea (*Vigna sinesis*), etc., not only fix atmospheric N in the soil but also provide considerable amount of organic matter when incorporated into the soil. According to Reddy *et al.*, (1986a) legumes like pigeonpea, showy crotalaria (*Crotalaria spectabilis*), hairy indigo (Indigofera hirsuta) and jointvetch (*Aeschynomene americana*), when grown as a green manure crops, reduced root-knot nematodes in the following crop. Hiremath and Patel (1998) reported that incorporation of the three green manure crops, *S. rostrata*, S. aculeata, or *C. juncea* was comparable with that of 100 kg N per ha for grain yield of 5 t per ha of rice during summer season.

4.4.1.11 Agronomy of certain green leaf manure shrubs/trees

Application of green leaves gathered from shrubs and trees growing in waste lands to the fields where crops are to be raised is known as green leaf manuring. *Cassia auriculata, Dodonia viscosa,*

Calotropis gigantea. etc., are common shrubs growing in the waste lands. Leguminous tree like *Pongamia glabra* and *Cassia siamea* can be planted in waste lands for augmenting the supply of green leaves.

Some of the most common shrubs and trees utilized for green leafy material is given below (Palaniappan and Annaduria, 2006):

(i) ***Azadirachta indica* (Neem):** It is profusely branching, large ever-green tree and gives plenty of foliage. It comes up in all types of soil. Trees are established by planting seedlings at a spacing of 5 to 6 m. One or two loppings in a year are taken in favourable seasons, each lopping weighing about 150 to 200 kg of green mutter.

(ii) ***Cassia auriculata***: It is very common plant in waste hills, hill slopes, plain sea shore, etc, almost in the wild condition. The plant is propagated by seeds. The seeds get dispersed and plants grow annually.

(iii) ***Derris indica* (Syn. *Pongamia glabra*)**: It is a leguminous, moderate sized evergreen tree. Trees are established by means of planting two to three months old seedlings, 4 to 5 m apart. Loppings may be taken once or twice a year. A tree yield approximately 100 to 150 kg of green material per lopping.

Mishra *et al.*, (1998) found that green leaf manuring with *Gliricidia, Ipomea* or *Pongamia* considerably reduced fertilizer needs of summer rice at Bhubneshwar.

(iv) **Glyricidia (*Glyricidia maculata*):** It is quick growing tree and often used for shade and green leaf manure in tea, coffee and cocoa plantation. It can be planted on alternate field bunds of wetland, 1 to 2 apart, or as a thick hedge by close planting in three to four rows at 0.5 m spacing or along field border as tall shrubs giving support to the fence line or along farm roads on sides for the production of green leaf. Glyricidia can be propagated by planting stem cutting or seedlings raised in nursery. The seedlings are transplanted when they are about 30 to 60 days old. Within two years after planting the plants are ready for lopping. Each plant gives 5 to 10 kg of green leaves annually.

4.4.1.12 Technique of green manuring in situ

The maximum benefit from green manure cannot be obtained without knowing (i) when the green manure crops should be sown ? (ii) when they should be buried in soil? and (iii) how much time should be given between the burying of green manure crop and sowing of the next crop ?

- **Time of sowing green manure crops in situ:** In India climate differs a great deal from region to region. The normal practice usually adopted is to begin sowing immediately after the first monsoon rains. In case of legume, the seeds at sowing time must be inoculated with specific bacterial culture of the right type and efficiency. This will increase the total nitrogen content of legumes.
- **The stage of green manure crop at burying (incorporation):** Based on experiments conducted, it can be generalized that a green manure crops take about six to eight weeks from the time of sowing to attain the flowering stage. The basic principle which governs the proper stage of turning under green manure crops, should aim at getting maximum succulent green matter at burying.
- **Time interval between burial of green manure crop and sowing of next crop:** The time interval or time lag that should be allowed for complete decomposition of the turned under green matter before planting the next crop depends upon weather conditions and nature of buried green material. Humid weather due to high rainfall and high temperature favours rapid decomposition. When the manure crop is woody, sufficient time should be allowed for its proper decomposition before planting next crop.

4.4.1.13 Green manure in cropping systems

In rice-wheat or maize-wheat rotation followed in Northern India there is a long period of 50 to 70 days between harvesting of wheat and transplanting of rice. Short duration grain legumes, such as green-gram (mungbean), black-gram (urdbean) and cowpea can be raised during this period and after picking pods, the residue can be incorporated into the soil.

Palaniappan (1997) reported that a short duration rainfed green manure crop can be grown in South India taking advantage of early monsoon rains in June which can be incorporated before planting the lowland rice under irrigation. Alternatively, a short duration grain legume can be grown before planting rice under irrigated condition, if farmers are unable to set apart separate land and time for green manure crop. After harvesting the pods for grain, the residue can be incorporated for green manure to rice crop. Cowpea was the best among the crops tested for grain yield, biomass, nitrogen added and rice yield.

Green manuring can be practiced in potato based cropping systems (e.g. Potato-wheat/sunflower). There is a long fallow period of 120 to 130 days after the harvest of wheat and before planting of potato crop in September. Green manure can also be included in oilseed based cropping systems, like Groundnut-*toria*/wheat.

Green manure can be intercropped with sugarcane and cotton crops and incorporated into soil after 50 to 60 days.

In rainfed areas, multipurpose trees such as *Leucaena leucocephala* and *Sesbania* can be raised on field boundaries for production of green manure, animal feeds, and fuel (Reddy, 2000).

The beneficial effect of green manuring varies from crop to crop. Rice is benefited most while increase in yield of cotton and sugarcane are relatively less. Reddy *et al.*, (1986 b) reported that *rye*, maize and wheat produced higher yield after green manured crops than after summer fallow.

4.4.1.14 Decomposition of green manure

The green manure incorporated in the soil undergoes a series of biochemical change before the nutrient content in the plants become available and humus is synthesized. The incorporated green manure in soil is acted upon by many types of microgranisms, such as bacteria, fungi, actinomycetes protozoa, worms and insect larvae resulting in formation of several intermediate and end products during the decomposition. The nutrients become available slowly and steadily for a prolonged period of time. The decomposition can be broadly of two types (a) Aerobic decomposition, and (B) Anaerobic decomposition influenced by factors, like (i) organism present (e.g., fungi or bacteria, autotrophic or hetrotrophic), (ii) temperature (optimum-30 to 35 C for normal decomposition, (iii) aertation (presence and absence of oxygen decides aerobic and anaerobic decompositions, respectively), (iv) moisture supply (optimum moisture is necessary for normal decomposition), and (v) soil factor (physical, chemical and biological properties of soil influence rate and type of decomposition).

A. **Aerobic decomposition:** Incorporated plant material made up of numerous compounds which can broadly be under three groups: (i) Carbon compound consisting of carbohydrates, fats, oils, organic acids, lignin and other cyclic organic compounds; (ii) Nitrogen compounds consisting of proteins, amino acids and other non-protein nitrogenous substances, and (iii) Mineral salts.

(i) **Carbon compounds**: The different carbon compounds, like glucose, starch, cellulose and hemicellulose are attacked by the microorganisms (e.g bacteria, fungi and actinomycates) to be converted finally to carbon dioxide and water.

$$\underset{\text{(Gulcose)}}{C_6H_{12}O_6} \xrightarrow{\text{Bacteria}} 6CO_2 \quad 6H_2O \quad \text{Energy}$$

(ii) **Nitrogen compounds:** The major nitrogenous compounds in plant material are proteins. The other nitrogenous compounds are urea, purine bases, lecithin, choline, cynamide, alkaloides, etc. In decomposition of protein, it is hydrolysed by proteolytic enzymes produced by microorganisms to polypeptides, amino acides and other nitrogen derivatives. These are further acted upon and ammonia is formed. Under aerobic condition, it is oxidized by the nitrifying bacteria to nitrate through process of aminization, ammonification and nitrification. Under anaerobic condition ammonia does not accumulate in the soil.

$$\text{Proteins} \xrightarrow{\text{Hydrolysis}} \text{Polypeptides, peptides \& peptones} \xrightarrow{\text{Aminization}} \text{Amino acids}$$

$$\text{Amino acids} \xrightarrow{\text{Ammonification}} \text{Ammonia} \xrightarrow{\text{Nitrification}} \text{Nitrite} \quad \text{Nitrate}$$

(iii) **Mineral salts:** Phosphours, potassium, calcium and magnesium, etc., are found in plant mainly in organic form and to some extent inorganic form. These are converted to more soluble form in the process of the decomposition and become readily available to plant.

B. **Anaerobic decomposition**: Under this condition, only those organisms which are capable of thriving in the absence of oxygen will develop to decompose the various constituents present in plant body.

(i) **Carbon compounds:** These are attacked by anaerobic bacteria and fungi and converted to methane, organic acids, alcohols and carbon dioxide. For example, if glucose is attacked by anaerobic bacteria and fungi, products like lictic acid alcohol, butyric acid and fumaric acid are produced.

$$C_6H_{12}O_6 \quad \text{Bacteria} \quad 2C_3H_6O_3 \text{ (Lactic acid)}$$

$$C_6H_{12}O_6 \quad \text{Bacteria} \quad 2C_2H_5OH \text{ (Ethyle alcohol)} \quad 2CO_2$$

$$C_6H_{12}O_6 \quad \text{Fungi} \quad 2C_3H_6O_3 \quad C_2H_5OH \quad CO_2$$

$$\underset{\text{(Ethyl alcohol)}}{2C_6H_{12}O_6} \quad C_2H_4O \quad 2CH_4$$

(ii) **Nitrogenous compounds:** Change in nitrogenous compounds are similar as in aerobic decomposition for conversion of proteins to ammonia. Under anaerobic decomposition, along with ammonia, various amines and mercaptans are also produced which cause putrefactive odour and hence, the decomposition is sometimes referred to a putrefaction. The end product ammonia further converted to gaseous nitrogen which may be lost to atmosphere.

(iii) **Mineral Salts:** As in the case of aerobic decomposition, the mineral constituents are also converted into more soluble forms in anaerobic decomposition.

Carbon-nitrogen ratio and decomposition: Materials with a very wide C : N ratio do not decompose rapidly, as the nitrogen contained in the green matter is not sufficient for microbial activity. Materials with very narrow C : N ratio decomposes more rapidly as excess nitrogen is available for the microbial activity. Only green matter with C : N ratio of 30:1 or lower will decompose in normal manner.

4.4.2 Inclusion of Legumes in Cropping Systems

Symbiotic nitrogen fixation by legumes in association with specific rhizobia needs no introduction. Legumes fix higher amounts of nitrogen (on an average 140 kg N per ha) compared to non-legumes (on an average 8 kg N per ha) (Burns and hardy, 1975). Legumes can fit well in different crop rotations either as sole (pure) crops or as inter/mixed crops with legumes or non-legume crops (Sekhon *et al.*, 1992). Grain legumes not only produce protein rich grains but also fix considerable among of nitrogen in the soil. Moreover, incorporation of its crop residue into the soil *in situ* helps greatly in improving physico-chemical properties of the soil.

Wherever it is not possible to include green manure in the cropping system, there is possibility of inclusion of legume in the system. Because of their shorter duration, small stature and erect growth habit, legumes could be easily accommodated as intercrops. Legumes of short (vigna group) or long (pigeonpea) duration group offer greater advantage as intercrops in cereals by providing additional profit to the farmers. Vigna group is harvested earlier and provides some nitrogen to the longer duration companion crop. Pigeonpea is of longer duration during the early growth phase and starts using the space effectively by putting additional growth after harvest of the early duration companion crop. These pulses provide nitrogen upto 50 kg per ha to the associated crops in vertisols. Higher benefits have been reported for crop combination of cowpea (fodder) + sorghum, green/black-gram + sunflower and pigeonpea + groundnut (Balasubrananiyan and Palaniappan, 2002).

Since initial crop growth upto 100 days after planting of sugarcane is very slow, this period can be utilized for growing legumes in inter-row spaces as intercrops for grain or fodder. Yadav (1980) reported that summer legume crop like black-gram and green-gram intercropped with spring planted sugarcane increased the cane yield markedly besides producing 0.4-0.5 t of pulse grain per ha.

Short duration grain legumes intercropped with cereals has been found to be beneficial in terms of yield and nitrogen economy, e.g., maize + blackgram/ greengram (Pathak *et al.*, 1968, 1970), maize + soybean (Srivastava and Soren, 1980), and pigeonpea + maize (Srivastava + Srivastava, 1991). Singh *et al.*, (1986) found that intercropping with soybean or urdbean (1:1) increased the productivity of maize crop by 15 to 20 per cent. Furthermore, the effect of intercropping maize + soybean or urdbean was equivalent to the addition of 80 kg N per ha. Verma *et al.*, (1980) reported that net return from pure stand of pigeonpea is always less than that obtained from pigeonpea + maize or pigeonpea + rice combinations.

Intercropping of pigeonpea with legumes, like greengram, blackgram, cowpeas and groundnut has proved successful without any adverse effect on pigeonpea (Giri and De, 1978; Saxena and Yadav, 1979). Similar finding was reported by Prasad and Srivastava (1991) in pigeonpea + Soybean intercropping system.

Inclusion of grain legume has been found to be advantageous for sustained yields of cropping systems under rainfed situations, e.g., cowpea-finger millet (Banglore) greengram-finger millet (Bhubaneshwar), sunhemp-wheat (Hoshiarpur) and blackgram-wheat (Rewa). The benefit of green manuring is felt only after continued application for 3 to 5 years. (Singh and Venkateshwaralu, 1985). Patel (1980) reported that pigeonpea intercropped with groundnut showed highly promising results in economising nitrogen equivalent to 90 kg per ha in following wheat crop, Rao (1974) compared two crop rotations, viz., maize-wheat-greengram and pigeonpea-wheat-greengram and observed that the latter proved more remunerative than the former under low fertility conditions. Maize- pigeonpea sequence has been found most promising at Dholi, Bihar (Roysharma *et al.*, 1980). Roy and Srivastava (1995) reported that among potato based cropping sequences, maize + groundnut-potato-greengram and maize + groundnut- potato + wheat-greengram crop sequences under irrigated condition were found to be promising in terms of yield equivalent and economic equivalent.

Inclusion of legumes in cropping systems increases soil fertility and consequently the productivity of succeeding cereal crops (Ahlawat *et al.*, 1981). Studying the effect of winter legumes on nitrogen economy of succeeding crop Srivastava and Srivastava (1993a) reported that productivity of rice increased by 34.6 per cent in sequence with chickpea (*cicer arietinum* L.) followed by peas (*Pisum sativum*) (16.3%). After three cycles of cropping chickpea-rice, lentil-rice and pea-rice showed positive trend in available soil nitrogen, whereas fallow-rice and wheat-rice exhibited negative trend. In another experiment on nitrogen economy and productivity of wheat succeeding grain legumes, Srivastava and Srivastava (1993b) found that groundnut-wheat rotation showed positive balance of available N in the soil and least drain of available P and K from the soil and thus, proved better than other rotations in maintaining soil fertility.

According to Dyer and Touchton (1990), inclusion of legumes in the cropping system can result reduction in nitrogen fertilizer requirements of the cotton crop. Baldock *et al.*, (1981) reported that alfalfa and soybean increased yield of the following maize and most of the increase was due to their nitrogen contribution. Where no fertilizer was used, maize yields on lands which had previously grown winged bean (*Psophocarpus tetragonobolus*) were 52 to 91 per cent higher than those where maize was the previous crop (Weil and Samarnayka, 1991) and the apparent nitrogen contribution by winged bean was found to be 70 kg N per ha.

Inclusion of legumes in the cropping systems is very important not only from the view point of nitrogen fixation but judicious utilization of soil nitrogen also. As nitrates can, sometimes, leach below the reach of grain crop roots and if a deep rooted legume (e.g., alfalfa) is sown in the following season, its roots can reach down and take up these nitrates, thereby preventing the wastage and pollution of groundwater (Rohweder *et al.*, 1988).

4.4.3 Crop Residue Management

4.4.3.1 Introduction

The importance of soil organic matter (SOM) in sustaining soil productivity has been appreciated since the down of human civilization. Important factors controlling SOM levels include climate, hydrology, parent material, soil fertility, biological activity, vegetation pattern and land use (Jenny, 1941). SOM is known to serve as soil conditioner, nutrient source, substrate for microbial activity, preserver of the environment and major determinant for sustaining or increasing agricultural productivity (Schnitzer, 1991). SOM is sensitive to human activities, viz, deforestation, biomass burning, land use changes and environmental pollution. India being a developing but also a highly populous nation of the world, could not escape the adverse effects of the menaces that are potential threat to soil health. Additionally, the unfavourable climatic conditions in the Indian Peninsula, have further enhanced the rate of decomposition of SOM and consequent depletion (Velayutham et al., 1999). The tremendous changes in land use during the last four decades in support of targeted yield have resulted in depletion of SOM level in Indian soils in general and in soils of Indo-Gangetic plains (IGP) in particular (Abrol and Gupta, 1998).

About 70 per cent area in the country is deficient in soil carbon, having less than 1 per cent organic carbon. Amount of SOM in most arid, semi-arid and subtropical soils is relatively low (0.5 to 2.0% and typically < 1%). Declining nutrients in soil due to exhaustive cropping system (e.g. rice-wheat) affect the sequestration of carbon in the soil and lead to decline in soil fertility. Therefore, the productivity of the system has shown stagnation, particularly of rice. In view of the high cost of inorganic fertilizers and the existing gap in additions of nutrients and crop removal, the recycling of available organic wastes has become a necessity in agricultural soils. The proper management of all type of residues not only augment the supply of plant nutrients but will also sustain crop production and maintain a cleaner environment on our planet.

4.4.3.2 Crop residues

Crop residues are renewable and readily available but are scattered organic resources. They are very often directly applied to the field since composting cannot be easily done. In addition to improving soil properties, conserving soil moisture and suppressing weeds, application of crop residues results in saving / substitution of chemical fertilizers to a great extent.

Crop residues are defined as the non-economic plant parts that are left in the field after harvest and remains that are generated from packing sheds or that are discarded during crop processing (Palaniappan and Annadurai, 2006).

Crop residues have at times regarded as waste materials that require disposal. Now there is an immense realisation that they are important natural resource of plant nutrients and not waste. These can be successfully be used for improving soil health and productivity (Beri *et al.*, 2002).

The forms of crop residues from various sources are listed below:

(i) Crop waste of cereals, pulses, and oilseeds (wheat, rice, pearlmillet, grain sorghum, greengram, blackgram, cowpea, pigeonpea, groundnut, linseed, and other crops).

(ii) Processing residues of sugarcane (bagasee), cotton (gin waste), stalk of maize, jute, topioca tree leaves, water haycinth and forest litter.

Crop residues, however, require preparation, usually consisting of size reduction and drying before they are suitable for use as a biomass feedstock for conversion to energy. Two principal conversion methods are used:

(i) Thermochemical, including direct combustion, pyrolysis, liquefaction and gasification with air or oxygen, and

(ii) Biological including anaerobic digestion and hydrolysis followed by fermentation.

Processing of crop residues produces a direct useable form of energy as hot combustion gas or an indirect energy form, such as liquid, solid or gaseous fuel.

4.4.3.3 Crop residue potential

After making use of economic parts of large number of crops grown annually in India, the remaining portion is mostly waste except for a few crops. The adoption of mechanized farming in many advanced regions in the country has resulted in leaving a sizeable amount of crop straw / stalk in field after harvesting. The potential of crop residues for recycling into valuable plant nutrients for sustained crop production is enormous. On national basis, not more than one-third of residues are available for utilization (Bhardwaj and Gaur, 1985) and in general, 50 per cent of the nutrients are mineralized in the soil on decomposition in cropping season. Table 26 and 27 provide data showing potential of crop residues of principal crops in India and average nutrient content of crop residues, respectively.

TABLE 26: Estimates of the available and realisable plant nutrient potential from the residues of principal crops in India

Crop	Residues yield (000' tonnes)	Nutrient content (%)			Total NPK potential (000' tones)
		N	P	K	
Rice	80744	0.61	0.09	1.15	1,493.8
Wheat	44987	0.48	0.07	0.98	688.3
Sorghum	11563	0.52	0.12	1.21	216.2
Maize	6219	0.58	0.09	1.25	119.4
Pearlmillet	8283	0.45	0.07	0.95	121.6
Barley	3180	0.52	0.08	1.25	58.8

Crop	Residues yield (000' tonnes)	Nutrient content (%)			Total NPK potential (000' tones)
		N	P	K	
Sugarcane	15645	0.45	0.08	1.20	270.7
Potato	5062	0.52	0.09	0.85	73.9
Groundnut	9580	1.65	0.02	1.23	277.3

Source: Bhardwaj and Gaur (1985).

Among various crop residues available in the country wheat straw and rice straw are mainly used as feed for cattle though these are poor in feeding quality as compared to green fodder. However, in the regions where mechanical harvesting is done, a sizeable quantity of both rice and wheat straw is left in the field. These are incorporated in the soil while preparing the field for next crop. Among different crop residues, sugarcane trash is available in large quantity and its management assumes special significance because it is hard to decompose and farmers burnt them in sugarcane fields (Modgal and Singh, 1990).

TABLE 27: Crop residue potential

Crops	Stubble added (kg/ha)	Addition of nutrients (kg/ha)			
		Organic	N	P	K
Rice	4200	1764	17.6	2.9	25.2
Sorghum	2889	462	6.1	2.6	95
Maize	667	93	0.6	0.2	2.7
Ragi	3111	899	43.5	3.8	20.5
Sesame	778	56	5.5	0.2	1.3
Cowpea	444	36	3.1	0.3	3.1
Lucerne	333	36	0.47	0.6	1.1

Source: Shanmuga Sundaram (1975).

Average nutrient content of certain common farm residues is given in Table 28.

TABLE 28: Average nutrient content of some common farm residues

Residue	Nutrient content (%)		
	N	P_2O_5	K_2O
FYM	0.80	0.41	0.74
Rice straw	1.59	1.34	3.37
Wheat straw	2.90	2.05	0.90
Water Hyacinth compost	2.00	1.00	2.30
Sugarcane trash	2.73	1.81	1.31

Source: Rani Perumal (1994)

4.4.3.4 Residue management practices

The C : N ratio of the organic materials determines the rate of decomposition and release of nutrients. The materials with very high C : N ratio decomposes very slowly like wheat straw, rice husk, rice straw and sugarcane trash, whereas crop residue of leguminous crops having relatively narrow C : N ratio decomposes faster. Decomposition of the organic material is an essential process for release of nutrients that can be used by the growing or succeeding crops.

The crop residues are managed in many ways like removal, *in situ* incorporation, and burning. The removed residues find their way in off-field uses like fodder, cardboard industry, surface mulch in no-till agriculture. Sidhu and Beri (1989) observed that burning of crop residues resulted in higher grain yield of rice and wheat (5.57 and 4.12 t per ha, respectively) than physical removal (5.53 and 4.025 t per ha) and incorporation (4.51 and 3.72 t per ha). The residues of both rice and wheat were managed which resulted in heavy loading of the soils with residues. The depression in yield of residues incorporation treatments was due to immobilization effect by wider C : N ratio (Toor and Beri, 1991). Application of irrigation after incorporation of residue and initial N application at sowing of wheat not only improve the soil moisture content which favourably affect the rice straw decomposition but also reduce the N immobilization and hence, improve the wheat yield over burning.

Effect of residue incorporation results in increasing grain yield. This has been attributed to (i) increase in N supply (ii) favourable N - release pattern (iii) increase in K supply (iv) increase in supply of other nutrients, and (v) improvement of soil structure.

Sidhu *et al.*, (1994) reported the beneficial effect of incorporation of chopped wheat straw on the yield of maize, groundnut and fieldpeas compared to residue removal. The beneficial effect was, however, much higher in sandy soil than that in sandy loam soil (*Bhatnagar* et al., 1987).

In Indo-Gangetic plains, the productivity is declining / stagnating due to the depletion of soil nutrients like C, N, P, S, Fe, Mn, and Zn. These soils require thorough incorporation of the crop residues so that after the decomposition and mineralisation, plants can use nutrients.

(i) **The rate of decomposition** of cereal residues, such as wheat and rice straw, and maize and pearlmillet stalk were always significantly higher than FYM in both normal and problem soils and cereal residues also differed among themselves in respect of rate of decomposition (Mudgal and Singh, 1990).

(ii) **The turnover of organic materials** in the soil are affected by the heterophic soil organisms including bacteria and fungi. Their requirement for energy is met through the oxidation of the carbonaceous material in the soil. The decomposition increases with a rise in temperature. It is further enhanced by adequate, although not excessive soil moisture and a good supply of oxygen. Decomposition proceeds under waterlogged conditions, although at a slower rate, is incomplete. Aerobic, and to a lesser extent anaerobic respiration releases the contained nitrogen in the form of NH_4.

(iii) **Immobilization of nitrogen** is not observed when the C : N ratio is brought down to 40 and the mineralization of nitrogen is satisfactory and steady. Addition of an inorganic source of nitrogen and phosphorous or organics comparatively richer in nitrogen like non-edible cakes and legumes can be advantageously utilized for lowering the C : N ratio of cereal straw/residues. Unless these are supplemented with external nitrogen and phosphorus during decomposition it may cause immobilization of N and for a longer period. Leguminous plant residues have low C : N ratio and, therefore, they degrade faster and help build good amount of available form of nutrients.

(iv) **Size of straw**: The decomposition is expected to be higher in case of chopped residues or of smaller size, because it provides microbes a greater surface area to act. Shorter cut of plant residues facilities uniform mixing.

(v) **Time of incorporation**: The effect of time of straw incorporation on grain yield depends upon temperature, cultivars and amount of straw incorporated. Sowing of wheat within one week of rice straw incorporation retarded the growth of wheat and reduced its yield continuously for 11 years (Sidhu and Beri, 1989; Beri *et al.*, 1995) due to fall in temperature after incorporation and little time available for decomposition. While in Australia (Bacon and Lewin, 1990), the residue of wheat incorporated 2 to 6 months before the transplanting of rice harnessed beneficial effect of residue incorporation on rice grain yield.

The residue incorporation at sowing may not yield short term beneficial results. The situation of temporarily immobilization of available N due to wide C : N ratio of crop residues can be averted if the residues are incorporated in the soil atleast 3 to 4 weeks before the crop is sown/planted. The best method of straw incorporation in rice is approximately one month before flooding for transplantation with supplemental N and P.

(vi) **Amount of residue**: Higher amount of residues, if incorporated, yields more microbial biomass that assimilates more N for immobilization. It also requires more N fertilizer, more time and oxygen to decompose. Sidhu and Beri (1989) found that residues of rice incorporated before wheat and residues of wheat incorporated before rice resulted reduced yields of both the crops. Application of 8 t per ha of wheat straw before rice depressed the rice yield while it improved the yield of following wheat. However, application of 8 t per ha of only rice straw of 4 t per ha each of rice straw before wheat and of wheat straw before rice sustained the yield of both the crops.

4.4.3.5 Crop residues and physical properties of soil

Appropriate management of crop residues retards degradation of soil physical properties and improves the soil. The well decomposed and easily decomposable crop residues generally show good effect in improving soil aggregation, while the materials relatively resistant to microbial decomposition, have a favourable effect on soil physical properties, such as porosity, bulk density and water holding capacity.

(i) **Soil structure:** Crop residues favours the formation of aggregates due to the addition of organic matter (Bhagat and Acharya, 1988). Soil structural stability increases due to straw addition and, better aggregate size distribution occurs due to a reduction in soil disturbance (Singh *et al.*, 1994).

(ii) **Bulk density and porosity:** The incorporation of straw along with application of farm yard manure (FYM) reduces the bulk density of soil and increases the porosity of the soil (Bhagat and Vera, 1991). Above ground biomass contributed directly to the reduction of machine-induced compaction effects. (Battikhi *et al.*, 1998). Leguminous crop residues, which were small in quantity, decomposed faster and acted as cementing agents of soil particles (Gardner *et al.*, 1994)

(iii) **Hydraulic properties:** According to Somani and Saxena (1981) the use of organic material in terms of crop residue hastens the hydraulic conductivity of the typical calcareous saline-alkali soils. While comparing the different methods of residue application, Bhagat and Verma (1991) found the hydraulic conductivity for a saturated soil (0-15 cm soil depth) was maximum for straw incorporation + FYM application as compared to straw burning, straw incorporation alone, straw mulching, and FYM. In the sandy clay loam soil, crop residue incorporation was more effective in reducing soil moisture stress than crop residue mulched (Fagi and De Datta, 1983).

(iv) **Soil temperature**: Mulching with plant residue raised the soil temperature in winter by 3°c during the early stage of soybean growth (Kitou and Yoshida, 1994). Crop residue also modify the soil temperature regime (Fortin et al., 1994) and hence crop growth. Bhagat and Verma (1991) reported that under straw mulch, minimum soil temperature increased (by 1.2 to 2.8°c) due to reduction in upward heat flux from soil and the decrease in maximum soil temperature (by 1 to 2°c) due to shading effect of mulch. Symbiotic parameters of biological nitrogen fixation were improved with mulching and were responsible for better yield of soybean (Goswami and Pareek, 1976) and pigeonpea (Sidhu *et al.*, 1988).

(v) **Air permeability:** For a given residue, the minimum-till treatment generally had higher permeability than either the conventional or no tillage treatment. The plots with wheat residue had consistently higher air permeability than those with fieldpea residue (Saxton *et al.*, 1993).

4.4.3.6 *Crop residues and chemical properties of soil*

Effect of crop residues on chemical properties of soil is discussed below:

(i) **Soil reaction:** Soil reaction seems to be influenced differently by soil situations. The capacity of soil organic matter to buffer soils against rapid changes in pH is well known. This property arises largely from the biological activity, involving the interaction of humic substances and clay producing colloidal complexes, which have negative surface charges and hold cations, such as H^+, Al^{3+}, Fe^{3+}, etc. As a result of this activity, acids and bases are being produced continuously in the soil so that without a buffering capacity the pH of the soil would change to the detriment of its biology. Increase in pH due to addition of coffee straw (Paes *et al.*, 1996) and field bean shoots (Yan *et al.*, 1997) has also been reported. Earlier Mishustin and Temtsev (1982) reported that incorporation of rice straw decreased the soil pH and was found to be very effective in alkali soils, while Debnath and Hazra (1972) observed an increase in pH due to incorporation of rice straw in laboratory studies. Such changes are expected, as the resistance to pH change is dependent upon cation exchange capacity and texture of soils. Soil pH change is positively correlated to alkalinity, or total N and water soluble organic N of incorporated organic materials. Sekhon and Bajwa (1993) also reported that incorporation of organic materials decreased the precipitation of calcium and carbonates, increased the removal of sodium in drainage water, decreased pH and ESP in the soil and improved crop yields.

(ii) **Organic carbon:** The major contribution of total build up of organic C in soil comes from the C of humic acid and humic fractions which appeared to have increased substantially due to the addition of organic material (Srivastava and Fokin, 1973). Intensive cropping leads to the exhaustion of soil fertility and thus, decrease in organic C and N contents of the soil causing decrease in production. However continuous organic matter application results in increased organic C and N contents of soil. Walia *et al*, (1998) reported increase of organic C by the use of organic matter in the form of rice straw. Earlier Pal (1969) also reported an enhanced organic C content of soil due to incorporation of maize stover. Also an application of wheat straw and rice husk markedly increased the K supplying capacity of soils (Somani and Saxena, 1975). Crop management systems that include rotations with high residue producing crops and maintenance of surface residue cover (Omay *et al.*, 1997) with reduced tillage (Havlin *et al.*, 1990) resulted in greater soil organic C and N, which improve soil productivity.

(iii) **Cation exchange capacity (CEC):** It has been estimated that 20 to 30 per cent of CEC of many soils is attributed directly to the soil organic (SOM) alone (Stevenson, 1982). SOM acts as a reservoir for plant nutrients and prevents leaching of elements, which is achieved by the considerable CEC of the organic material. Addition of crop residues increases the soil CEC (Paes *et al.*, 1966; Bayer and Mileniczuk, 1997).

4.4.3.7 Crop residues and biological properties of Soil

Crop residues provide energy for growth and activities of microbes and substrate for microbial biomass, and provide conditions for a source-sink of nutrients (Singh and Sekhon, 2002). The microorganisms constitute the driving force behind soil organic matter and help in biodegradation of organic matter, formulation of organochemicals, nitrogen fixation, phosphorus solubilisation and in increasing availability of nutrients required for both plant and microbial growth. The augmented microbial activity can tap the inert nitrogen gas from the atmosphere, reduce the leaching losses and regulate the supply of phosphorus and potassium. Mukherjee and Gaur (1988) reported that application of organics influenced bacterial population to a greater extent than fungi and actinomycetes.

Nitrogen fixation: Asymbiotic nitrogen fixation (Azotobactor and Anaerobes) were stimulated by addition of straw in the soil (Mukherjee and Gaur, 1988). Straw incorporation increases the nodule volume, fresh weight and dry weight nodule (Shivshankar *et al.*, 1976). The favourable effect of straw on the nodule growth was probably due to indirect effect of straw in decreasing mineral nitrogen content of soil by immobilization. This is turn must have provided a suitable environment for symbiotic nitrogen fixation (Shivashankar *et al.*, 1976, Hechman and Kruchinski, 1995).

Enzymes: Incorporation of straw at 2 and 4 t per ha improved the nitrogenese activity of soybean by 14 per cent and 197 per cent, respectively (Shivashankar *et al.*, 1976). Straw incorporation increased acetylene reductase activity more in dry season than in wet season due to lower soil temperature in wet season (Lodha *et al.*, 1986). Maize residue incorporation increased phosphatase activity (Singh, 1991). Goyal and Chander (1998) reported an increase in the microbial biomass, dehydrogenese and alkaline phosphates activities with the addition of wheat straw to sandy loam soil.

4.4.4 Farmyard Manure (FYM)

Applying FYM to enrich the soil fertility is an age old practice. It provides the plants with many macro and micronutrients.

The term FYM refers to the decomposed mixture of dung and urine of farm animals along with the litter (bedding material) and left over material from fodder fed to cattle. Well rotten FYM contains about 0.5 per cent N, 0.3 per cent P_2O_5 and 0.5 per cent K_2O.

Factors affecting composition of FYM are given below:

(i) **Source of manure:** Sheep and poultry manures are richer in N, P_2O_5 and K_2O than cow, horse and pig manures.

(ii) **Food of the animal:** The richer the food in proteins, the richer will be manure in nitrogen. Same holds good for phosphoric acid and potash.

(iii) **Age of the animal:** The manure of young animals is not so rich that of mature animals.

(iv) **Function of the animal**: Animals producing milk and wood absorb more nutrients from their food compared to animals that are working or at rest. Thus, the manure from bullocks will be richer in nutrients than that from cows producing milk.

(v) **Manner of storage:** Under ordinary conditions of storage, FYM suffers loss of nitrogen and if any drainage is allowed to escape from the manure heap, a loss of potash also occur.

(vi) **Nature of the litter:** The most common litters are straws of rice, wheat, sorghum, pearl millet, and maize. The nature will depend on nutrient content of these litters.

The results of chemical analysis of the permanent manurial trial in alluvial sandy loam soils at Kanpur, revealed that application of organic manure alone greatly increased the organic C and available P_2O_5. The nitrogen content increased under all treatments in comparison with control. Similar result was obtained from permanent manurial trial at Gurdaspur (Punjab). The organic manure and chemical fertilizers seem to have lowered the fulvic acid content of the soil. In similar another study, a continuous application of 75.5 and 18.9 t per ha of FYM to sandy loam soil of a permanent manurial experiment at Sabour (Bhagalpur) resulted in highest build up of humic substances in the soil. All treatments having higher and lower doses of the organic manure contributed to the content of the humic substances in the soils. The observed trends on soil properties in permanent manurial trial at Chinsurah (West Bengal) having soil of acidic nature were slightly different. Highest total carbon and fulvic acid carbon were found in the treatments involving FYM either with phosphatic or nitrogenous fertilizers. However, application of

FYM alone contributed conspicuously to the humic carbon content of the soil (Bhardwaj and Gaur, 1985).

Experiments conducted at different centers of CRIDA have shown that response to application of FYM differs with crops, soil and rainfall in dryland agriculture. At Kovilpatti (Tamilnadu), pigeonpea benefited from FYM application more than sorghum. At Hissar (Haryana) and Jodhpur (Rajasthan) combination of FYM and fertilizers were more advantageous.

4.4.5 Biofertilizer and Organic Fertilizer/Native Phosphate

The sustainable agriculture envisages primary emphasis on manipulation and management of biological and organic systems not only for maximum yield but also to stabilize the agro-system and to minimize industrial input demands. The effective use of biofertilizers along with other source of plant nutrients play a vital role in increasing crop productivity and sustained agriculture production.

4.4.5.1 Biofertilizers

Biologically active products containing selective strains of microorganisms which can contribute nutrients to plants through microbial activity are known as biofertilizers. In other words, biofertilizers are microbial cultures that fix atmospheric N for crop use and in addition, certain bacterial/ fungal cultures help in phosphate solubilisation of both native and applied sparingly soluble phosphates. They can provide an economically viable support to small and marginal farmers by partly replacing inorganic fertilizers use in crop production.

Role of biofertilizers: Biofertilizers are important for their following contribution as agricultural input:

(i) Biofertilizers are low cost, effective and renewable sources of plant nutrients to supplement chemical fertilizers.

(ii) They contribute plant nutrients through biological nitrogen fixation and solubilization of fixed phosphates.

(iii) They can enhance 10 to 20 per cent of crop yield.

(iv) They control and suppress soil-borne diseases as some of the inoculants produce antibiotics.

(v) They help in proliferation and survival of beneficial micro-organisms of soil.

(vi) They help in mineralization of plant nutrients, improve soil properties and sustain soil fertility.

(vii) They are eco-friendly and pollution free.

4.4.5.2 *Types of biofertilizers-factors affecting use and methods of application*

Broadly, the biofertilizers are classified into two types- the biological nitrogen fixing biofertilizers and the phosphate solublizing or mobilising biofertilizers Nitrogen fixing biofertilizers (NFB) consists of microorganisms which fix atmospheric nitrogen either symbiotically (e.g., NFB for legumes-Rhizobium) or asymbiotically (e.g., NFB for non-legumes - Azotobactor, Azospirillum, Azolla, Blue-Green Algae) into the plants, whereas phosphate mobilizing biofertilizers have the ability to solubilize (e.g., Bacillus, Pseudomonas, Fungus) fixed insoluble phosphates of soil and absorb (e.g. VA-Mycorrhiza) them. Common microorganisms used as biofertilizers are presented in Table 29.

(a) **Rhizobium** soil bacterial have the ability to fix atmosphere N in symbiotic association with host legumes and to some extent in certain non-legumes, such as Parasponia. These bacteria enter into the plant through root hairs, multiply there, form nodules and fix biological N in the nodules.

It is the first microbial fertilizer to be used on a large scale. There is specific *Rhizobium* for every legume. Inoculation with efficient *Rhizobium* strain is very essential for the nitrogen gains and better crop yields (Table 30). About 15-30 per cent yield increase in pulses can be achieved by *Rhizobia* inoculation in farmers fields. *Rhizobium* legume symbiosis can meet 80 per cent of N requirement of the legume crop. Contribution of Rhizobium in biological N fixation is presented in Table 31.

Rhizobium is recommended for (i) pulse legumes, such as, chickpea, pigeonpea, peas lentil, blackgram, and greengram, (ii) oilseed legumes, like soybean and groundnut, and (iii) fodder legumes like berseem and Lucerne.

Dwivedi (1981) found that growing of groundnut left the soil richer in nutrient status than a more competitive cereal crops, like pearlmillet. Numerous works have confirmed about residual benefits from legumes inoculated with rhizobium culture to the succeeding crop (Subba Rao and Tilak, 1977; Ahlawat *et al.*, 1985). Result of All India Coordinated Pulse

Improvement Project (AICPP) over years have proved yield advantage in pulse crops to the tune of 10 to 15 per cent due to *Rhizobium* inoculation (Ali, 1985). However, under adverse soil and climatic conditions there was no yield advantage due to inoculation in many places in drylands (Rao and Das, 1982). Inoculation resulted yield advantage in many places under irrigated conditions.

TABLE 29: Common microorganisms used as biofertilizers

Contribution plant nutrients	Microorganisms	Crop benefited
Nitrogen	1. Symbiotic	
	(a) Rhizobium (with legume)	*(i) Pulse legumes*: chickpea, pea, lentil, pigeonpea, greengram
		(ii) Oil legumes: groundnut, soybean
		(iii) Fodder legumes: berseem, lucerne, Rice,
	(b) Azolla (Fem-*Anabaena azollae* symbiosis)	Rice
	2. Associative symbiosis *Azospirillum*	Rice, sugarcane, maize, finger millet
	3. Non-symbiosis	
	a. Heterotrophs, e.g., Azotobacter	Vegetables crops, wheat, rice, and other commercial crops.
	b. Photo-autrotrophs e.g., Blue-greenalgae	Rice
Phosphorus	1. Phosphate solubilizers and mineralizers	
	a. Fungi: Aspergillus, *penicillum*	*For all crops*
	b. Bacteria: *Bacillus, Pseudomonas*	*For all crops*
	2. Phosphate absorber (Root fungus symbiosis)	
	a. Ectomycorrhiza: *Pisolithus, Rhizopogan*	For all crops
	b. Endomycorrhiza: *Glomus*, Gigaspora	For all crops

Source: Verma and Bhattacharya (1990).

TABLE 30: Group of legumes and their specific *rhizobia* constituting a cross-inoculation groups

Rhizobium species	Cross inoculation group	Legumes
R. *trifolii*	Clover group	Trifolium
R. *meliloti*	Alfalfa group	Melilotus, Medicago Trigonella

Rhizobium species	Cross inoculation group	Legumes
R. *phaseoli*	Bean group	Phaseolus
R. *lupini*	Lupine group	Lupinus, Ornithopus
R. *leguminorsarum*	Peas group	Pisum, Vicia, Lens
R. japonicum	Soybean group	Glycine
Rhizobium sp. (Mescellany)	Cowpea group (Mescellany)	Vigna, Arachis

Source: Singh and Dicit (1998)

Factors affecting use efficiency of the rhizobium:

(i) **Organic matter:** Favourable for the survival of rhizobium.

(ii) **Sunlight and temperature:** Cells of freshly inoculated rhizobium do not withstand direct sunlight and temperature beyond 40°C for a log period.

(iii) **Micro-nutrients:** Presence of cobalt and molybdenum is very useful for N-fixation.

(iv) **Nitrogen:** High dose of N (Mineral fertilizers) decreases the efficiency of N-fixation, whereas low amount of N enhances the symbiosis. In biological N-fixation, P, K, Ca and Mg also have positive effect on symbiosis.

(v) **Salinity:** The *Rhizobia* are highly sensitive to the salinity. High salinity greatly hampers the infection of root hairs.

TABLE 31: Contribution of Rhizobium in biological N-fixation

Name of the crop	Quantum of N-fixation (Kg N per ha)
1. Alfalfa	100 - 200
2. Pigeonpea	168 - 200
3. Clover	100 - 200
4. Chickpea	85 – 110
5. Cowpea	80 - 85
6. Groundnut	50 - 60
7. Lentil	90 – 100
8. Greengram/Blackgram	50 – 55
9. Peas	52 – 77
10. Soybean	60 - 80

Source: Bhattacharya and Mishra (1995)

The method of Rhizobium application: 10 kg of normal size seeds, such as greengram, blackgram, pigeonpea, cowpea, lintil and berseem is treated with 200 g of

Rhizobium inoculant by slurry method. The big size seeds, such as groundnut, chickpea, soybean, peas etc., require 400 to 500 g of inoculant for 10 to 12 kg of seeds. Slurry of required quantity of inoculant in sufficient water (400 to 500 ml water for 200 g of inoculant) is prepared and slurry is poured over the heap of seeds to be treated in shade and mixed homogenously. Treated seeds are spread over plastic sheet, clean floor or gunny bag for drying purpose. Seed is treated immediately or just two to three hours before sowing.

(b) **Azotobacter biofertilizers:** Azotobacter is chemoheterotroph organism of aerobic characters having the capability of fixation of dinitrogen (N_2) as non-symbiont. This organism widely occurs in the rhizosphere of many plants. Azotobacter fixes on an average 20 to 40 kg N per ha, which increases the yield of a variety of cereals, oilseeds, vegetables, fruit crops and cash crops. On the basis of morphological and physiological features species of genus Azotobacter are *A. chrooccum, A beijerinckii* and *A. vinelandii*.

Importance of azotobacter:

(i) It serves as a broad spectrum inoculant for various crops, like wheat, barley, maize, rice, sorghum, oats, sugarcane, sugarbeet, tobacco, cotton, potato, sunflower, mustard, sesamum, linseed, tea, coffee, brinjal, onion, cauliflower, cabbage, lady's finger, all forest fruits and flower plants.

(ii) It improves the percentage of seed germination, helps in root proliferation, accelerates plant growth and increases crop yield up to 10 to 15 per cent.

(iii) It fixes atmospheric N in soil and thus, reduces the use of chemical N fertilizer to the extent of 20 to 40 kg per ha.

(iv) It produces and deposits biologically active metabolites, like B-vitamins, indol acetic acid, gibberallins, which help in plant growth promotion.

(v) It synlhesizes antibiotic substances which control or supress various fungal, bacterial and viral plant diseases.

(vi) It cause mineralization of fixed phosphate of soil which increases uptake of P and plants. Besides this, it also increases the activity of beneficial rhizospheric bacteria, such as ammorifiers, nitrifiers, N-fixers, phosphate solubilizers and cellulose decomposers.

Factors affecting efficiency of azotobacter:

(i) **Oxygen:** It is essential for N-fixation.

(ii) **pH:** 7 to 7.6 pH has a very good response to N-fixation.

(iii) **Salinity:** It inhibits N - fixation.

(iv) **Minerals:** Ca, Nacl and combined N and trace elements are essential for optimum N-fixation.

(v) **Temperature:** Azotobactor is a mesophillic bacteria and very sensitive to temperature. It withstand the temperature range of 30 to 35°C.

Method of application: The incoulant may be used different ways depending upon the crop to be cultivated. The different methods of application are seed treatment, seedling treatment, setts/tuber treatment. For seed treatment, 10 kg medium size seeds, such as wheat, cotton, etc, is treated with 200 g of azotobactor inoculation. For seedling treatment, suspension of one kg in 10 to 15 l of water is sufficient for treatment of seedlings. Suspension of one kg of inoculant in 40 to 60 l of water is required for treatment of setts/tubers. Approximately 5 kg of inoculant mixed with 100 kg soil or FYM is required for short duration crops (less than six months) for on ha of land.

(c) **Azospirillum biofertilizers:** Azospirillum, an associative symbiotic N-fixing bacterium has a high N-fixing potential. It also produces growth promoting substances which favour better growth of plants. Azospirillum is found to be associated with forage grasses and roots and rhizoshere of many members of graminae, particularly in the tropics.

Usefulness of azospirillum inoculants in crops

(i) It helps to save 20 to 30 kg N per ha.

(ii) It helps to increase grain yield of rice, barley, pearlmillet, finger millet, sugarcane and sorghum. It is also useful for tea, coffee, vegetable crops, fodder crops, fruits and flower plants.

(iii) It produces Indole Acetic Acid (IAA), gibberellins, and cytokinins when inoculated on crop plants resulting in increased root development and enhanced seed germination.

(iv) It helps in maintaining soil fertility in semi-tropical and tropical regions.

Factors affecting efficiency of the azospirillum:

(i) **Oxygen:** Its presence in good amount enhances the N-fixation.

(ii) **pH:** The activity of nitrogenous enzyme during N-fixation is found to be optimal at pH 7.

(iii) **temperature:** The optimum temperature for better efficiency of azospirillum is 32°C to 35°C.

(iv) **Organic matter:** High percentage of soil organic matter flavours the growth and efficiency of Azospirillum.

Method of application: It is the same as described for Azotobactor inoculant.

(d) **Herbaspirillum:** a new nitrogen fixing associative gram negative bacteria is found associated with some cereals, grasses and sugarcane in Brazil. This shows similar growth habits of azospirilum. Herbaspirillum helps plants through its ability to fix atmospheres N, harmone production and faster nutrient uptake. Its testing in lowland rice has initially indicated 7 to 10 per cent increase in yield with reduced duration of rice crop by a week.

(e) **Blue-green algae (BGA) biofertilizer:** BGA or cyanobacteria is very useful source of organic N for lowland rice. It promotes the growth of rice crop by supplying fixed N through exudation and microbial degradation of dead algal cells. Basically, BGA is a photosynthetic N fixing agent. In Indian soils predominant genera are *Nostoc, Aulosira, Tolypothrix.* The other commonly known species are *calothrix* and *scytonema.*

Various studies conducted in the country have shown that BGA applications @ 10 kg per ha after one week of transplantation contribute 25 to 30 kg N per ha per season and increases grain yield by 10 to 15 per cent. Normally continuous inoculation for 3 to 4 consecutive cropping seasons results in an appreciable population build up without any further inoculation.

Factors affecting efficiency of BGA:

(i) **Water:** 5 to 15 cm standing water is needed for the proper growth or formation of algal mat for efficient N - fixation.

(ii) **Sunlight:** BGA is phototrophic organism and hence, sufficient period of sunlight is very important.

(iii) **Temperature:** High temperature range of 35 to 40°C is useful for growth and N-fixation by BGA.

(iv) **pH and Salinity:** BGA grows well in neutral to alkaline soils with capacity to extract sodium to some extant. BGA is tolerant to salt.

Method of application: Blue-Green Algae is applied at the rate of 10 kg per ha. The BGA flakes is crushed into powder which is broadcasted in the standing water of rice field soon after one week of transplantation. The field is kept waterlogged atleast for 3 to 4 days immediately after BGA application.

(f) **Azolla biofertilizer:** Azolla is an aquatic fern and this has utility as a N-fixer and green manure in rice cultivation. The leaf of azolla contains blue-green algal symbiont, i.e., *Anabaena azolla* which is responsible for N-fixation. Azolla has very fast growing character. It doubles its biomass in 3 to 5 days and fixes 40 to 60 kg N per ha.

Factors affecting efficiency of Azolla:

(i) **Water:** It is essential for growth and survival of azolla. Since Azolla is free floating fern, 10 to 15 cm deep water is required.

(ii) **pH:** The optimum pH of 5.0 to 7.2 is good for growth of Azolla. Its growth is found very good in slightly acidic to slightly alkaline soils.

(iii) **Temperature:** Temperature below 5°C and above 35°C has adverse effect.

(iv) **Light:** The longer period of light and optimum temperature support the growth and N-fixation of Azolla.

(v) **Minerals:** Except N, all macro and micro nutrients are essential for normal growth of Azolla.

(vi) **Combined Nitrogen:** Generally use of N-fertilizer affect adversely the growth of Azolla. However, use of low concentration of ammonia and nitrate promote the growth of Azolla.

Method of application: The limitation of Azolla is very high requirement of initial inoculum (500 to 1000 kg per ha). Hence, its transportation is difficult. Azolla cannot withslamd dessication, so it is not useful for upland rice. Pest like pyralis and snails affect the Azolla.

Azolla can be used in two ways for rice cultivation: (i) As green manure intercropped with rice in flooded condition. The inoculation rate is 10 to 15 t per ha applied one week after transplantation. 10 kg P_2O_5 per ha as superphosphate is needed for good growth of azolla. (ii) As dual cropping by applying 0.5 to 1.0 t per ha of fresh azolla.

(g) **Phosphate solubilizing biofertilizers:** In organic farming, phospho-microbes have a considerable potential in mobilisation of soil phosphates and in solubilisation of low grade phosphate rocks used in place of chemical fertilizers, thereby increasing crop yield (Tiwari *et al.*, 1989). Bacteria (*Bacillus megatherium, Bacillus circulans, Bacillus sublilis, Pseudomonas striata, P. rathoris)* and fungi (*Aspergillus awamori, Penicellium digitarium, Trichoderma sp.*) and yeast (*Schwannimyces accidentalis*) have the ability to solubilise or mobilise insoluble phosphorus to the soluble form and then make it available to plants. The efficient cultures have shown capacity to solubilize insoluble inorganic phosphate, such as rock phosphate, tricalcium phosphate, iron and aluminium phosphate by production of organic acids. Bagyaraj *et al.*, (2000) has also reported that bacteria and fungi posses the ability to bring insoluble P into soluble forms by secreting organic acids. Lowering of soil pH brings about dissolution of immobilized from the phosphate. Some of hydroxyacids may chelate with calcium and iron resulting in solubilisation of soil P and thereby its higher utilization. Dissolution of rock phosphate may be too low for rapidly growing crops (Le Mare, 1991), therefore, it is suggested to apply rock phosphate about a month before sowing an annual crop to allow more phosphate to dissolve.

Vasicular- arbuscular mycorrhiza (VAM): Effects of VAM on growth, yield and P uptake of several crops have been reported both in absence and presence of fertilizer P. These include sorghum, pearlmillet, fingermillet, groundnut, pigonpea, cowpea, cotton, chillies, chickpea and wheat. The beneficial effects, in general, with these crops is more pronounced at low levels of native or applied P (Krishnana and Dart, 1984). The VAM has also role in integrated VAM - Rhizobium inoculation in legumes and in increasing the efficiency of locally available rock phosphate VAM fungi not only improve plant yield but saves phosphatic fertilizer to the tune of 25 to 30 per cent. Recent studies have shown that VAM response were much higher in the second crop.

Factors affecting phosphate solubilising bacteria:

(i) **pH:** The optimum pH for maximum solubilization of inorganic phosphate has been found to be in neutral or acidic soils.

(ii) **Temperature:** The optimum activity of fungi is around 35°C, whereas that of Pseudomonas ranges between 29 to 39°C. The activity of these microorganisms is hampered at higher temperature (40°C or more).

(iii) **Nitrogen sources:** The phosphate solubilizing microorganisms solubilize a good amount of rock phosphate in presence of nitrogen sources.

Method of application: Doses and method of application for phosphate solubilizing biofertilizer are same as described for azotobacter inoculant.

4.4.6 Waste Recycling

Biological and industrial wastes for harnessing nutrient energy is of prime importance for increasing production and maintaining sustainable agriculture and environment. When these wastes are recycled as manure for crop production and are subjected to the degradation and assimilative capacity of soil, pollution of streams and rivers receiving these wastes is reduced to a large extent as compared to their direct disposal in water bodies. They are important source of plant nutrients and humus. They also protect the soil from erosion.

Bio-gas slurry: Cow dung is an input in biogas plant which has the dual advantage of providing both fuel (gas) and manure (slurry). The wet slurry should not be used as such, it should be dried for usage as manure. It contains about 1.8 per cent N, 1.10 per cent P_2O_5 and 1.50 per cent K_2O. Hence, its manurial value is superior to that of FYM.

Sugarcane trash compost: Sugarcane trash can be recycled after composting. Technology has been developed for production of enriched compost from sugarcane trash (Bhavanase *et al.*, 1995). The composted trash contains 1.09 per cent N with C : N ratio of 20:1 as compared to fresh sugarcane trash containing 0.36 per cent N with a wide C : N ratio of 122:1.

Vermicompost: It is an organic manure produced by the activity of earthworms. Vermicompost is a mixture of worm casts which are rich in nutrients (as compared to FYM), vitamins, growth harmones and immobilized micro-flora. It contains 1.60 per cent N, 5.04 per cent P_2O_5 and 0.80 per cent K_2O with small quantities of micronutrients. The C : N ratio of vermicompost is much lower (1:16) than that of FYM (1:30). The worm species more commonly used are *Lumbricus rubellus, Eisenia foetida, Perionyx excavatus, Eudritus eugeniae and P. arboricola.* Vermicompost can be used for horticultural crops / forest trees (100-200 g/tree), potted plants (50 g/pot) and field crops (2 t/ha).

Industrial Wastes: Among the industrial by-products, spent wash from distilleries and molasses and press mud from sugar industry have good manurial value. Well decomposed press mud may be applied at 10 t per ha. It enhances the activity of microbes, improves soil fertility and is also used as an amelioring agent in saline and sodic soils. Composted coir waste inoculated with *Pleurotus spp* and urea (5 kg/t) is recommended for irrigated and dry crops at 12.5 t per ha. Crops like groundnut and maize showed increased yield with the application of composted coir waste (Balasubrananiyan and Palaniappan, 2000). Sewage sludge, particularly from industrial cities, is contaminated with heavy metals and this poses hazard to plants, animals and human beings.

4.5 WEED MANAGEMENT

According to Hosmani and Meti (1993), non-chemical methods of weed management, e.g., crop and crop variety selection, methods and time of seedbed preparation, date of sowing, crop rotation, planting geometry and density, intercropping, mulching, irrigation management, and soil solarisation can also be employed to control weed successfully. More competitive crop varieties (Dhaliwal,1998), and higher seed rates (Sodhi and Dhaliwal, 1998) can smoother weeds. Inclusion of large leaf size and high leaf area index (LAI) as criteria for selecting high yielding genotypes with improved ability to suppress weeds should be feasible (Wortmann,

1993). The traditional tall varieties could also be used beneficially to tolerate weeds in organic systems (Cosser et al., 1997).

Narrow row sowing is advocated to suppress weeds and obtain higher rice yields (Bhan 1968). At low weed pressure, grain yields can be higher in narrow rows (Limon Orgega *et al.*, 1998). Similarly, bidirectional method of sowing of crops can suppress weeds. Better weed control in narrow row as well as bidirectional method of sowing is attributed to quicker canopy closure and greater light interception early in the season.

Intercropping of short duration crops in the long duration main crops can offer weed suppression apart from high total productivity per unit area per unit time (Singh and Sekhon, 1998). Ali (1988) reported that under rainfed conditions, cowpea, urdbean and mungbean intercropped with pigeonpea suppressed weed flora by 20-45 per cent over sole pigeonpea (unweeded).

The practice of soil solarisation can be used to control weeds in widely row sown crops. The practice involves covering of a moist soil with transparent plastic sheet, which causes the ground temperature to rise by about 5-10°C, which kills weed seeds (Singh and Sekhon, 2002). Weeds can also be controlled by allelopathic chemicals secreted by crops (Bhatia *et al*, 2002). California research has turned sunlight to kill weeds by developing a solar powered machine (Bud Houkstra, 1992). The machine is pulled by hand between rows of crops and concentrates sunshine into single band of withering light. The solar concentrator known as 'Fresnel lense' is an acrylic sheet made of an array of tiny lenses.

Biological methods of weed control are not popular among the farmers, yet, however, there is a good scope to employ insects, plant pathogens and other biological agents to control weeds (discussed in detail in Chapter I of this book). The merits of biological control are relative cheapness, comparatively long lasting effects and least harm on the environment and the non-target organisms. Some examples of successful biological control of weeds are:

WEED	INSECT
1. *Cyperus rotundas* (Nut sedge)	*Althesapaecuta cyperi* (Stem boring weevil)
2. *Eichhornia crissipes* (water hyacinth)	*Tilapia sp* (Herbivorus fish)
3. *Lantana camara* (Lantana)	*Teleonema scrupulosa* (Tinged bug)
4. *Parthenium hysterophorus*	*Zygorama bicolorata*
5. *Opuntia intermis/ O stricta* (Prickly pear cactus/ spiny prickly pear)	*Cactoblastis cactorum* (Pyralid moth)

6. Salvinia molesta (water fern)	Crylophagous salvinia (Curculionid weevil)

4.6 INSECT, PEST AND DISEASE MANAGEMENT

Prior to the introduction of crop protection chemicals, man evolved several cultural and physical manipulations of controlling pests (including insects and diseases) which reduced the damage potential to crops of essentially all pests (Upadhyay *et al.*, 1996). For achieving cultural control of insect pests. Atwal (1986) suggested some agricultural practices, like tilling and cultivating the soil; use of clean seed; regulating irrigations; use of resistant varities, manurings clean culture; cropping scheme and the trap crops; pruning and thinning; and adjusting time of sowing.

Intercropping reduces insect infestation. For examples sorghum (sorghum bicolor) grown as pure crop had significantly high stem borer (*Chilo partellus*) infestation (32.6%) while sorghum inter-cropped with lablab (*Lablab purpureus*) or cowpea (*Vigna anguiculata*) had low incidence of 9.2 and 16.4 per cent, respectively (Mahadevan and Chelliah; 1986). Similarly, pure crop of sesame (Sesamum indicum) recorded 23 per cent shoot-webber (*Antigastra catalaunalis*) damage compared with 7.8 and 12 per cent when intercropped with pearl millet (*Pennisetum glaucum*) and groundnut (*Arachis hypogaea*), respectively (Muralibaskaran *et al.*, 1991). Intercropping with non-host plants, like barley (*Hordeum vulgare*) and linseed (*Linum usitatissimum*) brought about considerable suppressing effect on the incidence of most of the insect pests on peas (*Pisum sativum*) through changed crop canopy and resultant change in microclimate (Prasad *et al.*, 1987).

Insect management can be achieved by manipulating time of sowing of crops. Sowing of Indian mustard (*Brassica juncea* subsp. *juncea*) is recommended around mid-October, as early sown crop may be attacked by painted bug (*Bagrada hilaris*) and in later sowing heavy infestation of mustard aphid (*Lipaphis erysimi*) may occur (Joshi *et al.*, 1989). In another study, the average number of mustard aphid (*Lipaphis erysimi*) on raya (*Brassica juncea*) increased tremendously when sowing is delayed (Vir *et al.*, 1990).

Planting geometry and planting density can influence insect infestation. Chickpea sown at a wider spacing (60 cm x 20 cm) had more infestation of black aphid (*Aphis craccivora*) than when sown at a closer spacing (30 cm x 10 cm) (Lal *et al.*, 1989).

As non-chemical methods for managing diseases (Mehrotra, 1987), plant pathogen can be avoided by proper selection of the field, by selection of disease escaping varieties and other planting materials, by manipulating sowing time, planting density, etc. Pathogen can be eradicated by roguing of infected plants and alternative hosts, with crop rotation, fallowing, flooding, etc. Non-chemical methods of disease management are cheap as well. For example, in case of wheat loose smut disease caused by *Ustilago tritici* can be very effectively controlled by seed soaking in ordinary water for 4 hours and then sun drying. This treatment, done on a hot sunny day during May or June, kills the disease causing organism and the seed so treated produce healthy plants when sown in the following *rabi* season (October to December sowing). Similarly, yellow ear rot (tundu) and ear cockle (mammi) of wheat caused by *Clavibactor, tritici* and *Anguina tritici* can be controlled by agitating seed in ordinary water for a few minutes, then skimming off floating ear-crockle galls with a sieve and then burning them (PAU, 1998). The non-chemical methods of insect pest and disease management has also been discussed in detail in Chapter I of this book.

Indigenous technical knowledge (ITK) is the knowledge that people in a given community have developed over time and continue to develop. Some examples of use of ITK in pest and disease control are application of *Neem* or *Karanj* powder in soil at the time of land preparation for control of termite aphids, jassids and white fly and use of *neem* oil or *neem* cake for control of early stem borer in rice and mosaic virus in tomato. Similarly, salt water is used for control of *Khaira* and blight disease in rice, wilt in pigeonpea and weed control in rice (Das *et al.*, 2006).

4.7 GLOBAL SCENARIO AND PROSPECTS OF ORGANIC FARMING

Today the developed countries like USA, Australia, UK and many other countries are practicing organic farming over a large acreage. The developing countries are also concerned over the ill-effect of overdose of chemical inputs used in agriculture. Though organic share of total food sales in different countries is only 0.2 per cent to 6.0 per cent (Lohr, 1998), the area under organic farming is gradually increasing. The worldwide sale of organic food is at present is estimated at about 11 billion dollars (Singhal, 1999). According to Willer and Yussefi (2002) the global area of land under certified organic farming is about 15.8 m ha, of which 45 per cent are located in Oceana (Australia, 7.7 m. ha.); 23.5 per cent in Europe (Itly having the highest area, nearly 1.0 m ha), 0.3 per cent in Asia and 0.1 per cent in Africa. In this data, large area of Australia (17.7 m. ha.) and Argentina (3 m ha) are not included because these organic land is extensive grassland. The number of certified organic farms in USA increased from 2,743 (0.47 m ha) in 1991 to 4,859 (0.91 m ha) in 1995 (Klonsky and Tourte, 1998).

More and more developing countries are also embarking in organic farming because of the following motivations (National Research Institute, 1998):

(i) Market demand,

(ii) Unavailability of chemical inputs (because of either its non-availability in market or poor purchasing power),

(iii) Environment concerns,

(iv) Social concerns, and

(v) Personal philosophy

Inclination of farmers towards organic farming is often based on combination of motivations rather than one single reason.

4.7.1 Organic Farming in USA:

In USA, not only on small farms but on large farms of 300-500 acres and even up to 1400 acres the farmers do not use synthetic chemical fertilizers or pesticides (USDA, 1980). A large number of farmers have shown that organic production of crops, like maize, soybean, wheat, cotton and others is feasible on a large scale (Brummer, 1998). About 30 million Americans are buying organic food regularly through 900 shops selling exclusively the organic products (Singhal, 1999). The examples of organic farming adopted by farmers in California are organic rice and organic grape and apple productions.

In organic rice production, two years rotation that alternatives rice with purple vetch (*Vicia benghalensis*) and fallow is practiced. The vetch supplies 120 to 130 Kg. N pe ha. In the spring of fallow year, vetch is chopped and disked along with decomposed rice straw leaving a mulch on the field. The field is shallow tilled to control weeds. Seeding of rice into mulch followed by intermittent flooding in the early stage of rice growth which also breaks life cycle of weeds, pests and pathogens and thus, eliminating the use of pesticides. This practice has the disadvantage of significantly lowering the yield which is compensated due to premium price farmers obtain from organic rice.

For the organic grape production, the management practices of farmers include application of 2000 t of composted steer manure per year, application of enzyme digested mixture of fish waste materials and no-tillage methods with perennial rye grass and cover crops chopped periodically for weed control. The yield of grapes obtained with 20 to 25 per cent premium price is much better than that in the conventional farm.

Organic apple growers grow a permanent legume or grass cover under the trees. *Vicia faba* is disked into the soil, thus incorporating organic matter and nitrogen into the soil and thus, improving the soil water holding capacity. Mechanical disking are the most common methods for weed control in organic orchards. The apple production is found to be 25 to 50 per cent lower than in conventional farms, but organic farmers usually obtain a premium price of their products.

4.7.2 Organic Farming in other Countries:

The countries having an active interest in organic farming are Switzerland, Belgium, Denmark, France, the Netherlands, Germany, China, Japan, Sweden, Canada, Austria and Itly. In the European Union, the present organic sale of 4 to 5 billion dollars is expected to touch 50 billion dollars in ten years (Singhal, 1999).

In UK, the number of organic farmers is increasing. The organisation involved in organic husbandry are the soils association, British organic farmers, Organic growers association, Elm research Centre, Organic farmers and growers limited and Bio-dynamic agricultural association (Marland, 1989). In U.K., sale of organic produce increased from 34 million pounds in 1987 (Marland, 1989) to 445 million pounds in 1997 (Lohr., 1998).

In Australia, area under organic production increased from 150 thousand ha. in 1990 to 335 thousand ha. in 1995 with corresponding increase in sale of organic products from 28 million to 60 million Australian dollars (Lyons, 1999).

In Russia, farmers in several regions adopt low external outputs in agriculture resulting lower yields in organic agriculture. The results obtained in state farms showed that row crops had lower yield while cereals and clover performed better. Disease posed no problems and hence, the quality of vegetables was better (Lampkin, 1990).

The new state policy in Cuba to develop organic agriculture increased focus on integrated nutrient and pest management techniques. The Cuban have pioneered the use of predatory ants to control weevil in sweet potato. Nitrogen fixing micro organisms are used to increase nitrogen availability to crops. Widespread use of bacteria *Bacillus* to promote the solublisation of phosphorus is also unique to Cuba. Vermi-composting is practiced widely.

The oldest organic and bio-dynamic coffee farm is situated in Finca Irlands, Southern Mexico. In order to improve nutrient recycling, the fertilization strategy adopted in the farm are (i) utilization of all residues generated, such as coffee berry pulp, cocoa husks, manure, urine and kitchen wastes, (ii) incorporation of biomass produced by deep rooting trees, and (iii) planting of shade trees, especially leguminous are considered important.

4.7.3 Organic Farming in India:

In developing countries, like India, sustainable agricultural system aims at increasing productivity with maintenance of environmental quality and ecological balance, while the developed countries are more concerned for proper care of environment and ecology with status quo in productivity. Farmers in India too believe that modern technology based agricultural practices are unsustainable.

The scientific concept of organic farming appears to occur in India about a hundred years back. From mid 80's onward experiments and demonstrations in organic farming have been conducted in various parts of India. Farmers are also getting convinced that this is the natural way to get out of the evils of chemicals. In Karnatka, eight acres farm of Narayan Reddy at Sorahunase village, 25 Kms from Banglore (Bangaluru) has become a global known votary of natural farming visited by many renowned agriculture experts. According to Reddy, the planting of marigolds in the fields could help to eliminate many pests. He has developed his own techniques to preserve seeds, destroy banana nematodes and check the larvae that infests cabbage and cauliflowers. Devangi Prafulla, the first farmer in Asia Pacific region to receive the FAO award in 1988 is another champion of natural farming in the state of Karnatka. He set the countrywide record raising the yield of sugarcane per ha without using fertilizers and pesticides. His farm at Hosahalli on the outskirt of Shimoga stands out as shinning example of the potentials of organic farming (Rao, 1990).

As reported by Palaniappan and Anadurai (1989), Mr. R. C. Kannan, an enterprising farmer in Tamilnadu raised Basmati rice under organic farming during the dry season (May-September). Only organics were applied to the fields and nimbicidine used for pest management. The crop gave an average grain yield of 4938 Kg per ha. In organic tea cultivation in Tirunelveli district in Tamilnadu, tea is grown in a system approaching an agro-forestry situation with shade trees of multipurpose type. They contribute to nutrient cycling, transport nutrients from below the tea's rooting zone, and catch leaching nutrients. Because of the addition of regular loppings, a mulch-layer is formed under the tea bushes. *Grevilla robusta* is used as a high shade tree planted at 10 m x 10 m and *Erythrina lithosperma* as a low shade trees planted at 2.5 m × 2 m. In between, *Gliciridia sepium* is grown.

Tribals of Jharkhand grow and /or protect medicinal herbal plants along with forest and agricultural crops without the use of any chemical either as fertilizer or as plant protection measure. Their cultivation practices are based on traditional organic farming followed by them since decades. Dey (2009) reported that the Birhors, a premative tribe have been using herbal plants to cure several diseases including that of blood dysentery, diabetes, leucorrhoea, malaria fever, scorpian bite, ulcers, etc. The Birhors have knowledge of medicinal plants for treating cancer and cardiac diseases. The National Institute of Science Communication and Information (NISCAIR), a unit of council of Scientific and Industrial Research (CSIR) of India have given approval for documentation, publication and patenting of these medicinal plants collected from Birhor dominated Chalkari village under Topchanchi block, about 42 Km from Dhanbad. Some of these herbal plants are listed below (Table 32):

TABLE 32: List of herbal plants of medicinal value used by Birhors, a primitive tribe of Jharkhand State.

Plants (local name)	Botanical Name	Medicinal Use
Jhagra He Baha	*Abrus Precatorious L*	Antiseptic, Headache
Bakas Pata	*Adhatoda vasica Nees*	Dry Cough
Sinjo	*Aegle marmelos Corr*	Sunstroke
Soso	*Anacardium occidentalis L*	Headache
Satmuli	*Asparagus voluble Will*	Blood dysentery, urinary disease
Lama Jang	*Bauhinia vahlii Wt*	Hypotension
Kamraj	*Byttneris herbacea Roxb*	Leucorrhoea
Tejraj	*Cissampelos Pereira L*	Stomach pain, Malaria
Bhojraj	*Cordea macleodil Hook*	Dysentery
Putari	*Croton oblongifolia Linn*	Eczema
Tiril Banda	*Dendrophthoe falcta Ettingh*	Skin diseases, contraceptive

Plants (local name)	Botanical Name	Medicinal Use
Haser	*Dioscorea bulbifera Linn*	Source of Carbohydrate
Doodhraj	*Gnaphalium luteo-album L*	Pregnancy, constipation, insomnla
Anantmul	*Hemidesmus indicus R. Br*	Diabetes
Huding Bhendra	*Jatropha curcas Linn*	Pyorrhea, gums and teeth
Marang Ponga	*Listsea monometallic Roxb*	Abscess
Tulsi Kana	*Lerilla ocimodies*	Keeps snakes away
Garki Jon	*Mallotus roxburghianus Lam*	Blisters and ulcers
Munga Aada	*Moringa Indica*	Keeps snakes away
Tamaku Kana	*Nicotiana tabacum*	Scorpion bite
Patol Kohada	*Pereira tuberose DC*	Urinary trouble
Ramdatun	*Smilax zeylancia Roxb*	Pyorrhea, Leucorrhoea, Spermatorrhoea
Sehda	*Sterblus asper Lour*	Stomach ailments
Talheja Dadu	*Sterculia urens Roxb*	Stomach burn, urinary tract infection
Rola Jon	*Terminalia chebule Retz*	Lungs diseases
Sinduar	*Vitex negundo L*	Body ache, headache

Mohan (1996) reported that Friends Rural Centre, Rasuila in Hosangabad district of Madhy Pradesh and farmers in Bijner district of Uttar Pradesh and Palghat district of Kerala are practicing organic farming. Farmers around Bhopal, Madhya Pradesh are cultivating organic farming over 10,000 ha. (Anonymous 2000). At Bolpur in West Bengal, the Tagore Society of Rural Development showed encouraging results by adoption of organic farming. Similarly, experiment in natural farming conducted by joint venture of ITTI, Delhi and Gandhi Peace Foundation also reported promising results.

Considering the merits of organic farming and organic produce by the use of organic manure, crop residues, green manures, legumes in cropping system, bio-fertilizers and rural industrial wastes; no one can deny the acceptance of this system. However, India's population is expected to be around 1.3 billion in 2020 AD and 1.5 billion in 2030 AD. It is but natural that farmers in India are seriously concerned for raising farm productivity to feed the increasing population. Hence, not the whole cultivated area should be expected to be brought under organic farming.

The critical analysis of available information shows that problems of degradation of soil health and ground water pollution are caused due to imbalance fertilizer use, inadequate use of organic manures and residues, high losses and low use efficiency, and that no verifiable evidence exists to prove fertilizer use per se as a cause of such problems. Thus, while responding to the alarm of ecological security raised on the basis of experiences of some highly industrialised countries in temprate and humid climates using excessive amounts of fertilizers, we should not loose the sight of our food security concerns (Paroda, 1998). The danger to food security can be attributed as the main reason for not popularising or supporting this system by the Government and policy makers on a large scale (Singh and Sekhon, 2002). Therefore, it would be wise to talk about sustainability at reasonably higher productivity levels in developing countries like ours, where sub-optimal fertilizer use continue to be major constraint to low crop yields. Our dependency on quick acting commercial fertilizers is not adequate (75.75 Kg per ha in 1996-97) (GOI, 1998) as compared to our neigbouring country China which consumes on an average of 370.7 Kg NPK per ha or any other countries in South-East Asian region such, as Korea (375.1 Kg. per ha) or Japan (375.0 Kg. per ha). Sekhon (1997) estimated that, to meet the food grain requirements of the country's population of 169.0 million in 2025, India needs to apply 30 million t of plant nutrients. The consumption of fertilizer in India (1997-98) stands at 17 mt. Hence, for a longer proportion of cultivated are of our country, integrated nutrient management appears to be the only answer at present which involves a combination of inorganic, organic and biofertilizers in such a proportion which may keep the environment healthy and make the soil capable of producing at the accelerated rate without being physically, chemically and biologically damaged.

REFERENCES

Abrol, I. P. and Gupta, R. K. 1998. Indo-Gangatic Plains - issues of changing land use. LUCC Newsletter, March, 1998, No. 3.

Ahlawat, I.P.S, Saraf, C.F. and Singh, A. 1985. Production potential of summer and rainy season pigeonpea inter-cropped with cowpea and greengram. Indian Journal of Agricultural Sciences, 55: 565-569.

Ahmad, S. and Srivastava, V.C. 1991. Assessment of competition and productivity evaluation under legume + cereal and legume + legume inter-cropping system. Indian Journal of Dryland Agriculture Research and Development, 6 (1 & 2): 60-66.

Ali, M. 1985. Soil fertility and fertilizer management for busting the productivity of pulses in arid and semi-arid regions of Northern India. Proceedings of SAI-NR Seminar, New Delhi.

Ali, M. 1988. Weed suppressing ability and productivity of short duration legumes inter-cropped with pigeonpea under rainfed conditions. Tropical Pest Management, 34:384-387.

Anonymous, 2000. Organic cotton. Pestology, 24 (9): 90.

Atwal, A.S. 1986. Agricultural Pests of India and South-East Asia, 2nd edition. Ludhiana: Kalyani Publishers.

Bagyaraj, D. J., Krishnaraj, P.U. and Khanuja, S.P.S. 2000. Mineral phosphate solublization: Agronomic implications mechanism and molecular genetics. Proceedings of the Indian National Science Academy, B66: 69-82.

Balaoybrananiyan, P. and Palaniappan, S.P. 2002. Principles and Practices of Agronomy, Agribios (India), Jodhpur.

Baldock, J.O., Higgs, R. L., Paulson, W.H., Jackobs, J.A. and Shrader, W.D. 1981. Legumes and mineral N effects on crop yields in several crop sequences in the upper Mississippi valley. Agronomy Journal, 73:885-890.

Bhattikhi, A. M., Suifan, M.S. and Al-Bakri, J.T. 1998. Effect of tillage and plant residue management practices on some physical properties of vertisols. Dirasat Agricultural Sciences, 25: 363-373.

Bayer, C. and Mielniczuk, J. 1997. Effect of soil preparation and cropping systems on soil chemical properties. Revizta Brasileria de ciencia do solo, 21: 105-112.

Beri, V., Sidhu, B. S., Bahl, G.S. and Bhatt, A. K. 1995. Nitrogen and phosphorus transformations as affected by crop residues management practices and their influence on crop yields. Soil use and Management, 11:51-54.

Beri, V., Sidhu, B. S., Bahl, G.S. 2002. Crop residue management, pp: 190-212. In: Recent Advances in Agronomy (eds. Singh, G., Kolar, J. S. and Sekhon, A.S.). Indian Society of Agronomy, New Delhi, India.

Bhagat, R. M. and Acharya, C. L. 1988. Soil water dynamics during wheat growth under different soil management practices. Journal of the Indian Society of Soil Science, 36:389-396.

Bhagat, R. M. and Verma, T. S. 1991. Impact of rice straw management on soil physical properties and wheat yield. Soil Science, 152: 108-115.

Bhan, V. M. 1968. Influence of row spacing on upland paddy. Il Riso, 17: 149-154.

Bhardwaj, K. K. R. and Gaur, A. C. 1985. Recycling of organic wastes. ICAR Publication, New Delhi. pp: 5.

Bhatia, R. K., Singh, S. and Mehra, S. P. 2002. Allelopathy in agriculture, pp: 358-388. In Recent Advances in Agronomy (eds. Singh, K, Kolar J. S. and Sekhon, H.S.D. Indian Society of Agronomy, IARI, New Delhi.

Bhatnagar, V. K., Chaudhary, T. N. and Sharma, B. R. 1987. Effect of tillage and residue management on properties of two coarsed textured soils and yield of irrigated wheat and groundnut. Soil and Tillage Research, 3:27-37.

Bhattacharya, P. and Mishra, U.C. 1995. A book on bio-fertilizer for extension worker. Publication of The National Bio-fertilizer Development Centre, Gaziabad, India.

Bhattacharya, P. and Chakrabory, G. 2005. Current status of organic farming in India and other countries. Indian Journal of Fertilizers, 1 (9): 111-123.

Brummer, E.C. 1998. Diversity, stability and sustainable American agriculture. Agronomy Journal, 90: 1-2.

Bud Houkstra 1992. Killer lense blasts California's weeds. New Scientist, June 13, No. 1825.

Burns, R. C. and Hardy, R.W.F. 1975. Nitrogen Fixation in Bacteria and Higher Plants. Springer-Verlag.

Clement, A., Ladha, J. K. and Chaliflour, F. P. 1998. Nitrogen dynimics of various green manure species and the relationship to lowland rice production. Agronomy Journal, 90:149-154.

Closser, N. D., Gooding, M. J. Thompson, A. J. and Froud-Williams, R. J. 1997. Competitive ability and tolerance of organically grown wheat cutlivars to natural weed infestations. Annals of Applied Biology, 139: 523-535.

Das, S. K., Das, P. and Roy, S. K. 2006. Indigenous Technical knowledge in Agriculture. Hand Book of Agriculture. Directorate of Information and Publication of Agriclture, ICAR, New Delhi. pp: 1162-1182.

Debnath, N.C. and Hajra, J. N. 1972. Rice straw incorporation - Lab. studies. Journal of Indian Society of Soil Science, 20: 95-99.

Dey, S. 2009. Patent for tribal cures, New Report of Hindustan Times (Ranchi, Octobr 26, 2009), pp: ole

Dhaliwal, B. K. 1998. Competition studies between spring barley (Hordeum vulgare L.) cultivars and wild oats (Avena fatua L.). Indian Journal of Ecology, 25:18-31.

Dwivedi, R. N. 1981. Fertilizer News, 26 (10): 28-34. (C. F. Modgal, S. C. and Singh, C.M. 1990. Crop Residue Management, pp:14. In Agronomic Research Towards Sustainable Agriculture. Indian Society of Agronomy Division of Agronomy, New Delhi.

Dyer, L. J. and Touchton, J. T. 1990. Utilizing legume cropping systems to reduce nitrogen fertilizer requirements for conservation............ corn. Agronomy Journal, 82:1123-1127.

Fagi, A. M. and De Datta, S. K. 1993. Physical properties of rainfed wetland rice soils as affected by cropping system and crop residue management. Field Crops Research, 6:189-204.

Fanlilanan, S. 1990. Safe, inexpensive, profitable and sensible. International Agricultural Development, March - April: 24.

Fortin, M. C., Pierce, F. J. and Edwards, M. 1994. Corn leaf area response to early season soil temperature under crop residues. Agronomy Journal, 86: 355 - 359.

Fukuoka, M. 1985. The natural way of farming: The theory and practices of green philosophy. Tokyo, Japan Publication.

Funderburk, J., Higley, L. and Buntin, G. D. 1993. Concepts and directions in arthropod pest management, Advances in Agronomy, 51: 125-172.

Gardner, W. K., Drendel, M. F. and McDonalt, G. K. 1994. Effect of subsurface drainage, cultivation and stubble retention on soil porosity and crop growth in a high rainfall area. Australian Journal of Experiment Agriculture, 34: 411-418.

George, T. Ladha, J. K. Garrity, D. P. and Buresh, R. J. 1994. Legumes as nitrate catch crops during the dry-to-wet transition in lowland rice cropping system. Agronomy Journal, 86: 267-273.

Giri, G. and De, R. 1978. Inter cropping of pigeonpea with other grain legumes under semi-arid rainfed conditions. Indian Journal of Agricultural Sciences, 48 (3): 659-665.

Gill, G. S. and Diggle, A. J. 1993. Herbicide resistance in annual ryegrass in the cropping belt of Western Australia: A case for integrated weed management. *In* Proceedings of International Symposium on Integrated Weed Management for sustainable Agriculture, 1:203-207, Hisar: Indian Society of Weed Science.

GOI. 1998. Agricultural Statistics at a Glance, Department of Economics and Statistics, Ministry of Agriculture, Government of India.

Goswami, K. P. and Pareek, R. P. 1979. Nitrogen economy by leguminous crops: Nitrogen Balance in soybean, wheat and gram. *In* Proceedings of National Symposium Hisar: Haryana Agricultural University.

Goswami, N. N. and Mishra, S. N. 1980. Annual Scientific Report. Division of Soil Science and Agriculture Chemistry, IARI, New Delhi.

Gould, W. D., Hagedorn, C. and Mc Cready, R. G. L. 1986. Urea transformations and fertilizer efficiency in soil. Advances in Agronomy, 40: 209-238.

Goyal, S. and Chander, K. 1998. Soil microbial biomass and enzyme activities as affected by incorporation of wheat straw in sandy loam. Environment and Ecology, 16:393-396.

Gunjal, S.S. 1991. Organic and Natural Farming: Principles of Organic Farming. Kisan World, June 1991: 10-12.

Handa, S. K. and Walia, S. 1996. Pesticide residues and its implication in integrated pest management. *In* IPM system in Agriculture, Vol. 1., Principles and Perspectives (eds. R. K. Upadhyay, K. G. Mukerji and R. L. Rajak), pp: 62-94, New Delhi: Aditya Books Private Limited.

Havlin, J. L., Kissel, D.E., Maddax, L. D., Classen, M. M. and Long, J. N. 1990. Crop rotation and tillage effect on soil organic carbon and nitrogen. Soil Science Society of America Journal, 54: 448-452.

Heekman, J. R. and Kruchinski, D. 1995. Soybean nodulation and nitrogen fixation on soil amended with plant residues. Biology and Fertility of Soils, 20: 284-288.

Heichel, G. H. and Barnes, D. K. 1984. Opportunities for meeting crop nitrogen needs from symbiotic nitrogen fixation. *In* Organic Farming: Current Technology and its role in a sustainable Agriculture (eds Bezdicek, D. F., Power, J. K., Kenney, D. R. and Wright, M. J.), pp: 49-59. Madison: American Society of Agronomy, Crop Science Society of America and Soil Science Society of America.

Hiremath, S. M. and Patel, Z. C. 1998. Effect of Winter green manuring and nitrogen on summer rice. Indian Journal of Agronomy, 43: 71-76.

Hosmani, M. M. and Meti, S. S. 1993. Non- chemical means of weed management in crop production. *In* Proceedings of International Symposium on Integrated Weed Management for Sustainable Agriculture, 1: 299-305. Hissar: Indian Society of Weed Science.

Jenny, H. 1941. Factors of Soil Formation. Mc Graw-Hill Book Co., New York, pp:281.

Jones, M. R. 1986. The effect of increased energy prices on the viability of organic agricultural system on arable farms in U. K. Agriculture, Ecosystems and Environment, 15: 221-229.

Joshi, M. L., Ahuja, D.B. and Mathur, B. N. 1989. Loss in seed yield by insect pests and their occurance on different dates of sowing in Indian mustard (*Brassica juncea* sub-sp. *juncea*). Indian Journal of Agricultural Sciences, 59: 166-168.

Katyal, J. C. 1977. Influence of organic matter on the chemical and electrochemical properties of some flooded soils. Soil Biology and Biochemistry, 9: 259 - 266.

Kitou, M and Yoshide, N. 1994. Mulching effects of plant residues on soybean growth and soil-chemical properties of soil. Soil Science and Plant Nutrition, 40: 220-221.

Klonsky, K. and Tourte, L. 1998. Organic agricultural production in United States: Debates and Directions. American Journal of Agricultural Economics, 80: 1119 - 1124.

Krishnan, K. R. and Dart, P. J. 1984. Effect of mycorrhizal inoculation and soluble P on growth and P uptake of pearlmillet. Plant and Soil, 81: 247-256.

Lal, S. S., Yadava, C. P. and Dias, C.A.R. 1989. Effect of planting density and chickpea cultivars on the infestations of the black aphid, *Aphis craccivora* koch. Madras Agricultural Journal, 76: 461 - 462.

Lampkin, N. 1990. Organic Farming. Farming Press books, TPSWICH, UK.

Le Mare, P.H. 1991. Rock phosphate in agriculture. Experimental Agriculture, 27: 413-422.

Limon-Ortega, A., Mason, S.C. and Martin, A. R. 1998. Production practices improve grain sorghum and pearl millet competitiveness with weeds. Agronomy Journal, 90:227-232.

Lohr, L. 1998. Implication of organic certification for market structure and trade. American Journal of Agricultural Economics, 80: 1125-1129.

Lyons, K. 1999. Corporate environmentalism and organic agriculture in Australia: The case of Uncle Tobys. Rural Sociology, 64: 251-265.

Mahadevan, N. R. and Chelliah, S. 1986. Influence of intercropping legumes with sorghum on the infestation of the stem borer, *Chilo partlellus* (Swinhoe) in Tamil Nadu, India. Tropical Pest Management, 32: 162 - 163.

Majumdar, D and Gupta, N. 2000. Nitrate pollution of groundwater and associated human health disorders. Indian Journal of Environmental Health, 42: 28-39.

Malik, R. P. and Singh, S. 1993. Evolving strategies for herbicide use in wheat: Resistance and integrated weed management. *In* Proceedings of International Symposium on Integrated Weed Management for Sustainable Agriculture, 1:225-238. Hisar: Indian Society of Weed Science.

Marland, A. 1989. An overview of organic farming in UK. Outlook on Agriculture, 18 (1): 24-27.

Mehrotra, R. S. 1987. Plant Pathology, New Delhi: Tata McGrew Hill Publishing Company Limited.

Mishra, I., Pradhan, L., Rath, B. S. and Tirpathi, R. K. 1998. Response of summer rice to different source of nitrogen. Indian Journal of Agronomy, 43: 60-63.

Mishustin, E. N. and Temsev, Y.T. 1982. An aerobic nitrogen fixation and plant nutrient. *In* Trans. 12th International Congress of Soil Sciences, New Delhi, 2: 48-53.

Modgal, S. C. and Singh, C. M. 1990. Crop residue management, pp:7-23. *In* Agronomic Research Towards Sustainable Agriculture. Indian Society of Agronomy, Division of Agronomy, IARI, New Delhi.

Mohan, B. 1996. A pioneer of natural farming. Yojana, 40(11), 40-42.

Moss, S. R. 1993. Testing Black-grass (*Alepoeursus myosuroides* Huds.) for resistance to chlorotoluron in the glass house. *In* Souvenir International Symposium on Integrated Weed Management for sustainable Agriculture, pp: 29-32. Hissar: Indian Society of Weed Science.

Moyer, J. R. 1995. Sulfonylurea herbicide effect on following crops. Weed technology, 9:373-379.

Mukherjee, D. and Gaur, A. C. 1988. A study on the influence of straw incorporation on soil organic matter maintenance, nutrient release and asymbiotic nitrogen fixation, Zbl. Bakt. Abt, 11: 135-663.

Murlibaskaran, R. K., Mahadevan, N. R. and Thangevelu, S. 1990. Influence of intercropping on infestation of shoot-webber (*Antigastra catalaunalis*) in sesame (*Sesamum indicum*). Indian Journal of Agricultural Sciences, 61: 440-442.

Omay, A. B., Rice, C.W, Maddux, I.D. and Gordon, W.B. 1997. Change in soil microbial and chemical properties under long term crop rotation and fertilization. Soil Science Society of America Journal, 61: 1672-1678.

Paes, J.M.V., Audreola, F., Brito, G. H. and Loures, E. G. 1996. Coffee straw decomposition in three different soils and its influence on CEC and pH. Revizta Ceres, 48:674-683.

Paroda, R. S. 1998. Natural resource management for sustainable agriculture: a new paradigm. pp.: 1-xxi. In first International Agronomy Congress (Nov. 25-27, 1998). Proce Indian Society of Agronomy, ICAR, New Delhi.

Pal, M. 1969. Indian Journal of Agronomy, 14: 272-279. (c.f. Modgal, S. C. and Sing, C.M. 1990. Crop Residue Management, pp:14) j. *In* Agronomic Research Towards Sustainable Agriculture. Indian Society of Agronomy. Division of Agronomy, IARI, New Delhi.

Palaniappan, S. P. 1997. Green manuring and biological nitrogen fixation: South Indian Perspective. In plant nutrient needs, supply, efficiency and policy issues: 2000-2025 (eds. Kanwar, J. S. and Katyal, J. C.). National Academy of Agricultural Science, New Delhi.

Palaniappan, S. P. and Annadurai, K. 2006. Organic Farming - Theory and Practices. Scientific Pubnshers (India), Jodhpur, Rajasthan, India.

Patel, Z. G. 1980. Effect of extra early variety of pigeonpea grown singly and intercropped with short duration grain and fodder legumes on the fertilizer nitrogen requirement of succeeding *rabi* cereal (wheat). Ph.D. Thesis, Division of Agronomy, IARI, New Delhi.

Pathak, R. D., Ghosh, T. K. and Srivastava, V. C. 1970. Response of maize to interseeding legumes and nitrogen fertilization. Agriculture and Agro Industries Journal, 3 (122): 33-35.

PAU. 1998. Package of Practices for *Rabi* crops of Punjab. Punjab Agricultural University, Ludhiana.

Pionka, H. B., Glotfelty, D.E., Lucas, AD. and Urban, J. B. 1988. Pesticide contamination of ground waters in the Mahantango Creek Watershed. Journal of Environmental Quality, 17:76-84.

Powles, S. B. 1993. Multiple herbicide resistance in annual ryegrass (*Lolium rigidum*): A driving force for integrated weed management. *In* Proceeding of International Symposium on Integrated Weed Management for Sustainable Agriculture, 1:189-194. Indian Society of Weed Science, Hisar, Haryana, India.

Prasad, D., Singh, K. M. Katiyar, P N. and Singh, R. N. 1987. Impact of intercropping on the plant growth, pest incidence and crop yield of pea, *Pisum sativum* Linn). Indian Journal of Entomology, 49: 153 - 172.

Prasad, K. and Srivastava, V. C. 1991. Pigeonpea (Cajanas cajan) and soybean (*Glycine max*) intercropping systems under rainfed situation. Indian Journal of Agricultural Sciences, 61 (4): 243-246.

Ramanathan, K. M. 2006. Organic farming for sustainability. Journal of the Indian Society of Soil Science, 54 (4): 418-425.

Ramesh, P, Singh, M. and Subba Rao, A. 2005. Organic farming: Its relevance to the Indian context. Current Science, 88(4): 561-68.

Rani Perumal, 1994. Residual effect of manures in organic farming. Tamil Nadu Agricultural University, Coimbatore, India.

Rao, J. V. 1974. Studies on fertilizer management of wheat in maize-wheat-green gram and arhar-greengram cropping systems Ph.D. Thesis, Division of Agronomy, IARI, New Delhi.

Rao, R. 1990. Back to natural farming. Current. September 1-7, pp:18.

Reddy, K. C., Soffes, A. R. and Prine, G. M. and Dunn, RA. 1986b. Tropical legumes for green manure: Nematode populations and their effects on succeeding crop yield. Agronomy Journal, 78: 5-10.

Reddy, K. C. Soffes, A. R. and Prine, G.M. 1986 b. Tropical legumes for green manure: Nitrogen production and the effects on succeeding crop yields. Agronomy Journal, 78:1-4.

Reddy, S. R. 2000: Principles of crop production. Kalyani Publishers, Hyderabad-Chennai, India.

Rohweder, D.A., Shrader, W.D. and Templeton (Jr.), W.C. 1977. Legumes: What is their place in today's agriculture. Crops and Soils, 29 (6): 11-14.

Roy, K and Srivastava, V.C. 1995. Productivity and economics of potato based cropping sequences. Paper presented in Group Meeting of Potato Worker held at UAS, Dharwad (Karnataka), Aug.30 - Sept 1, 1995.

Roysharma, R. P., Thakur, H. C. and Sharma, H. M. 1980. Pigeonpea as a *rabi* crop in India. Proceedings. International Workshop on Pigeonpea. ICRISAT, Patancheru, Andhra Pradesh, India, 1: 26-36.

Saxena, M.C. and Yadav, D.S. 1979. Parallel cropping with short duration pigeonpea under humid subtropical conditions at Pantnagar. Indian Journal of Agricultural Sciences, 49 (2): 95-99.

Saxton, K. E., Kenny, J. F. and Mccool, D. K. 1993. Air permeability to define frozen soil infiltration with various tillage and residue. Transactions of ASAE, 36: 1369-1375.

Schnitzer, M. 1991. Soil organic matter - the next 75 years. Soil Science, 151: 41-58.

Sekhon, G. S. 1997. Plant nutrient needs, supply, efficiency and policy issues: 2000-2025. (eds. Kanwar, J. S. and Kalyal, J. C.). National Academy of Agricultural Science, pp:78-90.

Sekhon, H. S. ingh, G. and Verma, M. M. 1992. Sustainable agriculture - Pulses, a must in different cropping systems. Progressive farming, 28 (1): 19-20.

Sekhon, B. S. and Bajwa, M. S. 1993. Effect of organic matter and gypsum in controlling soil sodicity in rice-wheat-maize system irrigated with sodic wasters. Agricultural Water Management, 24: 15-25.

Shanmuga Sundaram, V. S., Arunachalam, N., Ramasamy, C., Gopalsamy, A., Durairaj, N. and Morachan, Y. B. 1975. The role of crop residue to meet the fertilizer storage. Madras Agricultural Journal, 62 (6): 353 - 356.

Sharpley, A. N. and Menzel, R. G. 1987. The impact of soil and fertilizer phosphorus on the environment. Advances in Agronomy, 41: 297 - 324.

Sharpley, A. N. and Smith, S. J. 1989. Mineralization and leaching of phosphorus from soil incubated with surface applied and incorporated crop residue. Journal of Environmental quality, 18: 101-105.

Shivashankar, K., Vlassak, K. and Livens, J. 1976. A comparison of the effect of straw incorporation and CO_2 enrichment on growth and nitrogen fixation and yield of soybean. Journal of Agricultural Science, Cambridge, 87: 181-185.

Sidhu, B. S. and Beri, V. 1989. Effects of crop residue management on the yield of different crops and soil properties. Biological Wastes, 27: 15-27.

Sidhu, B.S., Beri, V. and Meelu, O. P. 1988. The effect of soil temperature modifications on the natural pigeonpea (*Cajanas cajan* L.) - Rhizobium symbiosis. Journal of Agricultural Science, Cambridge, 111: 395-397.

Sidhu, B. S. Goswami, K. P. and Pareek, R. P. 1994. Influence of paddy and wheat straw application on crop yields. Journal of Research (PAU), 31: 147 - 153.

Singh, B. and Dhaliwal, G. S. 1992. Environmental impact of pesticides. *In* changing Scenario of our Environment (Eds. Dhaliwal, B. S. and Hansra, B. S. and Jerath, N.), pp: 165-174. PAU, Ludhiana.

Singh, B., Chanasyk, D. S., McGill, W.B. and Nybrog, M.P.K. 1994. Residue and tillage management effects on soil properties of a typic cryoboroll under continuos barley. Soil and Tillage Research, 32: 117-133.

Singh, G. and Sekhon, H. S. 1998. Weed suppression and higher crop productivity in pigeonpea and mungbean intercropping. In proceedings of the 6th European Weed Research Society Mediterranean Symposium. Montpellier, France, pp: 97-98.

Singh, G. and Wright, D. 1999. Effects of herbicides on nodulation, symbiotic nitrogen fixation, growth and yield of pea (*Pisum sativum*). Journal of Agricultural Science, Cambridge, 133:21-30.

Singh, G. P. 1991. Effect of crop residue management on biological properties of soil. M.Sc. thesis. PAU, Ludhiana.

Singh, M. and Singh, R. K. 1975. Prospects of growing pulses under multiple and intercropping. Indian Journal of Genetics and Plant Breeding, 35: 216-220.

Singh, N. B., Singh, P. P. and Nair, K. P. P. 1986. Effect of legume intercropping on enrichment of soil nitrogen, bacterial activity and productivity of associated maize crops. Experimental Agriculture, 22: 339-344.

Singh, R. P. and Venkateswarlu, J. 1995. Report of CRIDA, Hyderabad, India.

Singh, S. P. 1983. Summer legume intercop effects on yield and nitrogen content of wheat in the succeeding season. Journal of Agriculture Science (Cambridge), 101:401-405.

Singh, T. and Dixit, S. K. 1998. Use of biofertilizers for higher crop yields. Directorate of Extension, Ministry of Agriculture, Government of India, New Delhi.

Singhal, V. 1999. Indian Agriculture 1999. New Delhi: Indian Economics Data Research Centre.

Sodhi, P. S. and Dhaliwal, B. K. 1998. Effect of crop density and cultivars on competitive interaction between wheat and old oats (*Avena ludoviciana* Dur.). Indian Journal of Ecology, 25: 138-145.

Somani, L. L. and Saxena, S. N. 1975. Effects of some organic matter sources on nutrient availability, humus build up, soil physical properties and wheat yield under field conditions. Annals of Arid zone, 14: 149 - 158.

Somani, L. L. and Saxena, S. N. 1981. Role of some sources of organic materials alone and in conjunction with inorganic amendments in reclamation of a typical calcareous saline-alkali soil of Rajasthan, Agrochemical, 25: 232-242.

Srivastava, G. P. and Srivastava, V.C. 1991. Production potential of pigeonpea based cropping system under different planting pattern. Indian Journal of Agronomy, 36 (2): 173-176.

Srivastava, G. P. and Srivastava, V.C. 1993a. Effect of winter legumes on nitrogen economy of succeeding rice (*Oryza sativa*). Indian Journal of Agricultural Sciences, 63(8): 501-503.

Srivastava, G. P. and Srivastava, V.C. 1993b. Nitrogen economy and productivity of wheat (Triticum aestivum) succeeding grain legumes. Indian Journal of Agricultural Sciences, 63(11): 894 - 698.

Srivastava, V. C. and Fokin, A.D. 1973. Study on sorption, Kinetics and transformation of certain fraction of labelled fulvic acid in padzolic soil. Izvestia (Journal of Moscow Agricultural Academy), 4: 114-119.

Srivastava, V. C. and Fokin, A.D. Study of hunification of 14C labelled plant material in soil. Izvestia, (Journal of Moscow Agricultural Academy), 4: 98 -105.

Steinhart, J. S. and Steinhart, C.E. 1974. Energy use in the US food system. Science, 184:307-316.

Tewari, G. C. and Sharma, P. 1996. Pesticide resistance problem and its management in India. In IPM system in Agriculture, Vol. 1 Principles and Perspectives (eds. Upadhyay, R. K., Mukerji, K. G. and Rajak, R. L.), pp:41-61, Aditya Books Private Limited, New Delhi.

Tiwari, K. N., Pathanak, A. N. and Ram, H. 1980. Journal of Indian Society of Soil Science, 28 (2): 162-169.

Tiwari, K. N., Sulewski, G. and Portch, S. 2005. Challenges of meeting nutrient needs in organic farming. Indian Journal of Fertilizers, 1 (4): 41-48.

Toor, G. and Beri, V. 1991. Extent of nitrogen fertilizer immobilized by the application of rice straw and its availability in soil. Bio-resource Technology, 37: 189-191.

Upadhyay, R. K., Mukerji, K. G. and Rajak, R. L. 1996. Integrated pest management. In IPM system in Agriculture, Vol 1, Principles and Perspectives (Eds Upadhyay, R. K., Mukerji, K. G. and Rajak, R. L.), pp:1-23, Aditya Books Private Limited, New Delhi.

USAD. 1980. Report and Recommendations on Organic Farming. United States Department of Agriculture, Washington DC., USA.

Velayutham, M., Pal, D. K. and Bhattacharya, T. 1999. Organic carbon stock in soils of Indian, pp:71-96. In Global Climatic Change and Tropical Ecosystems (eds. Lal, R., Kimble, J. M., Eswaran, N. and Stewart, B.A.), Lewis Publishers, Boca Rattan, Florida, U.S.A.

Verma, L. N. and Bhattacharya, P. 190. Role of biotechnology in supplying plant nutrients in the nineties. Fertilizer News, 35(12): 87-92.

Verma, U. K., Singh, R. K., Srivastava, V. C. and Mukerjee, A. K. 1980. Place of mixed crops of arhar in cropping systems of Bihar. Indian Journal of Genetics, 40A: 87-88.

Vir, S., Singh, M. P. and Henry, A. 1990. Yield loss in important cultivars of raya and effect of date of sowing on aphid infestation under arid climate of Rajasthan. Indian Journal of Entomology, 56: 541-546.

Walia, S. S., Brar, S. S. and Kler, D.S. 1998. Management of crop residues in rice-wheat system. Environment and ecology, 16: 730-731.

Weil, R. R. and Samaranayake, A. 1991. Effect of winged bean on a following maize crop. Experimental Agriculture, 27: 329 - 338.

Wortmann, C. S. 1993. Contribution of bean morphological characteristics to weed suppression. Agronomy Journal, 85: 840-843.

Wu, J., O'Donnell, A.G. and Syers, J. K. 1993. Microbial growth and sulphur immobilization following the incorporation of plant residues into soil. Soil Biology and Biochemistry, 25:1567-1573.

Yadav, R. L. 1993. Agronomy of sugarcane-principles and practices. International Book Distributing Company, Lucknow, India.

Yan, F., Schubert, S. and Mengel, K. 1997. Effect of soil moisture stress on plant growth and diseases in continuously cropped soybeans. Biology and Fertility of soils, 23:236-242.

Zaidi, N. W. and Singh, U. S. 2004. Mass multiplication of *T. harziaanum* on cow dung. Indian Phytopathology 57 (2): 189-192.

CHAPTER V

AGRICULTURE AND ENVIRONMENT

5.1 INTRODUCTION

Agriculture has tested the resiliency of nature since its inception. Its objectives generally have been to achieve maximum yield, operate with a maximum profit, minimise year to year instability in production and prevent long-term degradation of the productive capacity of the agricultural system. According to United Nation's population projection, the world's population is forecast to increase from 5.3 billion in 1990 to 8.50 billion in 2025 with 84 per cent of this growth expected in developing countries (Bos *et al.*, 1995). This can lead to a global food deficit of 700 million tons in 2025 if the present yield growth rate continues. By the year 2050, it is believed that an additional 4.3 billion people inhibit the earth.

The role of "Green Revolution" in addressing the problem of food security in India is known to every one. The term "green revolution" was coined by Dr. William Gaud of the United States in 1968 to mark the beginning of an era of high yields in major cereals, like wheat and rice. India had released a special stamp titled "The Wheat Revolution" in July 1968 to indicate that the stagnation in productivity in wheat fields has at long last ended, thanks to the semi-dwarf, non-lodging wheat varieties containing the 'Norin' genes for dwarfing introduced from the programme of Norman E. Borlaug in Mexico (Swaminathan, 1993). In the post-independence period, foodgrain production registered nearly four times increase from mere 54.9 tonnes in 1949-50 to 199.3 million tonnes in 1996-1997 with a compound growth rate of 2.51 per cent. India supports 16.8 per cent world's population, 4.2 per cent world's water resources and 2.3 per cent global land. Though foodgrains production today touched 220 million tonnes, the food demands also on account of our burgeoning population (projected to be 1.5 billion by 2050 AD and may outgrow China as the most populous country of the world) and improved incomes are pushing up the total and preferential demands for food and other items. India would need 325 million tonnes of food grains by 2025 and to achieve this target, production has to be increased by seven million tonnes every year over the next 15 years, without much increase in irrigated crop land.

The benefits of green revolution gained in sixties gradually started vanishing with the population growth competing with the agricultural productivity growth. Today the average productivity of most of the field crops particularly pulses and oilseeds is extremely low excluding wheat and to certain extent rice. The echo of sustainable agriculture has also become louder. We are also being warned to the ill-effects caused due to second generation problems of the green revolution. Today Indian agriculture is facing tremendous new challenges due to continued population growth; stagnation in farm level productivity in intensive farming areas; deforestation practices; land degradation; pollution of air, soil and water; hazardous waste and improper solid waste disposal; destruction and erosion of genetic resources (crop, animal, tree, fish, etc.) and globalisation. It is envisaged that if the current practices of natural resource utilization and low productivity of agriculture continue, a child born today will have less chance of getting adequate food to eat, space to live and pure air to breathe tomorrow.

In the fragile and marginal environment of the developing countries like India, sustainability of agriculture threatened by lack of the appropriate management of natural resources and the use of chemical inputs needed to optimise the production potential of the land to feed the people and to raise them above the poverty line.

There is great concern now about decline in soil fertility, decline in water table, rising salinity, resistance to many pesticides and degradation of irrigation water quality. Over-exploitation of groundwater is a serious problem in many intensively cultivated regions in India. The watertable in the north-western part of India is receding at an annual rate from 0.2 to 0.5 metre. Soil salinity and waterlogging have already spread to several parts of Indo-Gangetic plains. In Punjab, the mean fertilizer use in rice and wheat has become much higher than the recommended dose resulting in increased NO_3 concentration in groundwater (Aggarwal *et al.*,

2006). Farmers in Indo-Gangatic plains also have to apply more fertilizers to get the yield as they were getting with less fertilizer 20 to 30 years ago indicating declining trend in productivity.

The current issues bothering environmentalists and conservationists are resource degradation, depleting forest cover, loss of biodiversity and global warming. Human activities have increased CO_2 and many other green house gasses. This increase and mankind's continued activity has the potential to warm the earth's climate over the next century (IPCC, 1995). Burning of grassland located near forest by people to ensure better growth of grasses in the next season, sometimes, causes out of control fire which spreads to forest. The top layer of the soil gets burnt and hardens if the fire continue to rage. This makes the soil susceptible to erosion even in the mild showers. The quantity of rainfall and its occurance has become more uncertain. In certain places, climatic extremes, such as droughts, floods, rainfall distribution and snowmelt have increased. The glacier is melting with increased rate and the sea level has risen by 10 to 20 cm. The snow cover is gradually decreasing. The global mean annual temperature at the end of 20th century was almost 0.5 to 0.7 above that recorded at the end of 19th century. It is projected that the average temperature of the air would rise by 1.4 to 5.8°C depending upon the region over next 100 years. The likely consequences of such environmental change are more floods, frequent droughts, and forest fires, decrease in agricultural and aquatic productivity, displacement of coastal dwellers by sea-level rise and intense tropical cyclones and the degradation of mangroves.

According to Scherr and Yadav (1996), the current and future hot spots of severe environmental deterioration include:

- the river basins of the Indus, Tigris, and Eupharates in the Middle East (salinization);
- the foothill of the Himalayas (water erosion);
- the forest margins of the lower Amazon (overgrazing and nutrient loss);
- the periurban areas of Mexco City (pollution from agricultural chemicals);
- the sub-humid, semi-arid border zone of West Africa (migration from dryland degradation).

In order to avoid soil degradation and to be ecological sustainable, agricultural land-use should maintain an equilibrium between outputs and inputs of natural resources, and should maintain ecosystem flexibility and biodiversity. **Ecological sustainability** is one component of sustainable land use, which refers to the need to maintain the long-term production potential of the land. Other components of sustainable land-use refers to social, cultural and economic aspects, such as the ability of the land manager to adopt proper management systems and the equitable distribution of inputs and outputs of production.

For an ecological sustainable land-use, the exploitation rate of natural resources by human land-use should not exceed the carrying capacity. **The carrying capacity** is defined as the level of exploitation of renewable resources which imposes limit on a specific type of land-use, that can be sustained without causing land degradation within a given area.

In India, large scale degradation of natural resources has occured due to population induced deforestation, conversion of forest land to agriculture, intensive cropping, reduction in the frequency of fallowing and indiscriminate use of chemicals. Among the natural resources; soil, water and vegetation are the most crucial for the sustenance of agriculture. It is, therefore, a major challenge for agriculturists to produce enough food and fibre to satisfy the increasing population in an environmentally sustainable manner. The problem may get further complicated by on going globalisation process and economic reforms of World Trade Organisation facing many countries including India to make structural adjustments in agricultural sector.

Therefore, it is important to understand the key environment issues from crop production perspective and vice-versa in an effort to enhance and sustain agricultural production and hence, food security for increasing population under shrinking resources.

5.2 ENVIRONMENT AND PLANT GROWTH

5.2.1 Definition and Scope

The term 'environment' means surroundings. The various components of environment consists of the atmosphere, lithosphere, biosphere, and hydrosphere.

The environment may be defined as "the aggregate of all the external conditions and influences affecting the life and development of an organism." It is the entire web of biological and physical interactions, which characterise the relationship between life and the earth.

Life on earth is supported by the atmosphere, solar energy and our planet magnetic filed. Energy from the sun is absorbed by the atmosphere recycling water and other nutrients and working with electrical and magnetic forces to provide a moderate climate on the earth.

The environment protection act approved by the Government of India in 1986 defines "the environment as the sum total of water, air and land, the interrelationship among themselves with human beings and with other organisms and property."

The awareness about environmental issues all over the world was aroused by publication 'Silent Spring' in 1962 by the famous biologist and writer Rachel Carson who reported about the long-term effects of misusing chemical fertilizers and pesticides for increasing agricultural production. The directive principles of state policies in the Indian constitution approved in the year 1976 states that the 'state shall endeavour to protect and improve the environment and to safeguard the forests and wildlife in the country and to protect and improve the natural environment including the forests, lakes, rivers and wildlife and to have compassion for the living creatures.' The Stockholm Conference on Human Environment (Sweden, 1972), the United Nations Conference on Environment and Development (Rio de Janerio, 1992), The World Summit on Sustainable Development (Johannesburg, 2002) have made the world aware about the deteriorating environment on earth.

Environmental pollution is greatest threat to agriculture and mankind. **Pollution** may be defined as "an undesirable change in the physical, chemical and biological characteristics of air, water and land that harmfully affect life (human, animal, plant and other desirable species) and resources on earth." Pollution is different from **contamination** which means the presence of harmful organisms and their products in environment causing diseases or discomfort. **Pollutant** means any solid, liquid or gaseous substance present in such concentration as may be or tend to be injurious to environment.

In order to protect agriculture and environment and to promote sustainable economic development for better life on this earth, there is an urgent need to redress major environmental issues in agriculture, such as pollution of air, water and soil leading to global warming, greenhouse effect, climate change, acid precipitation (rain), ozone depletion, soil degradation, water scarcity due to falling groundwater table, shrinking wetlands, deforestation, deserfication, threat to biodiversity, depletion of energy resources and natural resource management.

5.2.2 Physical Environment

The crop environment constitutes both soil and aerial, as part of the plant inside the soil and other part which is exposed to the atmosphere. Crop's soil environment consists of several physical, chemical and biological changes which are amenable for manipulations through management practices. However, aerial environment cannot be altered easily. Aerial environment of a plant is influenced by climatic factors which are: (i) light, (ii) radiant energy, (iii) temperature, (iv) precipitation, (v) composition of the atmosphere, and (vi) wind; whereas soil environment influenced by edaphic factors are: (i) availability of soil moisture, (ii) mineral nutrient supply, (iii) soil reaction, (iv) composition of soil air, and (v) biotic factors.

The environment, in which crops are grown, can influence the rate of growth and development which depends primarily on crop's genetic make up, soil environment and climatic conditions. Growth may be expressed in terms of dry weight, length, height or diameter, whereas development is defined in terms of the progression through the morphological changes which occur during the growth of the crop and is more readily described qualitatively than quantitatively.

It is essential to define the physical environment of crop plant to augment yield per unit area per unit time. The importance of physical environment for plant growth can be understood by the fact that plants are rooted to their environment and unlike human being and animals, they cannot walkout of the hot sun, scorching winds or cold winter nights. They have to modify their anatomy and physiology to changing environment or else die. A living plant is affected by the flow of energy, water vapour, CO_2, O_2 and ions between the plant and the environment. It is clear that both aerial and soil environments are closely related and integrated by the plant. However, the quantities measured in one case do not describe the conditions in the other.

The plant parts absorb certain amount of incoming solar radiation, whereas some part is reflected and rest is transmitted to the ground. The plants also redistribute the absorbed heat in the form of re-radiation, convection, conduction and transpiration. These mechanisms help to keep thermal environment below lethal limits.

As regards incident solar radiation, it is not clear whether it will increase or decrease although variations are expected in different regions of the earth. But it is expected that higher amount of water vapours in the atmosphere will absorb incoming solar radiation resulting slight decrease in radiation by about one per cent or so (Hume and Cattle, 1990).

5.2.3 Energy Environment

Environment is generally described in terms of certain parameters like rainfall, sunlight, temperature, humidity, wind, etc. But it is not the temperature as such that is important but the radiant heat received. Meteorological parameters may describe properties of atmosphere but for crop growth, the energy environment has to be understood. The energy environment of plant can be described by the energy conservation equation (Gates, 1962).

$$\text{Net rate of incoming energy per unit area} = \text{Net rate of outgoing energy per unit area}$$

$$S+R+LE+G+C+P+M=O$$

Whereas,
S = Solar radiation flux
R = Thermal radiation flux-Ground, atmosphere and surrounding objects.
L = Latent heat of evaporation
E = Rate of evaporation
G = Sensible heat flux by conduction from ground to surface
C = Sensible heat flux by convection in air
P = Photosynthesis
M = Metabolism

The energy will flow to or from plant by radiative transfer or by conduction along a temperature gradient or by mass flow by convection. Similarly, water vapour, CO_2, O_2 will flow to or from plant along a concentration gradient.

Energy enters an ecosystem at the rate at which radiant energy is converted by photosynthetic and chemosynthetic activity of chiefly green plant to organic substances. The process of photosynthesis in green plants depends on quality of light.

Extremely highlight intensity may have an inhibitory effect on photosynthesis, a phenomenon called **solarization** (Wang, 1963).

5.2.4 Microclimate of crops

The energy for plant growth is essentially derived from sun. Plants grows under the climate of earth and as it grows, it produces in its neighbourhood its own climate, termed as microclimate. Microclimate is a dynamic reaction of a larger number of meteorological and biological processes going on at and near the earth surface.

With the growth of the plant and with increasing development of its neighbouring plants, the microclimate considerably changes. The thicker the plant population, the larger will be the crop cover and the greater will be the influences of microclimate. An understanding of microclimate requires a detailed study of the physical pressure of the mass exchanges (flux of radiation, sensible heat, latent heat and transfer of momentum) between the vegetation, foliage and lower layer of atmosphere.

The height at which the solar radiation is absorbed depends on the albedo of surface and mutual shading of the leaves. The net solar radiation incident on the crop is usually positive during the day time and as the short wave components are cut off at night, it becomes negative. Air and soil on the surface of earth begin to cool due to net radiation loss from the earth.

Solar radiation that changes the day and night temperature and wind velocity that controls the effectiveness of transfer of heat, water vapour and CO_2, etc. are important factors in controlling the climate. Friction between air and earth surface plays a decisive role that affects the various transport processes near the ground surface. Exchange of heat, water vapour and CO_2 take place in a boundary layer between the ground and the atmosphere. It is in this layer that large differences in the smallest space are encountered.

5.3 CLIMATIC FACTORS AND CROPS

A crop has to pass through different weather conditions during its life cycle. Hence, the success or failure of crop production and its qualitative and quantitative improvement are closely related to the prevailing climatic environment. The problem relating to agronomical practices like varietal selection, sowing schedule, crop geometry, timing of fertilizer application and irrigation schedule, etc. can be better solved by the knowledge about the relationship between climatic factors and crops.

5.3.1 Weather and Climate

Weather is defined as the atmospheric conditions at a given time and place. It is variation or conditions of lower layers of atmosphere. Weather changes from hour to hour, from day to day, and from year to year. Generally weather may be referred to dry weather, wet weather, humid weather, cloudy weather, hot or cold weather.

Climate is the average weather conditions over a given region, zone, state, country or continent as a whole during a long period of time. A long-term average of climatic parameters at a given place represent more accurate the climatic conditions of that place. The climate of a place and a region can be categorised as arid, humid, hot, cold, tropical, sub-tropical, temperate and polar climate.

The yield potential of a crop mainly depends on weather or climate. More than 50 per cent variation in crop yield is due to climatic differences. The most important climatic factors that influence growth, development and yield of crops are solar radiation, temperature and rainfall. Atmospheric humidity and wind velocity also influence crop growth to some extent.

5.3.2 Solar Energy

Solar energy is one of the most important factor for many vital processes of the plant. It is available to plant in two distinct form: (i) Solar radiation and (ii) Infrared or thermal radiation.

5.3.2.1 Radiant energy

(i) **Solar radiation:** It is the radiant energy from the sun measured as a total amount (direct solar plus sky radiation) expressed in cal per sq cm per minute measured by phyrhelometer. Almost 99 per cent of solar radiation is received between 0.15 and 4.0 micron (μ) wave length, of this 9.0 per cent is in the ultraviolet region (0.15 - 0.40μ) and nearly 45 per cent each is the visible (0.40 - 0.70μ) and infrared (0.75 - 4.0μ). The non-visible short wave length (0.0005 to 0.2 μ) are known as cosmic rays, x-rays and gamma rays and are obtained from radioactive substances. These substances (atleast up to 0.33μ) are almost all absorbed in upper layer of atmosphere by atomic oxygen or ozone, hence life can be sustained on earth, because very little of these radiation can be tolerated.

For photosynthesis, only visible part of the solar energy is important. The blue (0.40 to 0.50μ or 400 to 500 nm) and red (0.60 to 0.70μ or 600 to 700 nm) lights more strongly absorbed by chlorophyll are most useful for photosynthesis. Hence, use of solar energy in wavelength from 0.4 to 0.7μ for photosynthesis in green leaves often expressed to as **photosynthetically active radiation** (PAR). The ratio of PAR to total solar radiation is close to 0.50 in both tropics and temprate regions (Monleith, 1972). The most commonly used unit of solar radiation in agriculture is the calorie per square centemeter per day (cal/cm^2/day). Out of the net solar radiation, a small part is utilized as chemical energy in the process of photosynthesis and other part is stored within crop and soil. The magnitude of evapo-transpiration depends upon available heat energy within the crop environment. Longer waves in solar radiation band are largely 0.75 to 4.0μ and are called **near infrared radiation** (NIR).

(ii) **Infrared thermal radiation**: Earth also emits radiation of long wavelength (1.2 to 40.0 micron) which is called infrared thermal radiation. It is also called **terrestrial radiation**. Earth actually absorbs solar radiation which is in short waves and emits back in the form of long waves. Terrestrial radiation heat the atmosphere. The thermal radiation from earth occurs during both day and night as against only day time radiation from sun, i.e. solar radiation.

(iii) **Albedo or Reflected radiation**: It is the ratio of the amount of visible light reflected by a body to the amount of incident upon it. It is expressed in the percentage and measured with albedometer. Albedo of plant surface like rice, wheat, potatoes, mungbean and pearlmillet are 12, 23-25, 19, 26 and 24 per cent, respectively. Albedo of soil surface like black soil, grey soil and ploughed field (moist) are 8 to 14, 10 to 30 and 14 to 17, respectively depending upon the moisture content in soil (Reddy, 1999).

5.3.2.2 *Measurement of solar energy use*

Solar energy use: Photosynthetic efficiency (Eμ) of a crop can be estimated with the following formula (Yoshida, 1981):

$$E\mu = \frac{\text{Net gains of chemical energy}}{\text{Total incident solar energy}}$$

$$= \frac{K \quad W}{S} \quad \frac{K \quad W}{s \ T}$$

Where, K = heat of combustion (cal/g)

W = dry matter increase (g/m^2)

ΣS = total incident radiation (cal/m^2)

S = average daily incident radiation ($cal/m^2/day$)

T = number of days

The crop growth rate (CGR): It can also be estimated as:

$$CGR\ (g/m^2/day) = \frac{E\mu\mu\ S}{K}\ 10^4$$

Where, $E\mu$ = The highest photosynthesis efficiency (mean value in 3%)

S = The incident solar radiation of 400 cal per square cm per day

K = 3750 cal per g ; a value adopted for all plant parts

10^4 = cm to m conversion

Thus, an expected maximum crop growth rate under incident solar radiation of 400 cal per sq cm per day would be:

$$\text{Maximum CGR} = \frac{0.03 \times 400}{3750} \times 10^4$$

$$= 32 \text{ g per sq m per day.}$$

5.3.2.3 Solar radiation effect on Crops

Solar energy provides light for various growth and development functions and thermal energy for various physiological functions. Light is indispensable for the synthesis of the most important pigment of the plant, i.e. the green chlorophyll. The chlorophyll is capable of absorbing radiant energy and converting it into potential chemical energy of carbohydrates. Permeability of protoplasm, intake and loss of water, enzyme activity, respiration, photosynthesis, leaf expansion, growth of stem and shoot, flower initiation, ripening of fruits and regulating rate of transpiration by controlling opening and closing of stomata are all influenced by light directly or indirectly. The main effect of solar radiation are thermal effect, photosynthetic effect, photoperiodic effect and others as presented in Table 33.

TABLE 33: Spectral region and influence on plant life

Type of radiation	Spectral region (μ)	Solar energy (%)	Effect on plant life			
			Thermal	Photo-synthetic	Photo-periodic	Others
(i) Ultraviolet	0.3-0.4	0-4	IS	IS	Moderate	H
(ii) Photosynthetically active radiation (PAR)	0.4-0.7	20-46	S	S	S	NU
(iii) Near infrared radiation (NIR)	0.7-4.0	50-80	S	IS	S	
(iv) Large wave radiation	4.00-100.0	-	S	IS	S	

S = Significant; H = Harmful; IS = Insignificant; Nu = Nutrient uptake

Source: Reddy and Reddy (1995)

Radiation Up to 0.25μ or 250 nm (ultraviolet) is harmful to most plants. Solar radiation in the range of 0.30 to 0.55μ (300 to 550 nm) has photoperiodic effect, 0.40 to 0.70μ (400 to 700 nm) is the most effective in photosynthesis, and above 0.74μ (740 nm) has practically no effect on photosynthesis, its most effect is thermal.

Light affects the plants in four ways: intensity (quantity), wave length (quality) duration or day length (photoperiod) and direction.

The variation in light intensity are always accompanied with change in temperature and relative humidity. In photosynthesis, about one per cent of the light energy is converted into potential chemical energy. Very low light intensity reduces the rate of photosynthesis and may even result in the closing of stomata and thus, resulting in reduced vegetative growth of the plants. During cloudy weather, the production of dry matter is limited. Very high light intensity are detrimental to plants in many ways. It increases the rate of respiration disturbing photosynthesis–respiration balance. It also causes rapid loss of water resulting in the closure of stomata. The most harmful effect of high light intensity is the phenomenon of 'solarization' in which all the cell contents are oxidized by atmospheric oxygen. The oxidation is different from respiration and is termed 'photo-oxidation'. Depending upon the nature of the crops such as *heliophytes* (sun loving) and *sciophytes* (shade loving) the dry matter production is affected. Many species produce maximum dry matter under high light intensity if water is available in plenty. However, crop like betel, sunflower, buckwheat and tobacco grown during summer produce more dry matter if slightly shaded.

Duration of light is of considerable importance from the farmers' point of view in selecting the variety of a crop. The length of the day has greater influence than the intensity. The response of plants to relative length of day and night is known as **'photoperiodism'**. Plants, which develop and produce normally when the photoperiod is greater than a critical minimum, i.e. more than 12 hours of illumination, are called **'long day plants'**. Those plants which develop normally when the photoperiod is less than a critical maximum, i.e. less than 12 hours of illumination are called **'short day plants'**. Some plants which are found to be unaffected by photoperiods are called **'day neutral plants'**. Photoperiod has significant influence on plant character, like floral intimation, floral development, zhizome production, etc. If a **long day plant** is subjected to short day periods, the internodes may be shortened to give rosette appearance and flowering will not take place. Similarly, the growth of the plant becomes abnormal when a short day plant is subjected to long day period. Among the crop plants, soybean, maize, millets, etc., and among the horticultural crops strawberries, chrysanthemum are examples of short day plants. Crops like wheat, barley, sugarbeet and horticultural crops like lettuce, spinach and cruciferous plants are long day plant. Tomatoes, asparagus etc., are not sensitive to day length.

In the tropical and subtropical areas, crop production is all the year round. However, in the main crop growing period of these regions, the radiation is lower than that in temprate regions because of rains, cloudiness and shorter day length. Contrary to this in temprate region, the crops are cultivated with longer days resulting in higher productivity of crops.

Rice requires solar radiation of 300 cal per square cm per day at the reproductive stage. Low temperature coupled with bright sunshine during the ripening phase is highly conductive for the development of carbohydrates. Rice crop requires an optimum photoperiod of 10 to 10.5 hours per day for normal flowering particularly for season-bound rice varieties. The crop is sensitive to photoperiod only after the completion of its basic vegetative period.

Maize is a short day plant and its flowering is accelerated by shorter photoperiod.

Sunshine is very important for proper development of cotton particularly during the early growth and fullbloom. Jute, a short day plant, requires a critical photoperiod of 12.5 hours.

In sunflower, photosynthesis is markedly affected due to low light intensity which adversely affects seed yield. Soybean is a short day plant and highly photoperiodic. Low light intensity during early growth and flowering stage reduces fruit setting in groundnut.

Total amount of radiation reaching soil surface or crop is a function of the intensity and duration. The efficiency of light interception by a crop is solely affected by type, branching habit, age, plant population, density, height, leaf arrangement, leaf angle, tillering and orientation of different layers of plant leaves and ability of leaves to transmit light. The plant parts (e.g., shoots, roots and leaves) show different orientation to the direction of light. In temparate regions, the southern slopes show better growth of crops than the northern slopes due to the direction of light contributing more sunlight towards the southern side.

In an ideal situation for most effective use of radiation, the top leaves should be more angular or nearly vertical, about 80^0 being optimal angle to horizontal and down below this angle should be gradually less so that the bottom leaves will be placed in almost horizontal. The upper 50 per cent leaves need to be an angle of 60^0 or more.

Efficiency in radiation utilization varies with the phase of crop growth, usually falling after heading or flowering. In general, most C_4 plants and a few C_3 plants show high photosynthetic activity with increased photosynthetically active radiation (PAR). In C_3 plants, the initial product of carbon assimilation is the three-carbon compounds. The photo-respiration is high in these plants (e.g., rice, wheat, cotton, soybean, potatoes, etc.). In C_4 plants (e.g., maize, millet, sugarcane, etc.,), the primary product of carbon fixation is four-carbon compounds which may be malic acid or aspertic acid. In Crassulacean Acid Metabolism (CAM) plants (e.g., pineapple, sisal etc.), the stomata are open at night and large amount of CO_2 is fixed as malic acid which is stored in vacuoles. During day the stomata are closed and hence CO_2 cannot enter.

5.3.3 Temperature and Plant Growth

Temperature is a measure of intensity of heat energy. The temperature of a place is largely determined by its distance from the equator (latitude) and the altitude. Based on the above, the vegetation are classified as tropical (rainforest, desert and grassland), temperate (grassland and deciduous forest), taiga (coniferous forest), tundra (low shrubby growths lichens) and polar (snow ice). Radiation per unit area is more in the tropics than other parts of the earth because of vertical fall of sun-rays and hence, the tropic region become the heat source and polar region becomes the sink. In transporting the heath energy from tropical to polar region atmosphere acts like a heat engine. Moving away from the equator the temperature decreases towards the polar region. Besides the season, temperature is also influenced by distance of a place from sea, latitude, altitude, contact of different bodies (water, minerals and organic matter), local topographical features and cloudiness.

Based on the prevailing temperature conditions, the vegetation are classified into four classes: megatherms, mesotherms, microtherms and hekistotherms (Table 34)

TABLE 34: Classification of vegetation based on temperature

Class	Region	Temperature conditions	Type of vegetation	Common crops cultivated
Megatherms	Equatorial and tropical	High temperature throughout the year	Tropical rain forests	Tropical crops, like rubber, cassava, varieties of rice, etc.
Mesotherms	Tropical and sub-tropical	High temperature alternating with low temperature of winter	Tropical deciduous forests	Sub-tropical crops like maize, sorghum, etc.
Microtherms	Temperate and high altitude plants (upto 3658 m of tropical and sub-tropical)	Low temperature	Mixed coniferous forests	Temperate crops like wheat, oats, potato, etc.
Hekistotherms	Arctic and Alpine regions (above 4879 m in tropics and 3658 m in temperate regions)	Very low temperature	Alpine vegetation	Pines, spruce, etc.

5.3.3.1 *Role of temperature*

Temperature determines the distribution of biological forms including introduction of plant species over the earth's surface, length of growing period, magnitude of plant growth and conditions for all plant functions and energy processes. It directly influences biochemical reactions, uptake of CO_2, rate of photosynthesis, respiration, cell wall permeability, absorption of water and nutrients, transpiration, enzyme activity and protein coagulation. The influences are of vital concern in crop production.

Both air and soil temperatures are components of plant environment and with dirunal changes affect the micro-climatic temperature. Air temperature influences photosynthesis while the soil temperature controls the uptake of water and nutrients needed for photosynthesis and growth and development of plants.

Cardinal temperatures: Plant grow best under certain inter-related conditions of temperature, moisture and light. Every plant community has its own minimum, optimum and maximum temperature known as the cardinal points. For each species of plant there are maximum and minimum limits of temperature beyond which it does not grow and within this range there is optimum temperature at which the growth is maximum. The range of maximum growth for most agricultural plants is between 15 and 40°C (59 to 140°F). The cardinal temperature ranges for cool season crops (wheat, barley, oats, potatoes, etc.) are 0 to 5°C minimum, 30 to 38°C maximum and 20 to 25°C optimum. The cardinal temperature ranges for warm season crops (rice, maize, sorghum, pearl millet, sugarcane, groundnut, cowpeas, etc.) are 15 to 20°C minimum, 30 to 45°C maximum and 30 to 38°C optimum varying from crop to crop.

Both extremes of temperature beyond cardinal points results in many injuries on the crops and adverse effect on soil condition leading to reduction in crop yields. High temperature often stops growth, hinders fertilization and causes sterility in flowers. The plant faces incipient starvation due to high respiration rate. The photosynthetic activity appears to become near inactivated at excessively high temperature. The general effects of excessive heat are defoliation, premature dropping of fruits and in extreme cases death of plants.

Plants are injured due to cold temperature by chilling injury when temperature is above the freezing point for sometime. The effects of chilling are chlorotic condition or bands on leaves of sugarcane, sorghum and maize when exposed to 60 hours at 2 to 4°C. Frost and freezing temperatures, sometimes, cause considerable damage to oilseeds, grains, forage and horticulture crops. The physiological, biochemical and molecular changes are caused by low temperature stress (Bhattacharjee and and

Mukherjee (1994). Freezing injury is caused in plants generally growing in temperate region. Water is frozen into ice crystals in the intercellular space. The protoplasm of the cell is dehydrated and mechanical distortion takes place resulting in killing of the cells. Frost damage to potatoes and tea is common in Nilgris of Southern India. Crop suffers for want of oxygen as ice and snow form thick cover over the ground in temperate regions. Ice in contact with roots inhibits diffusion of CO_2 and respiratory products may become harmful to the plants.

Low night temperature may change the reaction of plant from photoperiodic to non-photoperiodic. Potato, a short day plant for tuber formation becomes insensitive to photoperiod when the night temperature is in the range of 10 to 14°C and produces as many tubers in eight hours day as in sixteen hour day (Went, 1957). Such information aids in locating regions for year long potato production (Venkatraman, 1968).

Analysis of climatological data can give idea of frequency, intensity and timing of frost occurrence in a region. Thus, knowledge of risk free growing season would help in selecting suitable crops and varieties. Planting dates may be so adjusted that chances of frost damage at sensitive growth stages (e.g., flowering) are avoided (Pitzalis and Pandolfi, 1986). Selection of late flowering varieties may be useful. Kurashi *et al.*, (1966) found that cytokinin prevented chilling injury to peas, while Ghosh and Chatterjee (1975) observed that spray of thiourea helped in preventing low temperature injury to rice seedlings. Foliar spray of sulphuric acid and dimethyl sulfoxide (SMSO) was found to protect potato crop from cold injury (Bhanwar Lal, 1975). In case of chickpea, cold tolerance could be increased with the spray of SMSO, sulphuric acid and glucose (Porwal *et al.*, 1986; Rao and Sahu, 1991).

5.3.3.2 Temperature requirement of certain crops

Crops can grow within certain limits of temperature. For each species and variety there are not only optimal temperature limits, but also optimal temperatures for different growth stages and functions, as well as lower and upper lethal limits.

(i) **Rice:** Rice is grown in the humid regions of tropics and sub-tropics and has wide adaptability. Most of rice varieties are thermo-sensitive and hence, their growth and development phases are greatly influenced by the temperature. Minimum, maximum and optimum temperatures for germination are 10 to 12°C, 36 to 38°C and 30 to 32°C, respectively. A mean temperature of 22°C is required during the growth. Optimum temperature for flowering and grain formation is approximately 23 and 21°C, respectively. Rice yield is reduced significantly if the night temperature falls below 15°C during vegetative phase due to the formation of sterile spikelets. If the temperature exceeds 35°C at anthesis for more than an hour spikelet sterility has been found to be more.

(ii) **Wheat:** The crop is cultivated during the cool season with temperature of less than 25°C. Minimum, optimum and maximum temperatures for the germination are 3.0 to 4.5°C, 20 to 25°C and 30 to 32°C, respectively. A mean daily temperature of 15 to 20°C is optimum for the growth, development and flowering of the wheat. Higher temperatures of about 30 to 35°C have generally detrimental effect on growth Temperature above 32°C prevailing 3 to 4 weeks after flowering may ripen the grain prematurely.

(iii) **Maize:** It is a warm weather crop. It requires an average temperature of about 24°C. Low temperature below 19°C and high temperature of 35°C can decrease crop yield (Chang, 1981). For germination, minimum, optimum and maximum temperatures are 8 to 10°C, 32 to 35°C and 40 to 44°C, respectively. Production of maize is not possible in areas where the mean temperature is less than 19°C.

(iv) **Sugarcane:** It is tropical crop grown in the area where mean monthly temperature remains above 22°C. It requires an optimum temperature of 30°C which is favourable for its best growth (Subbaraya and Kumar, 1980). Growth and yield are reduced in the sub-tropical region, such as north-western India where temperature falls below 0°C during winter season.

(v) **Cotton:** It is warm season crop. It requires 32 to 34°C during germination, 25 to 27°C during vegetative phase and 28 to 33°C during flowering period. Mean summer temperature of 21°C for commercial production with a frost free period of 200 days are favourable for cotton. Growth and development are adversely affected when temperature either falls below 15°C or exceeds 40°C

(vi) **Jute:** It is a sub-tropical crop requiring a temperature range of 15 to 38°C.

(vii) **Groundnut:** It is sub-tropical crop cultivated over wide range of temperature. Optimum temperature for growth ranges between 21 to 30°C. The crop requires a temperature range of

14 to 16°C for seed germination, 30 to 33°C during vegetative phase and 27 to 30°C during fruit setting.

(viii) **Soybean:** The crop is basically of temperate origin requiring a temperature range of 32 to 33°C. A day temperature of 25°C or less may delay the flowering.

(ix) **Sunflower:** It is a photo-insensitive and non-thermo-periodic temperate crop. Optimum temperature range for good growth is 25 to 27°C. The minimum temperature of -2 to -3°C and maximum temperature above 42°C are harmful for the growth of crop.

(x) **Seasmum:** It is tropical and sub-tropical crop and requires an optimum temperature of 25 to 27°C.

5.3.4 Atmospheric Humidity

Humidity of the air is water in the form of invisible water vapour present in the atmosphere At a particular temperature and pressure, atmosphere containing the maximum possible amount of water vapour is termed saturated. Inspite of constant water vapour, any increase in temperature will make the air unsaturated. In unsaturated condition, the water vapour content of air is usually expressed as **relative humidity** which is the ratio between the actual humidity present and the saturation humidity possible at that temperature. Apart from temperature and pressure relative humidity is also affected by wind, exposure, vegetation and water content in soil. The rate of transpiration is directly proportional to the difference between the amount of water vapour actually present and amount that would be needed to saturate the atmosphere. This difference is termed **vapour pressure deficit**. As compared to relative humidity vapour pressure, deficit is more indicative of the potential rate of evaporation.

Evapotranspiration of crop plant increases with temperature but decreases with high relative humidity affecting the quantity of irrigation water. Warm and humid weather favour the growth of many bacteria and fungi resulting in increased occurance of pests and diseases, for example blight diseases of potato and tea. Insect parasites, such as aphids and jassids thrive well under moist conditions.

5.3.5 Precipitation

Reaching of atmopsheric humidity to the ground either as rain or snow is known as precipitation. It occurs as rainfall when the temperature is high and as snowfall when temperature is low. As the temperature increases the water holding capacity of atmosphere increases. When the air is cooled, its water holding capacity decreases and the air becomes saturated, as a result the condensation takes place. Moisture content of the air is more above the ocean than over the land, consequently the rainfall is more over the oceans. When the dew point is above 0°C, the common form of condensation are precipitation, fog, cloud, rain, dew and mist. When the dew point is below 0°C, snow, frost and hail are formed. Vegetation in the desert areas receive precipitation in the form of dew.

Precipitation is the major source of soil moisture for crop growth. The amount of precipitation above the basic minimum is required to enable the crop to achieve maturity but yields are not always directly proportional to the amount of precipitation. The excess of optimum rainfall may cause reduced yields. The minimum amount of rainfall on a single occasion that ensures atleast some soil water storage is usually referred to us **effective rainfall**. To be reasonably safe from evaporation, the rain must penetrate to a depth of atleast 10 to 12 cm (Staple, 1964). For this amount, rain needed has been estimated as 15 to 20 mm for a single rainfall. According to Bagnold (1954) annual effective rainfall is usually about one-third of the total rainfall. The mean of the total rainfall received during the past 10 years is known as **decennial rainfall**. If these decennial averages are plotted on a graph, the increasing or decreasing trend of rainfall can be detected.

Effect of precipitation: The yearly preciptation, both in total amount and seasonal distribution greatly affects the choice of cultivated crops of a place. Low and ill distributed rainfall are common features of dryland farming wherein drought resistance crops, like sorghum, pearl millet and other minor millets are commonly grown. On the other hand in places of heavy rainfall areas, crops like rice is grown in flat areas while tea, coffees, rubber, etc. are grown on the slopes.

Water requirements of crop are different in different crop plants. Many grass species can grow under semi-arid conditions. There are also wet limit. Cotton cannot be grown commercially in the tropical and sub-tropical areas which have excessive rainfall during maturity period. In the dry farming agriculture, crops directly depend on the amount and distribution of rains. According to Mahi and Hundal (2002), during winter season weather systems moving across north-west India from west to east can cause rain which is very useful for *rabi* (winter) crops like wheat, barley, pulses and oilseeds. Under situation of delay in onset of monsoon (early season rainfall aberrations) sowing of crops get delayed and the practice of dry seeding and transplanting have been tried to compensate the delay in sowing time by a few days to even more than four weeks. Some of the crops like pearl millet and finger millet can be grown in nurseries and seedling can be transplanted after receipt of rains (Lahiri, 1978). Sowing of pearl millet beyond third week of July gives poor yield in the Indian arid zone, whereas legumes like cluster bean and moth bean give good yields under such conditions. In red soils of Telangana region, castor gives better yields than pigeonpea under late sown conditions. Alternative crops suitable for delayed sowing conditions in Ranchi (Jharkhand) are transplanted millet, pigeonpea and potato. Mungbean, urdbean and pearl millet have been found suitable for Rath Dhiansar (J&K), Varanasi (UP), Agra (UP), Jhansi (MP) and Rajkot (Gujarat). Crops suitable for delayed monsoon conditions have been identified for different dryland agro-climatic regions (Venkateswarlu *et al.*, 1991).

In-season drought (intermittent rainfall aberrations) is a potential stress for nearly all rainfed crops. Germination and crop establishment is one of the most sensitive period. Poor germination and seedling mortality may result in low and erratic crop stand. Gap filling by transplanting after the rains or re-sowing the crop after subsequent rain is an effective means to achieve desired plant stand. Moisture stress during crop growth stages of high leaf area indices will have the maximum decrease in yield. Timely weed control, use of mulches, pre-sowing seed treatment and use of anti-transpirants may sometimes give good result.

5.3.6 Wind Velocity

When the air moves in horizontal direction on the surface of the earth caused by the difference in pressure in two adjacent areas due to changes in temperature, it is called wind. Temperature and pressure gradients are the main cause for the movement of air resulting generation of wind. The various other factors on which the velocity of wind at a place depends are geographical situation, topography, altitude, distance from sea shore, flat plains, vegetation, etc. Grace (1977) found that the lowest wind speeds were generally at low altitude sites away from the coasts and highest in coastal areas and the high altitude mountainous areas.

The action of wind velocity on plants may be direct or indirect. Growth of plants, in general, seems to be inhibited by wind speeds above 10 km per hour (Went, 1957). A light or moderate wind is useful for CO_2 supply resulting in greater photosynthetic rates. Net assimilation rate (NAR) and relative growth rate (RGR), the indices of growth increased with increase in wind speed of 2.5 to 7.0 m/s in barley and rice. Nitrogen concentration in both barley and rice increases with increase in wind speed. Wind also alters the balance of harmones. Wind increases ethylene production in barley and rice. An important function of wind is to help in pollination of flowers and dispersion of seeds, fruits and micro-organisms. Wind is also responsible for causing rainfall to a very large extent. In India, the monsoon type of rainfall is largely determined by particular patterns of wind movement. Wind has also a powerful effect on the humidity of atmosphere.

The adverse effects of winds are drying of leaf tips, fruit drops, deformation of shoots and reduction in plant height due to reduction in cell expansion in absence of full turgidity (Reddy and Reddy, 1995). Wind causes mechanical lacerations and bruises on the tissues of crop plants. Lodging of crop plants in wheat, maize, sugarcane, rice, etc., are caused by violent winds. A very high velocity of wind (gale, blizzard, hurricane, cyclone, etc.) breaks dead and living branches of trees and even uproots them completely. The effect is considerably reduced in presence of windbreak. In bare deserts high velocity of wind causes constant soil erosion and this makes it difficult for plants to grow. Shelter belts of vegetation are raised to arrest the movement of soil by erosion. When the plant cover is not thick, strong winds remove the dry soil as to expose the roots of plants and kill them (Mavi, 1994). Indirectly, the velocity of wind affects the rate of transpiration in plants. Much damage is caused to crops by hot dry winds at the time of flowering. The bloom injury is caused by evaporation of secretions from the stigma. In arid and semi-arid areas, it induces the spread of sand over agricultural lands, roads and villages.

As recommended above, the planting of wind breaks which may be a hedge or a shelter made of artificial material, and shelterbelt which is a block of a single or multiple rows of tree with or without shrubs, would be very useful in regions susceptible to excessive winds. In addition to wind reduction, shelterbelts also modify the microclimate. In areas shaded by selterbelts the solar radiation is reduced. The relative humidity in shelterbelts is usually higher than in open. From studies conducted with *Acacia tortilis* shelterbelts, Ramkrishna and Shastri (1977) recorded a reduction in maximum daily soil temperature by 10 to 16°C at 0 to 5 cm depth and by 4 to 5°C at 30 cm depth, besides about 7 per cent increase in mean daily relative humidity around the area.

5.3.7 Composition of the atmosphere

The gaseous envelope of invisible film of air surrounding the earth is called atmosphere. Ninety-nine per cent of total mass of the atmosphere is within 40 km from the earth. However, atmosphere extends up to a height of 1600 km. Our current atmosphere is an oxidizing atmosphere, whereas the original atmosphere of the earth was a reducing atmosphere as it did not contain oxygen. The earth's atmosphere 4.6 billion years ago comprised of hydrogen and helium. Through the process of outpouring of gases from the earth's interior many other gases were ejected into the atmosphere which included carbon-di-oxide, nitrogen and water vapour. Plant, e.g., *cyanobacteria* produced most of the oxygen of our atmosphere. It occurred a period of millions of years, the atmosphere evolved to its present state.

The atmosphere contains a mixture of gases consisting of nitrogen (78.0%), oxygen (21.00%), argon (0.90%) and carbon-di-oxide (0.03%) by volume in a constant proportion. The remaining fractional per cent of air comprise of hydrogen, helium, krypton, methane, neon, nitrous axide, ozone, xenon, etc. The lower part of the atmosphere contains water vapour (0 to 4%) by volume. The lower atmosphere also contains dust particles, pollen, salt crystals, bacteria, etc. Most of the living organisms are present at the bottom of atmosphere. Variations from this may be found over industrial plants which release large quantities of gaseous fumes, such as sulphur-dioxide, carbon-dioxide and carbon-mono-oxide as well as water vaopur and mineral particles.

Photosynthesis is approximately proportional to the concentration of carbon dioxide (CO_2) in the air surrounding the foliage of the crop. The CO_2 which gets incorporated into the organic compounds of the plants returns to the atmosphere by respiratory breakdown of these compounds and by the death, decay and combustion of plants. The CO_2 content dissolved in water is fairly high which meets the need of aquatic plants. In soil, excess CO_2 causes poor root respiration affecting adversely on the metabolic activities of roots in plants.

Atmospheric oxygen (O_2) is vital to life and normal growth of all living organisms including plants. The amount of O_2 is normally constant in the air because plants give off oxygen during photosynthesis.

The nitrogen (N) of the atmosphere is made available to the soil by lightening, rainfall and nitrogen fixing micro-organisms. The nitrogen in soil is made available to the crops by the activity of nitrifying bacteria. The plant tissues containing N are consumed by animals and part of this nitrogen is again returned to the soil in the form of animal excreta or dead animal tissues.

The emission of six major gases in the atmosphere responsible for global warming are carbon dioxide (CO_2), methane (CH_4), nitrous oxide (N_2O), hydrofluo-carbon (HFC_3), perfluora-carbon (PFCs) and sulphur hexafluoride (SHFs). Carbon dioxide concentration is increasing at the rate of about 1.5 ppm per year (Keeling *et al.*, 1984). Ozone depletion may lead to increase UV radiation with adverse impact on earth's environment, and mankind. Certain gases such as sulphur-dioxide (SO_2), carbon mono-dioxide (CO) and hydrofluoric acid (HF), when released into air in sufficient quantities, are toxic to plants. Acid rain is often due to relatively high concentration of sulphur di-oxide and sulphates resulting in damaging effect on crop and soil. Injury to vegetation by fluoride released during the manufacture of metallic aluminium and the production of phosphate fertilizers has been reported. Near about the cement factories, the fall out of fine cement particles also causes damage to the crops.

5.4 EDAPHIC FACTORS AND CROPS

The nature of soil is complex and dynamic. It provides physical anchorage to plants and acts as a store house for water and mineral nutrients. Principal factors characterising soil-climate are soil air, soil temperature and soil moisture, soil reaction, mineral nutrient supply and biotic factors. They have also pronounced effect on growth and development of crop plants.

5.4.1 Soil air

Soil air is continuation of atmospheric air and there is continuous exchange of gases between atmosphere and soil air. The movement of soil air is controlled by temperature, water and diffusion. Mass flow resulting in response of differences in the partial pressure are the main processes in exchange of air between soil and atmosphere. Nitrogen, oxygen, and carbon dioxide are the important gases in both atmosphere and soil air. Usually the total amount of carbon dioxide plus oxygen in soil air remains equal to that of atmosphere. However, the carbon dioxide content is higher and oxygen is less compared to atmospheric air. Soil structure has profound influence on oxygen diffusion and soil aeration. As a rule higher the bulk density more compact the soil and smaller the amount of pore space.

Oxygen is needed for the respiration of roots. In well aerated soils, carbon dioxide released by respiration of roots and micro-organisms is freely exchanged with the air above the soil. Soil air is useful in increasing the nutrient availability of the soil by breaking down the insoluble minerals into soluble forms by decomposing plant and animal remains, and by bringing about nitrifying and nitrogen fixing processes of bacteria. Free living bacteria, like *azotobacter*, symbiotic bacteria, like *rhizobium*, blue-green algae, etc. fix a good amount of nitrogen in soil. The decomposition of protein content of the dead plants and animals by micro-organisms also add to the nitrogen pool of the soil.

Restricted soil aeration is the greatest limiting factor in the development of an extensive root system due to the accumulation of carbon dioxide. It impairs the essential process of an established root system which retards both water and nutrient absorption. Restricted aeration prevents the orderly function of essential biological processes associated with good soil fertility. Germination is inhibited in the absence of oxygen. Decomposition of organic matter is reduced and change in microflora occurs. Symbiotic nitrogen fixation is strongly affected by soil aeration.

Crops like potatoes, cotton, legumes, linseed, safflower and tea are sensitive to oxygen supply and hence, need higher level of oxygen in soil air as compared to other crops. Rice being the semi-acquatic plant can tolerate very low level of oxygen in the soil. Other cereals are intermediate in oxygen requirements.

5.4.2 Soil Temperature

The energy received from the sun is absorbed by the soil which is converted into heat. Soil temperature is one of the most important edaphic factors that controls the physico-chemical processes and micro-biological activities involved in the crop production.

The temperature of soil in the field is dependent upon the net amount of heat absorbed by soil and thermal properties of soil, such as heat capacity and thermal conductivity. **Specific heat capacity** is defined as the amount of heat required to raise the temperature of a unit mass or unit volume of soil by one degree. The specific heat of water is 1.0. All other constituent of soil have much lower specific heats. The specific heat of soil increases with increase in moisture content. The specific heat is 0.33 cal per g when the soil moisture content is 20 per cent. Flow of the temperature in soil is expressed as **thermal conductivity**. It increases with water content and decreases with porosity. Tillage improves prosity and decreases thermal conductivity resulting in higher surface temperature.

Movement of heat in soils can be expressed as:

$$\frac{Q}{At} = q = -Kc\,\frac{t}{x}$$

Where,

q = amount of heat flowing

A = Area

T = Time

q = flux of heat

Kc = the thermal conductivity of soil and refers to the ability of soil to transmit or conduct heat.

$\frac{Q}{At}$ = the rte of heat flow per unit area.

Factors affecting soil temperature are (i) the angle with which soil faces the sun and (ii) insulation by air, water vapour, clouds, dust, snow, plants and mulch. Factors affecting amount of heat entering the soil are latitude, soil slope and distribution of vegetation and water bodies on the land.

Soil temperature affects seed germination, root and shoot growth, absorption and transportation of water and nutrients by plants, decomposition of organic matter, microbial and enzymic activities and weathering process in soil. Soil temperature first affects plant growth particularly underground portion during germination which is a slow process in cold soil. The germination becomes more rapid as the temperature of the soil rises upto a certain optimum temperature. A cold soil is not conductive to rapid growth of the most crops. The same is true in condition of extreme warmth. The maximum metabolic activities of the plants and the maximum absorption of water by the roots take place generally between 20 to 30°C. Lowering of temperature below 20°C causes an appreciable reduction in rate of absorption of water by most crops. Soil temperature controls the micro-biological activity and process involved in the availability of nutrients to plants. Nitrification does not begin until the soil reaches a temperature of about 5°C (40°F).

Crops like maize and sorghum need a high soil temperature for active seedling growth. Wheat seeds put forth radicle 6 days after sowing at 5°C, 3 days after sowing at 10°C and within 2 days after sowing at 18°C. The optimum soil temperature for germination for different crops are 20°C for wheat, barely and oats, 31°C for rice, 20°C for maize, 25°C for sorghum, 29°C for soybean, 33 to 35°C for cotton, and 29°C for tobacco (Venkataraman and Krishnan, 1992). Mulches, irrigation and windbreaks can alter the soil temperature to a limited extent to meet the specific crop requirements.

5.4.3 Soil Moisture

Water is indispensable for any form of life, some of which contains more than 90 per cent. It is an essential part of protoplasm and is an important ingradient in photosynthesis. It is required for translocation of nutrients and dissipation of heat. It gives turgidity to plants and keeps them erect.

Water that enters the soil either remains in the pores and percolates through them to lower depths. The character of the soil pore space determines to a considerable extent the deposition of infiltrated water. Water retention and movement, therefore, constitute two important phases of soil-moisture relationship (Baver, 1956). Soil moisture is in highly dynamic condition and is held up by the soil particles by attractive forces of adhesion and cohesion. Soil water is depleted due to evaporation from soil surface, transpiration through the plant and deep percolation into the soil beyond the root zone.

As water availability to crops reduced gradually, plants are subjected to moisture stress causing reduction in yield. Increase in water increases plant nutrient availability and mobility, and biological activities including enhanced mineralisation. Soil water is controlled by rainfall, irrigation and drainage.

Soil structure is purposely destroyed in rice cultivation by puddling to avoid percolation losses of water. The pH of most acid and alkaline soils converges between 6 and 7 within 2 to 3 weeks after flooding. The redox potential of water and top layers of submerged soil is +3 to +5 volts and lower layers have a stable redox potential of 0.3 to -0.3 volts. The tendency of a given system to oxidise and reduce the susceptible substances (sulphate, nitrate in flooded soil) is known as redox potential. It is positive in aerable soils and negative in highly reduced soils.

5.4.4 Soil Reaction

Soils may be neutral, acidic or alkaline depending upon their content of the basic salts and acidic components. For the growth and development of most crops neutral soils are best. Availability of plant nutrients depends on soil reaction. The most convenient method of expressing the relationship between H^+ and OH- is pH. pH influences availability of plant nutrients through its influence on decomposition, cation exchange capacity and solubility of materials. Low pH beyond particular limit (soil acidity with high alumina content) is not only toxic to the plants but also interferes with the absorption of several nutrients. Similarly high soil pH (high alkalinity due to excess of cations like Na, K, Ca and Mg) also adversely affects crop growth.

pH is defined as the logrithm of the reciprocal of the hydrogen ion (H^+) concentration of a solution in g per litre, usually written as:

$$pH = \log \frac{1}{(H\ )}$$

At neutrality, the hydrogen-ion (H^+) concentration has been determined to be 0.0000001 or 1 x 10^{-7}g of hydrogen per litre of solution. Substituting this concentration into the formula, we get

$$\begin{aligned} pH &= \log \frac{1}{0.0000001} \\ &= \log 10{,}000{,}000 \\ &= 7 \end{aligned}$$

At a pH of 6, there would be 0.000001 g of active hydrogen or 10 times the concentration of H^+ compared to pH 7 (0.0000001 g of H per litre).

Soil vary considerably in the degree of acidity or alkalinity or in reaction. Acid soils with high iron, aluminum and manganese and alkaline soil with high calcium decrease the availability of phosphorus due to its fixation in soil as insoluble phosphate. Availability of phosphorous is high within a pH range of 6.5 to 7.5 due to higher solubility of phosphorous compounds. At low pH, phosphorus is precipitated as Fe and Al phosphates while at high pH, less soluble calcium phosphate is formed. Decomposition of organic matter by micro-organisms and activities of nitrifying and nitrogen fixing bacteria are considerably reduced at pH below 6 and fastest between 6 and 8. Metallic cations, such as Fe, Mn, Ca, Zn and B precipitate at high pH and hence, their availability is less in alkaline soils. At lower pH, potassium compounds are converted into non-exchangeable forms and thus, their availability is reduced. Availability of molybdenum declines with decrease in soil pH. Ammonium fertilizers left on soil surface are subjected to loss by volalisation of pH more than 7.

The range of reaction under which crops grow satisfactorily is fairly wide provided there is sufficient available plant food and humus in soil and drainage is satisfactory. Most of the common crops, such as cereals and many grasses can be grown successfully on soil pH ranging from moderately acidic to mildly alkaline (pH 5.5 to 7.5). Potatoes grow well in strongly acidic to neutral soil (pH 5.1 to 7.0). Some of the grasses, tea, coffee and fruit trees (e.g. blueberries) thrive well on strongly acidic soil (pH 5.1 to 5.5). Lucern, sweet clover, cabbage and cauliflower thrive under slightly acidic to alkaline conditions (pH 6.1 to 7.8).

5.4.5 Mineral Nutrient Supply

The mineral content of the soil is derived from the weathering of the rocks and minerals and consists of particles of different sizes. The principal minerals occurring in earth's crust are feldspar (almunium silicate with varying amount of silicates of K, Na and Ca), quartz (silicon-di-oxide), mica (silicate of Al and K with silicates of Fe, Mg and Na), limestone (calcium carbonate with magnesium carbonate it occurs in dolomite), hornblend (silicate of ca, mg and Fe with sodium silicate), augite (silicates of Ca, Mg, Fe and Na in varying porportion), Olivine (ferro-magnesium silicate) and serpentine (hydrated magnesium silicate), clays (hydrated aluminium silicate) containing varying amounts Ca, Mg, Na and Fe and other minerals, such as tourmaline (borosilicate of Al with alkaline metals and Fe or Mg), rutile (titanium oxide), zircon (zirconium silicate), glauconite (hydrated silicate and Fe and K), magnetite and Hematite.

Chemical composition of soils vary similar to that of the rocks from which they originate. Thus, it evident that compounds of Si, Ca, Mg, Fe, K, Na and Al form the principal chemical constituents of soils. Besides these, soil contains small quantities of number of other mineral elements like B, Mn, Mo, Zn, Cu, Co, I and F which are known to be essential for the well being of plants in very small quantities. Such elements are called trace or micro nutrients.

Besides inorganic substances, soil contains organic matter in amounts ranging from less than 1.0 per cent in arid sandy soils to as much as 90 per cent in peaty soils. The organic matter of soil is a source of mineral nutrients essential for the crop growth. During decomposition, the humus is mineralised. As a source of plant nutrient, it contains 95 per cent of the total nitrogen, 50 to 60 per cent of total phosphorus and 10 to 20 per cent of total sulphur. It increases water holding capacity due to its content of organic colloids. It can hold by adsorption a great amount of minerals particularly the cations in the ionic form, thus, increasing the exchange capacity of soils.

Soil capacity to make available plant nutrients has profound influence on crop production. Nutrient requirements of a particular crop can be assessed by soil analysis for determining the suitability of soil for a particular crop and in formulating more precisely the fertilizer requirements. However, any crop can be grown on any soil with great variation in final economic yield and net return. To meet the crop needs for optimum production, soil can be improved considerably by soil and water conservation improving nutrient and water retentive capacity, modifying soil reaction by use of amendments and judicious application of organic matter and fertilizers.

5.4.6 Biotic Factors

The soil biological environment is composed of soil organisms belonging to both plant kingdom (Flora) and animal kingdom (fauna). The soil organisms consist of micro-organisms (microscopic) and macro-organism (developed). Bacteria, fungi, actinomycetes and algae constitute the soil microflora whereas macroflora of soil consists of roots of higher plants. Protozoa and nematodes are important soil microfauna and earthworms, moles, ants, etc. are the macrofauna.

Raw organic matter in the soil cannot be used by plants directly as food. It is broken down first into humus and then into simpler products by different kinds of micro-organisms before it can be utilized by the plants. Sugars, starches and proteins are broken down first, then cellulose and fatty substances (lipoids) and lastly lignin or woody substances. Decomposition of organic matter primarily forms part of the feeding and growth processes of these microscopic plants and animals.

Rhizosphere provides environment around root surface of higher plants for rapid proliferation of micro-organisms due to exudation of energy rich compounds. Bacteria mostly colonise on the rhizosphere compared to the rest of the soil (Reddy and Reddy, 1995). Bacteria are the smallest and simplest forms of plant life. They are single celled organisms, most of which are either spherical or rod shaped. The most important species of soil bacteria are Rhizobium, Clostridium, Nitrobactors and Nitrosomonas. These bacteria are heterotrophic, their population depends on soil organic matter. The important functions carried out by specific bacteria are nitrogen fixation, mineralization of organic matter releasing nutrients like N, P, K, S and Fe. Bacteria are also involved in decomposition of cellulose and oxidising sulphur and manganese.

The number of fungi is less than bacteria in soil. Fungi dominates is acid soils since bacteria and actinomycetes are inactive. The important genera of fungi are *Aspergillus*, *Fusarium*, *Trichoderma*, Mucor and *Penecellium*, etc. The fungi act as scavengers decomposing several types of materials in soil. Fungi decompose lignin which bacteria cannot. Certain fungi feed on protozoa and nematodes and thus, maintaining microflora equilibrium in the soil. Mycorrhiza, another soil fungi solubilizes phosphates in soil and makes available to plants on which they live.

Actinomycetes are thread like organisms and intermediate in size between bacteria and fungi. They are aerobic and heterotrophic and are less active in acid soils. The optimum temperature range for mesophiles are in the range of 28 to 37°C. Thermophiles are present in manure and compost pits. They are capable of decomposing more resistant organic matter.

Algae are autotrophic and contain pigment to fix carbon-di-oxide. Both micro and macro organisms are present in this group: (i) Cyanophyta (blue green algae) occur mostly in tropical regions, (ii) Chlorophyta (grass green algae) exists in grassland of temperate region, (iii) Yanthophyta (yellow green algae) and (iv) Bacillariophyta (golden green algae). The last two groups are less abundant. The most common genera are *Anabaena* and *Nostoc*. They fix atmospheric nitrogen and are used as biofertilizers.

Soil macro-fauna like earthworms, moles, ants, etc. consume plant residues and help in decomposition of organic matter. Protozoa (microfauna) feed on bacteria and actinomycetes. Nematodes (microfauna) are parasitic on crop plants and some live on organic matter. Bacteria and algae as biofertilizers has been dealt in details in Chapter IV of this book.

5.5 ENVIRONMENTAL ISSUES IN AGRICULTURE

5.5.1 Air Pollution

The increase in industrial activity that has taken place rapidly during the last 100 years contributed significantly for the deterioration of environment quality by releasing toxic gases harmful for both human and plant population. The major constituents of pollution gases are carbon-dioxide (CO_2), sulphur dioxide (SO_2) and nitrous oxide (N_2O). USA followed by USSR (erstwhile) and China are found to be on the top of the list of countries that are emitting 48.8 per cent of the total CO_2 emissions in the world. India too has contributed its share of adding CO_2 into atmosphere by moving to a sixth place in the world during 1991 from thirteenth position in 1951. These will have indirect effect by way of increase in air temperature which may affect plant growth as well as their water needs. Crop plants are very sensitive to air pollution as compared to human and animal populations as plants remain stationary and continuously get exposed to environmental stresses (Ramakrishna *et al.*, 2000).

The gaseous pollutants diffuse into plants through stomata. Little amounts are also absorbed from cuticle itself. Injury to vegetation commences with foliar lesion, phytomass and impairment in reproduction and regeneration. This is followed by changes in canopy structure, viz., open shrunken asymmetrical canopy and defoliated branches. Further impact are gradual elimination and disappearance of plant species, and injury to ground cover of herbs and shrubs.

Pollutants discolour foliage, and irreversibly change the chlorophyll molecule into photo–synthetically inactive pigment incapable of making any food for the plant causing reduction in root and shoot lengths, number of leaves per plants, biomass productivity and yield. Thus, air pollution in the cities and surrounding countryside could have significant impact on agricultural production.

Absorption of pollution is maximum in the plants that are cultivated under optimum moisture and fertility conditions due to maximum stomatal opening. However, the absorption ratio comes down drastically with the closure of stomata during night-time. With the sufficient supply of moisture during day-time and high relative humidity conditions, the stomata will open and absorption of pollutants will be maximum. The effect of SO_2 absorption reduces the level of chlorophyll pigment in the cell and affects the photosynthetic ability of plant and hence, reducing the production potential of the plants. Similar findings was reported by Verma and Agarwal (1995) in wheat and Verma *et al.*, (1991) in fruit plants. Injury due to oxides of nitrogen is similar to that of SO_2.

Sensitivity and tolerance of plants depend on the geographical location, climate, growth stage and maturation. The most sensitive ones are fodder plants, moderately sensitive are wheat, leafy vegetables, beans, strawberries, roses etc. Least sensitive are root crops and cabbage (Padmanabha Murthry, 1999).

Vegetation is a potential sink for suspended particulate matter (SPM). SPM in atmosphere vary in size from few microns to several hundred microns. The origin could be either from industrial or vehicular sources or from natural sources like bare soil surfaces. The SPM gets deposited on the leaf surfaces, which are greasy, cool and moist, and prevents the re-entry of dust into the atmosphere. However, the dust deposited on the crop plants alters the energy and moisture balances of plants by way of reduced albedo and transpiration values. It was reported that cement dust settled cotton plants reduced the height, stem width and chlorophyll content. Similar reports were made in respect of wheat and soybean plants (Chitralekha and Dakasina Murthy, 1997). Studies by Seema *et al.*, (1996) on rice showed the yield loss was nearly to an extent of 29 per cent.

5.5.1.1 Depletion of ozone layer

Ozone layer in the stratosphere protects the earth by filtering out ultraviolet-B (UV-B) radiation from the sun which is known to be most damaging as compared to other kinds of radiation, i.e. UV-A and UV-C. The UV-B radiation has disastrous effects on life of the earth (i.e. human, animal and vegetation) and the environment by causing DNA damage resulting in mutation and diseases like skin cancer, eye cataracts in human and animals and reduction in yield of crops. It reduces the reproduction of phytoplankton, the microorganism at the bottom of marine food chain which is vital to the survival of marine animals. Experiment conducted by Pal *et al.*, (1997) revealed that mungbean (a dichotyledonous C_3 plant) was sensitive to ambient UV-B levels (22.8 per sq cm per nm) in terms of leaf area development, plant height and net photosynethsis, while maize (C_4) was unaffected.

Ozone, a triatomic form of oxygen derives its name from the Greek word 'ozein' meaning smell as it has a pungent smell in the concentrated form. Stratosphere is a layer of atmosphere which lies above tropopause. It lies beyond the height of 8 to 18 km extending up to 50 km depending on latitude. But unlike to troposphere (lower layer of atmosphere extending 8 to 18 km from the earth), stratosphere absorbs heat from the sun in the form of ultraviolet radiation. It does so since oxygen seeping up from the troposphere below absorbs energy from the sun. The solar radiation breaks up the two atoms of oxygen in a molecule apart, leaving each one free to link up with another molecule of diatomic oxygen and forming two molecules of ozone. Thus, in the process of formation of ozone layer, the harmful ultraviolet radiation is filtered which otherwise makes life impossible on the earth. Because the stratosphere is warmer at higher attitude, hot air no longer rises there, since the air above is hotter than the air below, convection cannot occur. The stratosphere acts as a lid on the troposphere keeping convection and weather below the tropopause (Reddy and Reddy, 1995). However, excessive ozone levels in the troposphere have been responsible for death of plants through reaction with chlorophyll. Ozone is by-product of photochemical smog and reacts with hydrocarbons to form peroxynitrates that damage sensitive biological tissues in the eyes, nasal passage, throat and lungs. It is also associated with increase in respiratory disorders, such as asthma. The gas lowers human resistance to cold and pneumonia and make eyes itch, burn and water.

Depletion of ozone in the stratosphere is threat to mankind. There will be adverse effect on humans, animals and crops and will bring about unpredictable change in weather and climate. The result of monitoring of this important gas of stratosphere indicate that record ozone depletions have occured in 1995 to 1997. The ozone hole reached a record 10 million square miles in 1996. European commission's Environment and Climate Programme reported in May, 1997 that "for the third winter in succession, occurance of low temperatures in Arctic's lower stratosphere coupled with the presence of man-made pollutants has resulted in up to 40 per cent ozone depletion in the Northern Hemisphere (Ramkrishna, *et al.*, 2000). Ozone concentration of the stratosphere is declining at nearly half a per cent per year. This is due to release into the atmosphere of chemicals such as chlorofluorocarbon (CFCs), nitrous oxide and methene used as refrigerants, cleaning fluids and aerosol propellants, etc. Being inert, these chemical reach the atmosphere virtually intact. Reaching the stratosphere of the atmosphere they stay there for many years. The ultraviolet rays of sun breaks them apart and release chlorine which destroy ozone.

For every one molecule of chlorine for chlorofluorocarbon, about 1,00,000 molecules of ozone are removed from ozone layer. The environment need has always been and continues to be is to stop using all ozone depleting substances as quickly as possible.

5.5.1.2 Greenhouse effect and global warming

The gases like carbon dioxide, methane and nitrous oxide (along with water vapour comprising one per cent of atmosphere) are called greenhouse gases as working principle is same as that which occurs in greenhouse. Emission of greenhouse gases forms a thick cover in the atmosphere trapping the heat emitted by earth. It leads to warming of the planet and change in weather pattern around the world. It is estimated that nitrous oxide (N_2O) contribute 5 to 7 per cent of the accumulated greenhouse effect caused by human activities. Methane (CH_4) is produced during microbial decomposition of organic materials in the absence oxygen. Two major sources are lowland rice production and natural wetlands. Rice paddies (lowland) contribute 28 per cent of world's methane emissions.

The amount of CO_2 emitted are much larger than those of CH_4 and N_2O, whereas human activity contributes less than 5 per cent of CO_2 emitted, it accounts for 68 per cent of the CH_4 and 40 per cent of the N_2O. The effectiveness of each greenhouse gas to trap heat in the atmosphere varies. Taking the global warming potential (GWP) of CO_2 as 1, that of CH_4 is 21 and that of N_2O is 310 (Isherwood, 1998). **Global warming potential** (GWP) is an index defined as the cumulative radiative forcing between the present and some chosen later time 'horizon' caused by a unit mass of gas emitted now.

The atmospheric concentration of carbon dioxide (sources: fossil fuel combusion and deforestation), nitrous oxide (source: fossil fuel, fertilizers and biomass burning) and methane (source: rice fields, swamps, marshes, fermentation, wetlands, livestock and gas leaks) are 351, 0.31 and 1.68 ppm, respectively. The consumption of fossil fuels all over the world has gone up to several thousands folds since the 1850s and at the same time millions of hectares of tropical forests which are efficient sink for carbon-dioxide have been denuded of trees. CO_2 has increased from 250 ppm (1940) to approximately 350 to 370 ppm now due to industrialization and urbanization and it may reach 450 ppm in next 40 years (2050). Taking into account the projected concentration of greenhouse gases there could be doubling of CO_2 by as early as 2030. It has been projected that if the concentration of greenhouse gases continues to increase at the present rate, the global temperature may increase by 1.5 to 4.5°C by 2050 (Parry and Duinker, 1990).

Among the greenhouse gases, carbon dioxide and water vapour play a vital role in maintaining the delicate balance that determines the temperature on the surface of the earth. The two gases are transparent to visible solar radiation and hence, allow sunlight to pass through, but they prevent infrared radiation from escaping and thus, effectively trap solar heat. This produces greenhouse effect and heats up the atmosphere. The all India mean annual surface temperature derived from 73 stations across India shows a significant warming 0.4°C per 100 years which is comparable to a global mean trend of 0.3°C per 100 years (Hingane *et al.*, 1985). It has been estimated that earth's mean temperature would have been as low as -23°C without carbon dioxide rather than +15°C as it is today.

The carbon di-oxide beyond optimal level will lead to over heating of the atmosphere which causes climatic change. A rise in temperature both in terms of intensity and rate would lead to melting of ice caps and glaciers resulting in floods, thermal expansion of ocean and rise of sea level may lead to permanent inundation of resource rich and thickly populated coastal areas and disappearance of many islands in Pacific, Caribbean and the Indian ocean. Direction of wind will change resulting change in climate of many countries. Frequently of storms, hurricanes, floods, droughts, forest fire, etc. is likely to increase. Quality of water of sea will deteriorate affecting ground water of land. Desertification will increase in countries located on both sides of equator. Increase in temperatures can create favourable conditions for the growth of insect, pests, and diseases, which can have a devastating effect on world's agriculture. Ozone layer will break at many places which will allow penetration of ultra-violet rays on earth adversely affecting all forms of lives.

Environmental effects of greenhouse gases and global warming led to the International Communities who met in Kyoto (Japan) in 1997 and Copenhagen in 2009 to request all the countries to start reducing carbon dioxide emissions immediately and prepare for an orderly phase out of fossil fuels. The rich countries were held historically responsible for greenhouse gas emission and were asked to reduce emission by 5 per cent of its 1990 levels. It has been proved beyond doubt that instead of depending mainly on fossil fuels, systems can be developed which use the renewable energy. The advent of portable hydrocarbon air-conditioners, evaporative water coolers, etc has shown the potential to reduce the energy consumption as well as greenhouse gas emissions.

5.5.1.3 Climate change

Global climate change is one of the greatest threats to the mankind. It can cause disastrous effects on ecosystems with rapid loss of biodiversity, decline of natural resources, change in monsoon pattern and global rainfall distribution. The natural greenhouse effect keeps the planet 33°C warmer than it would otherwise be, allowing the earth to sustain life. Human activities including industrial revolution and population explosion have substantially contributed to increased burning of fossil fuels, thereby increasing the concentration of carbon dioxide (CO_2) and emission of other greenhouse gases like methane (CH_4), nitrous oxide (N_2O), hydrofluor-carbons (HFCs), perfluorcarbons (PFCs) and sulphur hexafluorides (HSFs) responsible for global warming. The indiscriminate use of man made gases like chlorofluorcarbon (CFC) have altered the radiation balance of the earth's atmosphere resulting in increased temperatures. The average temperatures are expected to increase more near the poles than at equator. The shift in climatic zone will be more pronounced in the higher latitudes. In mid-latitudes the shift is expected to 200 to 300 km for every increase in 1°C (IUCC, 1992).

There is no ambiguity in the data that since 1860, concentration of carbon dioxide in the atmosphere has risen 30 per cent, i.e. from 280 ppm to 360 ppm. It is now projected that South Asia might have an increase in temperature from 0.1 to 0.3°C by 2010 and 0.4 to 2.4°C by 2070. The precipitation is expected to remain the same, upto 2010 and may change by 0 to 10 per cent by 2070 (IPCC, 1998). The general circulation models (GCM) indicate that doubling of CO_2 concentration by 2030 will increase average global temperature by 1 to 3°C. The sea level are also expected to rise between 15 to 94 cm over the next century, thus, increasing the vulnerability of coastal low lying areas. Similarly decrease in stratospheric ozone has been observed between 1969 and 1988 in the range of 3.4 to 5.1 per cent between 30 N and 64 N. Ozone depletion may lead to increase in ultraviolet (UV) radiation with adverse impact on earth's environment.

Climate change models predict in general an increase in temperature and variation in precipitation and radiation levels which are likely to affect the crop production. Such climatic changes has been ascribed to increased levels of greenhouse gases. (Hundal and Abrol, 1991).

Millions of Indians are already bearing the burnt of climate change resulting in wayward rainfall, ravaging floods and lost of livelihoods. The two cyclones that hit the Orissa in quick succession in 1999 ravaged large areas of agricultural land and forests and changed the local climate. There was flood in Bihar and drought in parts of Chattishgarh and Jharkhand. Changes in rainfall pattern have been reported from many parts of the country. Geography coupled with high levels of poverty, population density and greater dependence on agriculture in Asia have made this part of the world quite vulnerable and would be one of the areas to be hit by the global crisis. Food security and health are already high on worry list of South Asia. The high temperature that come with climate change move crops from their optimal range to lower yields. A recent study on vulnerability of Indian agriculture to climate change confirms that an increase in temperature by one degree celsius may reduce yields of wheat, soybean, mustard, groundnut and potato by 3 to 7 per cent. Much higher losses are expected at higher temperature. Productivity of most crops is expected to decrease only marginally by 2020 but 10 to 40 per cent by 2100. It also predicts rising sea and river water temperature to affect the breeding and migration of fish, increased water, shelter and energy requirement for livestock and loss of 1.5 million tonnes of milk by 2020. Our glaciers are melting faster. Studies show that extremes in rainfall and temperature are increasing, the date of onset of monsoon might get more variable in the future, the intensity of storms would increase by at least 10 per cent. While no will be left untouched by these events, the maximum costs will be paid by those displaced by natural calamities like floods and droughts, and the human activities which are aiding these near apocalyptic phenomena, like mining, industrialization, deforestation and land degradation.

Agriculture has both a cause and effect relationship to climate change. The change caused by green house gases (GHG) will effect the plant metabolic activity and hence, the production directly. Reports indicate that contribution of methane emission from submerged rice paddies and mechanization in agriculture towards emission of green house gases. Govinda Rao *et al.*, (1994) in their study on implication o climatic change for India concluded a decrease in cereal production and increase in tropical cyclonic activity posing serious flood damages along their trajectories.

Similarly, increase in rainfall by 5 to 15 per cent in South Asian region was predicted by Moughton *et al.*, (1990) due to enhancement of concentration of greenhouse gases. According to Gadgil (1996) it is not clear that over the Indian region whether temperature and rainfall will simultaneously increase.

Green plants need CO_2 for promoting plant growth and agricultural productivity without increasing the water demand for crop transpiration (Goudriaan and Unsworth, 1990). Doubling of CO_2 concentration may increase the photosynethic rates by as much as 30 to 100 per cent. C_3 Plants such as wheat, rice and soybean will become more water efficient as they grow quickly. However, the response in C_4 plants such as maize, sorghum, sugarcane, millets etc. may not be spectacular (IUCC, 1992). Moreover, expected climatic warming may have adverse effects on agriculture, partly offsetting the positive direct CO_2 effects. Model predicts that the global temperature may rise more than 4.5°C by the year 2100. This projected change is larger than any climatic change experienced over the last 10,000 years. Climatic zones and thus, ecosystems and agricultural zones could shift towards the poles by 150-550 km in the mid-latitude regions. Forests, deserts, rangelands, sand and other unmanned ecosystems may face new climatic stresses. Consequently, many will decline or fragment or some flora and fauna species will become extinct. Beneficial and detrimental effects of climatic change will not be evenly distributed over the world. Cool and temperate climatic zones will benefit but in the tropics, a further increase in temperature will be undesirable. While increase in rainfall and concentration of CO_2 generally lead to an increase in yield, the impact of an increase in temperature is generally negative in the tropics.

In a study on the effect of increased CO_2 and temperature on Indian Agriculture, Sinha and Aggarwal (1993) predicted a shift in wheat production belt. Earlier More and Sadaphal (1981) reported decrease of wheat yield by 4000 kg per ha for a unit increase of 1°C maximum temperature and 0.5 hours sunshine. In similar studies, Ramkrishna *et al.*, (1996) have estimated a fall in productivity of pearl millet in Rajasthan by 10 to 15 per cent due to increase in temperature by 2°C. Rao *et al.*, (1992) observed the positive effects of increased CO_2 on the sorghum productivity. Studies of Lal *et al.*, (1998) using validated wheat and rice models, showed higher yields of both crops under elevated CO_2 levels (28 and 15%, respectively for a doubling CO_2). A 2 to 3°C rise in air temperature nearly cancels out the positive effect of elevated CO_2 on the wheat and rice yields.

Several studies with crop growth models have predicted alteration in mean crop yields as a result of the changed global warming (Cooter, 1990), and such changes are predicted to cause great impact on the productivity of different crops and to have important economic implications. Impact to various climatic changes on wheat (Porter and Miglietta, 1991); soybean and maize (Sinclair and Rawlins, 1993); wheat rice and maize (Hundal and Kaur, 1995a, 1995b and 1995c); groundnut (Kaur and Hundal, 1995), maize (Muckow and Sinclair, 1991); wheat, maize and soybean (Adams *et al.*, 1990) have been reported. Rosenwig and Parry (1994) estimated that the net effect of climate change is to reduce global cereal production up to 5 per cent.

5.5.1.4 Acid Rain

The phenomenon occurs when suspended air borne pollutants, sulphur di-oxide and nitrogen oxide from the burning of fossil fuels, such as, petrol, diesel, and coal react with the rain droplets to form acid like H_2SO_4 and these droplets fall on the ground as acid rain. According to Isherwood (1998), global emission of ammonia from N-fertilizers is estimated at 4.5 million tonnes. Ammonia can react in the atmosphere in a such a way as to contribute to acid rains and its direct deposition in soil can favour acidification through nitrification. It can cause considerable damage to the living and non-living things on earth. The classic examples are the threat posed on the "Taj Mahal" by the acid rains due to the oil refinery at Mathura (Uttar Pradesh) and tanneries in around Agra (Uttar Prasesh). It can carrode buildings, monuments and be hazardous to human health. Acid rain mobilizes heavy metals, like cadmium, mercury, nickel, etc., in soils and that are slowly leached to contaminate the surface water bodies.

Acid rains cause considerable damage to agriculture. Some of the effects that acid rain can have on plant and soil include increased leaching of inorganic nutrients, damage to leaves at pH below 3.5, reduced microbial activity, reduced availability of soil nitrogen, etc. Kulshrestha *et al.*, (1996) observed that the average pH of the rain water at New Delhi was 5.7 and its chemical composition was dominated by NH_4 and SO_4. The acidity of rain water was due to SO_4 rather than NO_3 and it was mainly neutralzed by NH_4 and Ca. Singh and Agarwal (1996) while studying the response of wheat to simulated acid rain, have reported that above and below ground biomass decreased significantly in the plants treated with acidic precipitation. Relative to control, the number of grains per plant and yield per square metre declined significantly under all acid rain treatments. Verma (1992) reported that rapid decrease in rain pH in India is caused by rapid increase in nitrate radicals formed on oxidation of nitrogen oxides. The increase in nitrogen oxides is caused by pollution from the rising number of vehicles in the country.

5.5.2 Water Crises and Pollution

The challenges of growing water scarcity is becoming increasingly serious due to wasteful use. Siltation of reservoirs is causing significant loss of the created potential at the rate of about 1 lakh ha annually. Most irrigation projects in India are operating at a low efficiency in the range of 30 to 40 per cent, losing rest of the water during conveyance from the reservoirs to the fields. Water gets misused by farmers for various reasons including lack of realisation of the cost of water, lack of concern for damage and deprivation to the downstream farmers from the overuse of water in the upstream, disregard to ill consequence of waterlogging and salinisation, and lack of access to water management technologies which could enhance the efficiency of irrigation and water use. In tube-well irrigated area, there has been an overexploitation of ground water leading to decline in water table in many areas of the North-Western Plains. Overdraft of ground water at rates higher than recharge has caused increase in pumping cost due to lowered water table. Although the country is endowed with fairly rich rain water resource (about 115 cm per annum), the same is not harvested for agricultural purposes.

A critical appraisal of existing rain water availability shows: (i) India receives 400 million ha metre of rain water annually (ii) Nearly 160 million ha m falls on agricultural land (iii) about 24 million ha m equivalent rain water is available for harvesting in small scale water harvesting structures (v) Nearly 186 million ha m goes to rivers as run-off and (vi) About one-fourth of total annual rainfall is received before or after the cropping season (Paroda, 1998).

The other important concern related to water crises are limited scope for expansion of irrigation potential, and increasing pollution of water resources through industrial effluents and ecological problem owing to floods and wetlands. Some important inorganic pollutants entering the water bodies are mercury through industrial effluents, cadmium through industrial discharge (e.g., electroplating, welding, refining, fertilizer and pesticide industries), lead through industrial discharge (e.g. battery, lead pipes industries), fluoride through aluminum and fertilizers plants, and arsenic through pesticides industrial discharge. Mercury contaminated water is poisonous. Cadmium accumulates in liver and kidney leading to carcinogenesis. Leads affect the central nervous system and abnormalities in fertility and pregnency. Fluoride at higher concentration (more than 1.5 ppm) causes fluorosis affecting bones and teeths. Arsenic poisoning results in melanosis leading to skin lesions, leprosy and gangrenous ulcers.

The other pollutants responsible for contamination of water bodies are (i) discharge from municipalities, industrial complexes, domestic sewade / wastes and / or rural runoff adding organic and metals to surface water; (ii) surface runoff containing pesticides, herbicides, insecticides and fertilizers from agricultural farms; (iii) commercial toxic compounds like mine tailings, dyes, acidic materials, etc., (iv) other materials like dead organic matter.

The in-plant treatment of industrial waste water is not cost effective and hence, often disposed off directly in surface water causing pollution of water bodies. Such polluted water when used for irrigation can be harmful to crops.

The crisis is not only limited to agriculture but fresh water for drinking is also becoming a scarce resource. The per capita comfortable level of 2100 cubic metre per year is estimated to reduce to the stress level of 1700 per cubic metre per year by 2030 and to the scarcity level a few years later.

The holy Ganges has become so polluted that bathing and drinking its water has become health hazards. The river flows through 29 cities and about 48 towns and nearly 1 billion litre of mostly untreated row sewage enters the river every day. The other sources of pollution of this river is leather industries near Kanpur (UP) from which chromium and other chemicals leak into the river, domestic and industrial wastes, sewage and solid garbage, agriculture runoff containing pesticides and fertilizers, floating of partially burnt or unburnt corpses, defection and mass bathing by man and animals.

The arsenic contamination of ground water in Karimganj, Dhemaji and Dhubri districts of Assam and belt of West Bengal extending up to Bangladesh is serious threat to the people in region. A long exposure to arsenic contaminated water causes lung cancer, skin diseases, gastro-intestinal, neurological, cardiovascular and respiratory problems. Among the several examples of water pollution in India, few worth mentioning are textile dye contamination of groundwater and water of Nogyal, Bhavani, and Cauvery in Tirupur and Coimbtore.

Industrial pollution has threatened the natural habitats of pearls in pearl banks of Tuticorin coast in Gulf of Mannar. The Chennai coastal water has showed high levels of pesticides like DDT, lindane, endosulphan and heptachlor. The city of Mumbai generates nearly 2,250 million litres of sewage per day that flows directly into the Arbian sea. The beaches are choked with solid wastes and raw sewage. Dissolved oxygen levels in water are seriously low along the polluted part of the coast. In Punjab, high levels of pesticides has been reported in vegetables and groundwater. The persistant of pesticide - a non biodegradable toxins in environment for a long period and their bio-accumulation in the food chain causes severe toxicity.

5.5.3 Soil Degradation and Pollution

The soil is uppermost layer of lithosphere. It is an indispensible resources on which we depend for our food, fodder, fuel, fibre and energy.

India has total geographical area of 329 million ha, out of which 187 million ha representing 57 per cent presently suffer from various kinds of degradation. While water erosion is the major cause contributing to loss of valuable top soil (132 million ha) and terrain deformation (16.4 million ha), wind erosion is dominant in western region resulting in 13 million ha degraded land. Soil deteriorates due to chemical process is estimated to affect about 14 million ha, which includes accumulation of salts, i.e. salinisation (10 million ha) and loss of organic matter and plant nutrients (4 million ha).

The soil degradation can be natural or man made. Natural degradation occur in nature due to the action of wind and water. The degradation caused by man-made activities include deforestation, inappropriate cultivation, over extraction of ground-water, overgrazing, improper waste disposal and mining of resources, etc. Soil erosion leads to loss of vital plant nutrients from the soil and development of salinity and alkalinity leading to desertification, lowering of water table and decline in crop production. It must be understood that it takes almost 1000 years to form an inch of soil from parent rocks but quite often that is lost through soil erosion in less than a single year of intense.

The major cause of soil pollution are heavy metals from applied nutrients, irrigation with poor quality water and presence of pesticide residues in soil. A gradual decline in native soil fertility in intensively cultivated soil with inorganic fertilizer source is being observed particularly by the farmers in Indo-Gangetic plains. Increasing yields associated with increasing cropping intensity has resulted in reduced organic source of nutrients in many parts of India. Consequently the farmers have to apply more fertilizers to get the same yield. The micro-nutrients such as sulphur, zinc and boron have also started showing deficiencies.

Many fertilizers, particularly phosphatic fertilizers contain varying amounts of aresenic, cadmium, chromium, and mercury, etc. These potential toxic elements accumulate in soil, damage beneficial soil microflora, adversely affect yield and quality of crops and enter into food chain of human and livestock causing serious health problems. Pesticides that are used to kill microorganisms also kills the seeds. The fruits borne by pesticides fed plants or trees are poisonous.

5.5.4 Pesticide Residues

Incidence of pests and diseases is the manifestation of environmental condition. Nearly 10 per cent of agricultural production annually is affected by pests and diseases. As control measure two-third of nearly 60,000 tonnes of pesticide entering the Indian environment each year is used in agriculture and remaining one-third is used in public health programmes. BHC, DDT, malathion accounts for the bulk. Part of applied pesticides ultimately finds its way into the soil, water and food chain. Long-term effects of pesticide residues in the human body include carcinogenecity, reduced life span and fertility, increased cholesterol, high infant mortality, and varied metabolic and genetic disorders.

Highly persistant pesticides (half-life above 6 months) are DDT, HCH, Aldrin and Endosulfan. Pesticides with half life up to 6 months are Atrazine, Mouron, Linuron, Simazine, Terbecil and Trifluralin are moderately persistent. Slightly persistent (half life 2 to 6 weeks) pesticides are Dicamba, Dalapon, EPTC, TCA, whereas non-persistent pesticides (half life 1 to 2 weeks) are 2, 4-D, Diquat and Endothal (Aggarwal, *et al*, 2006). Chlorinated hydrocarbon insecticides can persists for a period of 16 months to 5 years; Urea, tiazines and picloram herbicides can persists for 3 to 18 months; phenoxy and nitrite herbicides for 1 to 6 months; phosphetic insecticides for 1 to 12 weeks, benzoic acid and amide herbicides for 3 to 12 months; carbamate and aliphatic acid for 3 to 5 months. Out of all chemical produced and released into the environment by human beings, some toxic organic substances (that include many pesticides, industrial, chemical and chemical by-products) are termed Persistent Organic Pollutants (POPs) that make them an urgent global environmental health problem (Sharma, 2006).

Pesticide persistance is usually specified in residual half-life (RL_{50}), which is a measure of the time in days required to reduce pesticide residues to half of its initial deposits. The amount of pesticide lost per unit time is proportional to the total amount present in a particular time. Degradation of this type is called "first-order degradation" and can be represented mathematically as:

$$\frac{dN}{dt} = \lambda N$$

Where as λ = decay constant

N = mass present at time t

Once released in environment, pesticides may be broken down or degraded by the action of sunlight (photolysis), water (hydrolysis) and other chemicals or microorganisms, such as bacteria, fungi (microbial degradation). This degradation process usually leads to formulation of less harmful products, but in some instances can produce more toxic products.

5.5.5 Loss of Biodiversity

A growing worldwide realisation that biodiversity is fundamental to agricultural production and food security, as well as a valuable ingredient of environmental conservation has been clearly observed.

Biodiversity or **biological diversity** refers to the variety of living organisms in a particular habital or geographical area. It is measured as the numbers of species or sub-species of plants, animals, and microorganisms. It is widely recognised that maintaining biodiversity is important in agriculture and bioproductive systems. The major storehouse of biodiversity is now under threat. With more than half the world's species found in closed tropical rain forest, deforestation will be the single greatest cause of species extinction in the next 50 years. Some of the steps which require urgent attention are (i) restriction on felling of tress, (ii) soil and water conservation, and (iii) checking forest degradation.

Agricultural biodiversity or **Agro-biodiversity**, is defined as the variety and variability of animals, plants and microorganisms used directly or indirectly for food, fibre, fodder, fuel and other aspects of agriculture. It forms the basis of food and livelihood security to billion of people and the development of all food production including for industrial agriculture and for the bio-technical industries.

The predominant pattern of agricultural growth have eroded biodiversity in agro-ecosystems including plant genetic resources, livestock, insects and soil organisms. Monoculture, adoption of limited number of crop varieties, continuous use of fertilizers and pesticide, mechanization and diversification of limited crops and farm animals in last few decades has gradually led to the genetic and species erosion in some agricultural lands. There are also decline in the population of several indigenous cattle, organic matter decomposition (soil biota) and land-races of aromatic and scented traditional rice, minor millets and several kind of indigenous vegetables.

The erosion of biodiversity has caused economic losses, jeopardizing productivity and food security, and leading to broader social costs. It has been noted with concern that the diversity is being lost from the 'natural' habitats due to the expansion of agricultural production to frontier areas and also from the agricultural fields due to the adoption of improved varieties and other technologies by the farmers. Genetic resources of actual or potential value to food and agriculture are also being lost at an alarming rate, due to habitual destruction, land degradation, over-exploitation of water resources, industrial agriculture, forestry practices and urban expansion (Gautam and Kochhar, 2000).

The major causes of the losses of agricultural biodiversity (Sur *et al.*, 2002) are:

(a) The rapid expansion of industrial and green revolution agriculture, intensive livestock production, fisheries and aquaculture (some production systems using genetically modified varieties and breeds), cultivating few crop varieties in monoculture and rearing a limited number of domestic animal breeds, fish and aquatic species.

(b) Globalisation of the food system and marketing, and the extension of industrial patenting and other intellectual property systems to living organisms which have led to the widespread cultivation of fewer varieties and rearing of few breeds for a more uniform, less diverse but more competitive.

As a consequence there has been:

(i) Marginaliation of small-scale, diverse food production systems that conserve farmer's varieties of crops and breeds of domestic animals, which form the genetic pool for food and agriculture in the future.

(ii) Reduced integration of livestock in arable production, which reduces the diversity of uses for which livestock are needed.

(iii) Reduced use of natural fisheries techniques, that conserve and develop aquatic biodiversity.

The main cause of genetic erosion in crops is the replacement of local varieties by improved or exotic varieties and species.

5.6 MITIGATING POLLUTION IN CROP ENVIRONMENT

5.6.1 Reducing Greenhouse Gases from Agricultural Source

Carbon dioxide (CO_2), methane (CH_4) and nitrous oxide (N_2O), the three most important greenhouse gases absorb solar radiation rather than allowing the heat to be radiated away from the earth. India's per capita emission of greenhouse gases is 1.1 tones only as compared to 20.1 tons for the United States of America (USA). Contribution of Indian agriculture is only 3.2 per cent of world total methane emission of which 25 per cent is from agricultural soils. Nitrous oxide is miniscule in contribution of Indian agriculture. Methane and nitrous oxide from Indian agricultural soils are responsible only 0.23 and 0.1 per cent, respectively of the global warming caused by World's methane and nitrous oxide emissions.

5.6.1.1 Nitrous Oxide

The soil is considered to be contributing 65 per cent of total nitrous oxide emissions. The major sources are soil cultivation, fertilizer and manure application and burning organic material and fossil fuels. It represents loss of nitrogen and nitrogen use efficiency.

Deep placement of fertilizer in reduced zone, use of slow released N fertilizers, banding fertilizer within the crop rows and thus, by improving uptake of N hold the key to reduce nitrous oxide emissions.

5.6.1.2 Methane

Primary source of methane from agriculture include animal digestive processes, wetlands, rice cultivation as well as manure storage and handling. Methane is about 20 times more effective than CO_2 as a heat trapping gas.

Aerating manure during composting and use of solid rather than liquid manure may reduce methane but promote nitrous oxide formation.

In rice cultivation, methane emission can be reduced by adopting short term drainage for frequent aeration.

Because most of methane is produced in the rumen of livestock by fermentation, practices that speed the passage of feed from rumen can also reduce methane formation. For this, strategies that reduce feed input per unit of product output may be adopted.

5.6.1.3 Carbon dioxide:

For mitigating carbon dioxide, carbon can be sequestered in soils by increasing carbon inputs and/or decreasing their decomposition which can be achieved by reduced tillage and greater allocation of land to agroforestry and biofuel crops.

5.6.1.4 Conventional fossil fuel

In order to reduce emission of greenhouse gases in the environment, conventional fossil fuels may be replaced by renewable energy sources. Biofuel including woodfuel, charcoal, biogas, ethnol produced from agri-resources, agricultural wastes, crop residues and energy crops have been considered as possible sources. Agri-resources for ethnol production include biomass consisting of sugars, starch, cellulose and ligno-cellulosic residues.

Biodiesel for transportation, tractors and pumps: Biodiesel consisting of the fatty acid esters of simple alcohol is a potential replacement for a portion of the diesel fuel. It is a renewable resource, which reduces HC, CO, and CO_2 exhaust emissions, and is non-toxic and biodegradable. Among different promising biodiesel plants, a non-edible oilseed plant known as jatropha or *ratanjyot* (*Jatropha curcus* - a plant of Latin-American origin) is prominant.

Jatropha oil has been used as a diesel substitute in several countries including Austria and Nicaragua. Its seeds contain about 35 per cent of non-edible oil. Jatropha is drought-resistant perennial shrubs and can be grown easily on marginal soils. It is hardy which grows quickly and can live upto 75 yeas. About one-third of the energy in the fruit of jatropha can be extracted as an oil with similar energy value to diesel fuel. Jatropha oil can be used directly in diesel engines added to diesel fuel as an extender or trans-esterised to a biodiesel fuel. Other promising biodiesel plants are *Karanj (Pongamia pinnata)* with an oil content of 25 to 28 per cent and *Simaruba glauca* with an oil content of 55 to 65 per cent.

5.6.1.5 Degradation of heavy metals by microbes and plants

Microbes are generally useful for assisting in reclamation of sites with heavy metals which cannot be destroyed unlike organic contaminants which can be degraded. Several fungi are also good in the accumulation of heavy metals like cadmium, copper, mercury, lead, zinc and others. *Rhizopus arrhizus* is for example, useful in treating uranium and thorium.

Higher terrestrial plants (belonging to the family *Brassicaceae, Asteraceae, Cunouniaceae, Cyperaceae, Caryphyllaceal* and *Fabaceae)* can also be used for remediation of metal polluted soils. This process is called phytoremediation which include phytostabilization and phytodecontamination. However, the reason of unpopularity of this process is slow growth of the plants and low bioavailability of metals.

Certain biotechnological approaches, such as ecological improvement of rhizospher by rhizosphere technology and use of genetic engineering tools for improvement of plant for bioremedian are also worth-noting.

5.6.1.6 Degradation of hazardous chemical by microorganism

The fungi are unique among microorganisms which have ability to transform a wide variety of hazardous chemicals as they secrete a variety of extra-cellular enzymes that facilitate decomposition of some pollutants. *Phanerochaete chrysosporium, P. sordida* and *Trametes hirsuta* have been shown to degrade *PCP* (Pentachlorophenyl) and creosote in soil. *P. chrysosporium* also degrade a number of toxic xenobiotics. The basidiomycetous fungus *Pleurotus ostreatus* is known to produce fungal laccase, which helps in the degradation of atrazine and organophosphorous insecticide. Bacteria are also good degraders of toxic pesticides, such as halocarbons. *Pseudomonas putida* degrade polychlorinated biphenyls, tricholraethyelene and some other aromatic hydrocarbons. Methanogenic bacterial consortium degrades perchloro-ethane. Mono and dichloro-benzenes can be degraded aerobically by various *Pseudomonas* and *Alcaligenes* strains.

5.6.2 Measures to Reduce Air Pollution

5.6.2.1 Farming systems approach

Agriculture is also a contributor to global air pollution. According to Conway and Pretty (1991), a farming systems approach can contribute to the development of a low-polluting agriculture in two ways. First, by identifying and strengthening such on-farm resources as nitrogen-fixing organisms and the natural enemies of pests and diseases, it can reduce reliance on the inputs of polluting agrochemicals. Second, through an analysis of whole farm systems, it can identify those elements and processes that contribute to efficiency in the use of resources, and hence, reduce farm wastes and losses of nutrients. It can be argued that a farming systems approach is the only way to achieve significant reduction in pollution arising from agriculture. Individually, narrowly focused approaches are at best likely to substitute one form of pollution for another.

5.6.2.2 Tree Plantation

The greatest ecological potential of tree plantation seems to be buffering and maintaining sustainable environment and production capacity of agricultural systems. The woody plants contribute to both protective functions (enhancing ecological sustainability) and productive functions (taking the place of chemical fertilizers). There should be proper conservation, protection and preservation of forests. Afforestation should be at mass level. There should be green belt of trees around the cities, villages, industrial areas and other pollution areas. Trees are the best controller of air pollution. It is reported that money plant, *Coleus blumeri, Phaseolus vulgaris* and *Ficus* absorb carbon monoxide and *Punus* species and *Juniperus* metabolize nitrozen oxide emissions.

5.6.2.3 Conservation of energy

Air pollution can be reduced to a considerable extent by promoting energy efficiency and conservation, cutting down on wasteful energy consumption activities by using safer and cheaper alternative energy source (Biogas, solar power, hydrogen gas, gastrol, etc.)

Use of CFC free refrigerators, air conditioners, sprays, perfumes and deodorants should be encouraged. Fossil fuel based public transport system should be phased out by CNG gas based transport systems. Use of catalytic converter in vehicle should be promoted to reduce emission of carbon-mono-oxide in air. Unnecessary burnings and use of polystyrene cups and plastic carry bags should be avoided. Use of jute or coir bags for shopping should be popularised.

5.6.3 Control of Water Pollution

The effects of water pollution are not only devastating to human beings but also to animals, fish and birds. Polluted water is unsuitable for drinking, agriculture, recreation, and industry. Contaminated water destroys aquatic life, reduces reproductive ability and is hazards to life on earth. Pollution of water can be either prevented or controlled by the following measures:

(i) To adopt integrated nutrient management by incorporating use of bio-fertilizers, organic manures and biopesticides since a large amount of chemical fertilizers used are washed into the waterways resulting in contamination of both surface and ground water.

(ii) To adopt integrated pest management by minimizing the use of chemical pesticides including herbicides and insecticides. These chemicals pollute the water and are toxic to all forms of life.

(iii) Agroforestry, social forestry and afforestation can contribute to land rehabilitation and increase production by minimizing flood, soil erosion and runoff.

(iv) Proper treatment of waste waters before discharge into natural water bodies is an essential pre-requisite. Polluted water can be reclaimed by sewerage treatment plants to change them in non-toxic forms. The treated water can be reused in the factories and even for irrigation but only after water analysis to judge its suitability.

(v) Water haycinth can purify water polluted by biological and chemical wastes. It can also filter out heavy metals as well as other toxic substances found in industrial waste waters.

(vi) Oxidation ponds have been found to be useful in removing low levels of radioactive wastes.

5.6.4 Control of Soil Pollution

Following measures may be adopted to control soil pollution:

(i) To adopt soil and water conservation measures including contour farming and strip cropping.

(ii) To adopt integrated nutrient management and integrated pest management techniques with minimum use of chemical fertilizer, pesticides and herbicides.

(iii) To use land according to land capability classes and adopt cropping on watershed basis.

(iv) Proper forest management and adoption of agro and social forestry and afforestation on large scale.

(v) Controlled grazing.

(vi) Improvement in mining techniques to reduce spread of mine's dust.

(vii) Improved disposal techniques for dumping industrial wastes.

(viii) For solid waste management, application for techniques to ensure orderly execution of collection, processing and disposal.

5.6.5 Conservation of Biodiversity

The conflicts between agriculture and biodiversity can be overcome with sustainable farming systems, changes in agricultural policies and institutional mechanism including linkages. Integration of agricultural practices with biodiversity maintenance can have multiple ecological and socio-economic benefits, particularly ensuring food security (Gautam and Kochhar, 2000). Though no single path can be laid yet various policies and practices can be devised to promote and use biodiversity in agriculture (Sur, *et al.*, 2002). These include:

(i) Diverisified farming operations with viable alternatives, like mixed and multiple cropping systems.

(ii) Emphasis should be laid on crop mosaic system where a farmer has a choice to make from a wide arry of varieties available before him.

(iii) Promote policies that encourage market diversification through developing new market and by popularising forgotten crops for international markets.

(iv) Tap indigenous technical knowledge of the area that emphasised biodiversity.

(v) Rapid agro-diversity assessment are required to study the impact introduction of a technology in relation to changes in parameters, like varietal density, land use pattern and environment.

(vi) Adoption of soil and water conservation practices is necessary to provide conductive nutrient and moisture conditions.

(vii) Inventories of the resources using photo-interpretation and remote sensing techniques need to be prepared so that base lines are available for future comparisons.

Pathak and Hegde (1996) reported that agroforestry had potential to conserve and utilize biodiversity. Single species plantation though provide wood at short rotation have been found to be deplete biodiversity compared to mixed culture of forests and agriculture including homegardens in tropical humid climates. Kang *et al.*, (1990) demonstrated that under agroforestry system, earthworm population was larger than in control showing the complementarity of the system and its role in encouraging soil biodiversity.

5.7 PROTECTED AGRICULTURE

Sustainability in agricultural productivity is possible only if the impacts of the variability in the input resources are minimised through efficient management/ manipulation. Agro-climatic resources are one of those important components which need to be efficiently managed as their influence on crop growth and agricultural productivity are quite large. For providing much needed protection to agriculture, efforts have been made to develop technologies which may help to ameliorate the impacts of adverse weather condition on crop growth and yield. Some of these weather based strategies include: (a) Drought proofing, (b) Irrigation and rain water management, (c) Response farming, (d) Microclimatic manipulation, to name a few (Ramkrishna *et al.*, 2000)

5.7.1 Drought Proofing

In India, drought is a recurring phenomena in some parts or other. Hence, there is a need to develop infrastructure, like (i) an early warning/monitoring system to generate information on onset, duration, intensity and termination of drought conditions, (ii) Research to make a reliable assessment on the impact of drought, (iii) creation of buffer stock for foodgrains and fodder (iv) providing economic relief through creation of durable assets and employment to reduce dependence on agriculture in drought prone areas, and (v) coordination mechanism of activities in implementing drought management.

Drought proofing through judicious and integrated management of natural resources is necessary in highly drought prone areas. The principle and practices useful in agricultural drought management are directed towards efficient use of available rain water for agricultural production as given below (Ramkrishna *et al.*, 2000):

(i) Increasing the infiltration of rain water into the soil through tillage, land configuration and residue management,

(ii) Storing runoff water for later use,

(iii) Reducing deep percolation of water through deep rooted crops or through soil amendments using moisture barriers, like asphalt,

(iv) Reducing evaporation by maintaining a mulch over the surface, limiting tillage and achieving timely rapid or uniform crop establishment to shade the soil surface,

(v) Reducing water use by eradication of non-crop plants, like weed, and

(vi) Maximization of crop yields by growing well adopted crop genotypes, timely crop establishment and good agronomic practices.

5.7.2 Irrigation and Rainwater management

This has already been discussed in Chapter-I.

5.7.3 Response Farming

Based on the analysis of weather data from Africa, Middle East and India Stewart (1988) found that there is a relationship between the time the rainfall season begins (date of onset) and the rainfall amount and duration thereafter. He suggested use of response farming concept to stabilised crop production and minimum risk. He suggested that if off-season rainfall is below normal, the farmer would prepare his land to retain all the rainfall, as the rainfall is more likely to be below normal. The farmer may need less fertilizer for top dressing. In case the off season rainfall is above normal, farmer may expect a wet year and plan their strategy accordingly. Broad rainfall predictions before start of the season can provide usefull guide to farmers to finetune their management practices depending on the actual weather conditions as the season progress. Analysis of 28 years (1957-84) Hyderabad data by Stewart (1988) showed that when off season rainfall were low (16 years), 75 per cent of following rains were below the median and when off season rains were high (12 years), 83 per cent of following monsoon rains were above the median. Based on this analysis of weather data, he suggested the use of response farming concept.

Response farming is a mean of coping with seasonal variability in climatic elements in a production system through identification and quantification of the season variability and the related risks and addressing the same at farm level through development of appropriate alternate strategies to minimise the productivity and economic losses. Earlier the concept was developed and tested by Stewart (1987) at Kenya by analysing the relationship between onset of rainy season versus duration of rainy season and expected quantum of rainfall. The methodology aims to accomplish improved prediction of expected behaviour of the climatic element (e.g. rainfall) in the approaching cropping season enabling improved decisions at the field level. For example, the method demonstrates the possibility to utilize the date of onset to act as a rainfall (amount, duration, average rainfall per day) predictor for the remainder of the season.

5.7.4 Manipulation of Micro-climate

The adverse effect of climate can be reduced to limited extent by adopting the following measures:

- Use of surface mulch can improve the soil thermal regime and minimise the splash erosion by reducing rain drop impact Mulch also conserves moisture in soil enabling the plant to have extended period of moisture availability.
- In absence of adequate rainfall, irrigation can make crops remunerative

- Provision of adequate drainage leads to profitable crop production in lowlying areas.
- Adverse effect of high wind velocity can be reduced by windbreaks and shelterbelts.
- Frost damage can be minimised by mulching.
- Heat trapping can increase the growing periods in temprate climate. Gomes (1998) reported that use of non-woven row covers, a very thin permeable fabric can raise the temperature of the soil and air during the day, diminish heat loss during the night and raise minimum temperature. Its permeability to air, water and light ensures a uniform heat and water transfer and creates balanced micro-climate under the cover.
- Afforestation helps in microclimate amelioration. Trees due to their canopy architecture, shade and density help in reducing temperature extremes of environment and surface soil due to reduction in radiation. A decline of 0.1 to 0.7°C was evident beneath the canopy cover of *A. tortilis* than that of surrounding open area at 07 hrs (Ramkrishna and Shastri, 1977). Around 14 hrs, this decline was of the order of 0.6°C to 2.0°C. Similarly, it was found during monsoon that the soil temperature beneath the tree cover was lower by as much as 10°C to 16°C in top soil zone and 4°C to 5°C at 30 cm depth when compared to open field conditions. Dela Cruz and Luna (1992) found decrease in light intensity under *Acacia auriculiformis* plantation with higher relative humidity. This favoured soil moisture conservation and better soil organic matter content due to more favourable activity of soil microbes. They also reported 2.4 to 5.9°C lower temperature under tree canopies compared to grassland during the day time. Gill (1985) also reported microclimate modifications in *Eucalyptus* and *Acacia* plantations raised on sodic soils. Gupta and Rao (1994) found that afforestation / agroforestry on degraded lands has the largest potential for carbon sequestration.

Carbon sequestration is necessary for sustainable development and arresting global warming and climate change. Agroforestry has potential to moderate climatic change predictions by storing the fixed carbon for long time in wood and thus, halting the depletion of ozone layer.

REFERENCES

Adams, R. M., Rosenzweig, C., Pert, R. M., Ritchie, J. T., McCarl, B.A., Glyer, J.D., Curry, R. B., Jones, J.W., Boote, K. J. and Allen Jr., L. H. 1990. Global Climate Change and U.S. Agriculture, Nature, 345: 219-224.

Aggarwal, P.K. 1993. Agro-ecological zoning using crop growth simulation models: Characterisation of wheat environment of India. *In* Systems Approaches for Agricultural Development *eds.* De Varies, F.P., Teng P. and Metaselaar, K.), pp:97-109. Klwer Academic Publishers, London.

Aggarwal, P.K., Choudhary, A., Joshi, H. C., Singh, S.D., Gupta, N. and Pathak, H. 2006. Agriculture and Environment. Hand Book of Agriculture, pp:55-71. ICAR, New Delhi.

Bagnold, R.A. 1954. The physiological aspects of deserts. *In*: Biology of deserts (*ed.* Coundsley-Thompson, J.L). Institute of Biology, London.

Bhanwar Lal. 1975. Effect of different methods of planting with or without mulching and use of chemicals to mitigate cold injury on potato production. M.Sc. Thesis, Univ. of Udaipur.

Bhattarcharjee, S. and Mukherjee, A. K. 1994. Biochemical and molecular changes during cold stress in higher plants with special reference to protein metabolism. Advances in Plant Science Research, 1:117-141.

Baver, L. D. 1956. Soil Physics (3rd edition), Asia Publishing House, Newyork, USA.

Bos, E., Vu, M. Y., Massiah, E. and Bulatao, R.A. 1995 World Population Projection, 1994-95 edition: Estimates and Projection with Related Demographic Statistics, John Hopkins University Press, Baltimore, MD., USA.

Chang, J. H. 1981. Corn yield in relation to photoperiod, night temperature and solar radiation. Agriculture Meteorology, 24:252-262.

Chitralekha, R and Dakshinamoorthy, M. 1997. Effect of certain kiln dust pollution on soil properties and on sorghum plants. Madras Agricultural Journal, 1997 (Publ. 1998), 84 (11-12): 671 - 674.

Corrway, G. R. and Preety, J. N. 1991. Pollution and farming systems. J. of the Asian Farming Systems Association, I (i): 29-53.

Cooler, E.J. 1990. The impact of climate change on continuous corn production in the Southern USA. Climate Change, 16:53-82.

Dela Cruz, L.U. and Luna, A.C. 1992. Effect of *Acacia auriculiformis* and *Gamelina arborea* on soil and microclimate of degraded grassland in *Nueva* Ecuja, Phillipines. Paper presented at an International Symposium on "Rehabilitation of Degraded Grassland" held during Sept. 15-20, 1992 in Tsukuba, Japan.

Gates, D. M. 1962. Energy exchange in biosphere. Harper and Row, New York, USA.

Gates, D. M. 1965. Energy, plants and ecology. Ecology, 46 (1):13.

Gates, D.M. 1996. Spectral distribution of solar radiations at the earth's surface, Science, 151:523-529.

Gautam, P. L. and Kochher, S. 2000. Agrobiodiversity Management. *In.* Natural Resource Management for Agricultural Production in India. (*eds.* Yadav, J.P.S. and Singh, G.B.) pp: 483-550. International Conference on Managing Natural Resource for sustainable Agricultural Production in the 21st Century, Feb. 14-18, 2000, New Delhi, India.

Ghosh B. and Chatterjee, S. K. 1975. Repairing of low temperature induced injury in rice seedlings by thiourea, cystine and potassium thiocynate. Geobios (Jodhpur), 2:31-32.

Gill, H. S. 1985. Studies on the evaluation of selected tree species for their tolerance to sodicity and mechanical impedence in a highly sodic soil with particular reference to root growth behaviour. Ph.D. Thesis. Krushetra University.

Gomes, R. 1998. FAO's experience in the production of agro-meterological information to the user community. *In*: User requirements for agrometerological services (*eds*. Shivakumar, M.V.K., De, U.S., Sinha Ray, K. K. and Rajeevan, M. J.). Proceedings of an International Workshop, 10-14 November, 1997, Pune, India.

Goudriaan, J. and Unsworth, M. M. 1990. Implication of increasing carbon dioxide and climate change for agricultural productivity and water resources. *In* impact of carbon dioxide, trace cases and climatic change in global agriculture. American Society of Agronomy (ASA). Special Publication No. 53, pp:111-130, Madison, WI, USA: ASA.

Grace, J. 1977. Plant response to Wind. Academic Press, London.

Gupta, R. K. and Rao, D. L. N. 1994. The potential of agroforestry for sequestering carbon in wasteland and mitigating climatic change. Pages 90-95 in Agroforestry systems for sustainable land use (*eds*. Singh, P., Pathak, P. S., and Roy, M.H.). Oxford and IBH Publishing Co. Pvt. Ltd., New Delhi.

Hingane, L. S., Rupa Kumar, K. and Ramna Murty, B. V. 1985. Long-term trends of surface air temperature in India. Journal of Climatology, 5:521-528.

Hoffman, G. J, Howell, T.A. and Soloman, K. H. 1990. Management of farm irrigation systems. American Society of Agricultural Engineers, pp:1040, St. Joseph, Michigan, USA.

Hume, C. J. and Cattle, H. 1990. The greenhouse effect: Meteorological mechanisms and models. Outlook on Agriculture, 19:17-23.

Hundal, S. S. and Abrol, I. P. 1991. Perspective of greenhouse gases in climatic change and plant productivity. *In*. Proceedings of Indo-US Workshop on Impact of global climatic changes on photosynthesis and plant productivity (*eds*. Abrol, Y. P., Gananam, A., Govindjee, Ort, D. R., Teramura, A. H. and Wattal, P. N.), pp:767-779, New Delhi: Oxford and IBH Publishing Co. Pvt. Ltd.

Hundal, S. S. and Kaur, P. 1995a. Effect of environmental stresses on potential production of major cereal crops in Punjab. *In*. Proceedings of International Conference on Sustainable Agriculture and Environment, 11-13 January 1995, HAU, Hisar, Haryana, India.

Hundal, S. S. and Kaur, P. 1995b. Effect of CO_2 concentration and global warming on productivity of rice. *In*. Proceedings, National Symposium on Agriculture in Relation to Environment, 16-18 January, 1995, IARI, New Delhi.

Hundal, S. S. and Kaur, P. 1995c. Effect of climatic changes on rice production in Punjab, India, International Rice Research Notes, 20 (1):32.

Hundal, S. S. and Kaur, P. 1996. Climate change and its impact on crop productivity in Punjab, India. *In* climatic variability and Agriculture (*eds*. Abrol, Y. P., Gadgil, S. S. and Pant, G.B.), pp:578-600. New Delhi: Narosa Publishing House.

IUCC. 1992. Information unit of climatic change. Fact sheet No. 101. Impacts of climate change. UNEP, Polaisdes Nations, CH 1211, Geneva.

IPCC. 1995. International Panel on Climate Change, Working Group I. Climate change, the IPCC scientific assessment. (*eds*. Houghton, J. T. and Emphraums, J.J.) Cambridge University Press, Cambridge, U. K.

IPCC. 1998. The Regional Impact of Climate Change - An assessment of vulnerability (*eds*. Watson, R., Zinyowera, T. and Moss, R), Cambridge University Press, U. K.

Isherwood, K. F. 1998. Fertilizers and the Environment. *In* Proceeding of First International Agronomy Congress held at New Delhi on No. 23-27, 1998 (*eds*.

Singh P, Prasad R and Ahlawat, I.P.S.). First Printed October 2000. Indian Society of Agronomy, ICAR, New Delhi.

Kang, B. T., Reynolds, L. and Atta Krqash, A. N. 1990. Alley Cropping. Advances in Agronomy, 43:315-359.

Kaur, P. and Hundal, S. S. 1995. Effect of climatic changes on simulated growth and yield of groundnut (*Arachis hypogaea*) in Punjan, India. *In* Proceedings of Indo-German Conference on Impact of Modern Agriculture on Environment, Vol. 3 (*eds.* Arora, S. K, Behl, R. K., Tauro, P. and Joshi, U. N.), pp:15-19, 1-3 December, 1993, HAU, Hissar.

Keeling, C. D., Carter, A. F. and Mook, W.G. 1984. Seasonal, latitudinal, and secular variations in the abundance and isotopic ratio of atmospheric CO_2. Journal of Geophysics Research, 89:4615-4628.

Kulshrestha, U. C., Sarkar, A. K., Srivastava, S. S. and Parashar, D.C. 1996. Investigation into atmospheric deposition through precipitation studies at New Delhi (India). Atmospheric Environment, 30(24):4149-4154.

Kurashi, S. Tejoka T., Ushijima, T. and Tojaki, T. 1966. Effects of cylokinins on frost hardiness. Plant cell physiology, 7:705-706.

Lahiri, A. N. 1978, Soil-water-plant relations and production optimisation from arid areas. *In* Arid zone Research and Development (*ed.* H. S. Mann) pp:223-228, Jodhpur: Scientific publishers.

Lal, M., Singh, K. K., Rathore, L. S., Srinivasan, G and Saseendran, S.A. 1998. Vulnerability of rice and wheat yields in NW India to future changes in climate. Agricultural and Forest Meteorology, 1998, 89(2): 101-114.

Mahi, G. H. and Hundal, S. S. 2002. Climate and Agriculture. *In*: Recent Advances in Agronomy (*eds.* Singh, G., Kalar, J. S. and Sekhon, H.S.). Indian Society of Agronomy, IARI, New Delhi.

Mavi, H. S. 1994. Introduction of Agrometeorology,. Oxford & IBH Publishing Co. Pvt. Ltd., New Delhi.

Monteith, J. L. 1972. Solar radiation and productivity in tropical ecosystems. Journal of Applied Ecology, 9: 747-766.

Morey, D. K. and Sadaphal, M. N. 1981. Effect of weather elements on yield of wheat at Delhi, Panjabrao Krishi Vidyapeeth Research Journal, 1:81-83.

Muchow, R. C. and Sinclair, T. R. 1991. Water deficit effect on maize yields modelled under current and "greenhouse" climates, Agronomy Journal, 84:1052-1059.

Padmanabha Murthy, 1999. Impact of pollution on plant growth and yield. Journal of Agrometeorology, 1(1):1-5.

Pal, M., Sharma, A., Abrol, Y. P. and Sengupta, U. K. 1997. Exclusion of UV-B radiation from normal solar expectrum on the growth of mungbean and maize. Agriculture, Ecosystems and Environment, 61(1):29-34.

Paroda, R. S. 1998. Plenary Lecture on Natural Resource Management for Sustainable Agriculture: A New Paradigm. *In.* Proceedings of First International Agronomy Congress held at New Delhi on November 23-27, 1998 (*eds.* Singh, P., Prasad, R. and Ahalawat, I.P.S.). First printed October 2000. Indian Society of Agronomy, ICAR, New Delhi, India.

Parry, M. L. and Duinker, P. N. 1990. The potential effects of climatic change on agriculture. *In.* Inter-Government panel on climate change, pp:2-45. Geneva: WMO and UNEP.

Pathak, P.S. and Hegde, N. G. 1996. Agroforestry and biodiversity work intended to benefit India. Diversity, 12(3):47-48.

Pitzalis, M. and Pandolfi, A. M., 1986. Adverse meteorological conditions and methods of protection: frost. Italia Agricola, 123(4):69-812.

Porter, J. R. and Miglitta, F. 1991. Modelling the effect of CO_2- induced climatic change on cereal crops. *In.* Proceedings of Indo-US Workshop on Impact of Global climate changes on photosynthesis and Plant Productivity (*eds.* Abrol, Y. P., Gananam, A., Govindjee, Ort., D. R., Teramura, A. H. and Wattal, P. N.), pp:495-507. New Delhi: Oxford and IBH Publishing Co. Pvt. Ltd.

Porwal, B. L., Singh, H. G. and Mathur, P. N. 1986. Metabolic changes associated with chemical cryoprotection in gram (*Cicer arietinum* L.). Biochemical and Physiological Pflanzen, 181:659-664.

Ramana Rao, B. V. and Srinivas Rao, M. 1996. Weather effects of Insect Pests, pp:280-294. *In.* Climate variability and Agriculture (*eds.* Abrol, Y. P., Sulochana, Gadgil and Pant, G. B.). Narosa Publishing House, New Delhi, India.

Ramakrishna, Y. S. and Shastri, A.S.R. 1977. Micro-cliamte under *Acacia tortitis* plantation. Annual Progress Report (Jodhpur: CAZRI), pp: 69-70.

Ramakrishna, Y. S., Rao, G. G. S. N, Kesava Rao, A.V.R. and Vijay Kumar, P. 2000. Weather resource management. *In.* Natural Resource Management for Agricultural Production in India (*eds.* Yadav, J. S. P. and Singh, G. B.). International conference held on Feb 14-18, 2000, New Delhi, India, pp:245-370.

Ramkrishna, Y. S., Victor, U.S. and Ramana, Rao. 1996. Moisture regime variability and its impact on agriculture in the Indian Arid zone. *In* Climate variability and Agriculture (*eds.* Abrol, Y.P., Gadgil, S. and Pant, G.B.), Narosa Publishing House, New Delhi.

Reddy, S. R. 2000. Principles of Crop Production, Kalyani Publishers, Ludhiana, pp:18.

Reddy, T. Y. and Reddy, G.H.S. 1995. Principles of Agronomy (second revised edition). Kalyani Publishers, Ludhiana, India.

Rao, D. G., Katyal, J. C., Sinha, S. K. and Srinivas, K. 1992. Impacts of climate change on sorghum productivity in India: simulation study. Climatic change and agriculture: Analysis of potential international impacts. Proceedings of a symposium sponsored by the American Society of Agronomy in Minneapolis, MN, 4-5 Nov. 1992. ASA-special-Publication 1995, No. 59, 325-337.

Rao, S.S. and Sahu, M. P. 1991. Effect of Sulphur and foliar applied chemicals on cold tolerance in chickpea (*cicer arietinum* L). Journal of Agronomy and Crop Science, 167:320-325.

Rosennzweig, C and Parry, K. L. 1994. Potential impact of climate change on world food supply. Nature, 367:133-138.

Scherr, S. and Yadav, S. Land degradation in the developing world: Implications for Food, Agriculture and Environment to 2020. 2020: Vision for Food, Agriculture and the Environment Discussion Paper 14. Washington, D.S.: International Food Policy Research Institute.

Seema, P., Mukundan, K and Vishwanath, S. 1996. Impact of Cement kiln dust on rural economy: a case study. Indian Journal of Agricultural Economics. 51(3):407-411.

Sharma, K. K. 2006. Pesticide Residue. *In*: Handbook of Agriculture, pp:587-615. ICAR, New Delhi, India.

Sinclair, T. R. and Rawlins, S. L. 1993. Intern-seasonal variation in soybean and maize yields under global environmental change. Agronomy Journal, 85:406-409.

Singh, A. and Agarwal, M. 1996. Responsde of two cultivars of *Triticum aestivium* L. to simulated acid rain. Environmental Pollution, 91(2): 161-167.

Subbaryya, I. K. and Kumar, R. 1980. Crop-weather relationship of sugarcane and yield producing in north-west Andhra Pradesh, India. Agriculture meteorology, 21:265-279.

Staple, W. J. 1964. Dryland agriculture and water conservation. *In*: Research on water. ASA special publication. Soil Science Society of America. Madison, Washington, USA.

Stewart, J. I. 1987. Proceedings INRAN-ICRISAT-TROP SOILS-USAID-USDA Workshop on Soil, Crop and Water Management Systems for Rainfed Agriculture in Sudano-Shelian zone, Niamey, Nigger, 11-16, pp:225-264.

Stewart, J. I. 1988. Development of management strategies for minimising the impact of seasonal rainfall variation. *In*: Drought Research priorities for the Dryland Tropics (*eds*. Bidinger, F.R. and Johnsen, C.), pp:131-150. Patancheru, Andhra Pradesh: ICRISAT.

Sur, H. S., Grewal, S. S. and Hadda, M. S. 2002. Biodiversity and environmental conservation. *In* Recent Advances in Agronomy (*eds*. Singh, G., Kolar, J. S. and Sekhon, H. S.), pp:543-569. Indian Society of Agronomy, New Delhi, India.

Swaminathan, M. S. (ed.). 1993. Wheat Revolution: A dialogue. MacMillan India Ltd. Madras, India, pp:164.

Venkataraman, S. 1968. Climatic consideration in cropping pattern. Proceedings of the symposium on new cropping pattern in India. ICAR, New Delhi, India.

Venkataraman, S. and Krishnan, A. 1992. Crop and Weather, ICAR, New Delhi, India.

Venkateswarlu, J., Murty, T. V. and Padmanabhan, M. V. 1991. Contingent crop production strategy in rainfed areas under different weather conditions, pp:76, Hyderabad Central Research Institute for Dryland Agriculture.

Verma, G. S. 1992. Rain acidification in India caused by nitrates. Atmosfera, 5(4):233-239.

Verma, K. S., Kang, M.S. and Sharma, A. K. 1991. Injurious effect of some abnoxious gases on guava. Plant Disease Research, 6(2):140-141.

Verma, M. and Agrawa, M. 1995. Effect of sulphur-dioxide on biomass and yield of wheat cultivars grown at different fertility regimes. New Agriculturist, 6(1):37-39.

Wang, J. Y. 1963. Agricultural meteorology. Pacemaker Press, Milwaukee, Wisconsin, USA.

Went, F.W. 1957. Experimental control of plant growth. Chronica Botanica, 17. Ronald Press Co., New York.

Yoshida, S. 1981. Fundamental of rice crop science, IRRI, Los Banos, Philippines.

Subject Index

—A—

Aalbizia lebbeck, 167, 181
Abelmoschus esculentus, 150
Acacia spp, 148, 149, 161, 190
- *albida*, 146, 178
- *arabica*, 190
- *auriculiformis*, 171, 330
- *catechu*, 148, 174, 176
- *leucophplea*, 167
- *nilotica*, 147, 149, 165, 167, 170, 172, 173, 176, 177, 178
- *radiana*, 176
- *senegal*, 88, 146, 165, 166, 178
- *tortilis*, 165, 176, 181

Acid rain, 314
Administrative manageability, 10
Adoptability, 195
Advantages of GM, 223
Aegle marmelos, 172
Aerobic decomposition, 232
Aerva pseudomentosa, 177
Agaricus gisporus, 75
Age of the animal, 246
Aggressivity, 185
Agricultural biodiversity, 320
Agricultural resource determinants, 104
Agricultural sustainability, 12
Agriculture and environment, 275
Agriculture with sericulture, 150
Agri-horticulture, 136
Agri-silviculture, 136, 202
- system, 145

Agri-silvi-fishery, 150
Agri-silvi-pastoral system, 149
Agro-biodiversity, 320
Agroforestry systems
- classification, 144

Agroforestry, 80, 128
- benefits, 138
- concept, 131
- design criteria, 195
- for soil-conservation, 148
- in problematic soils, 173
- research, 195
- system, 154

Ailanthes excelsa, 145
Air permeability, 243
Air pollution, 306
Albedo, 284
Albizia spp, 152, 161, 190
- *lebbeck*, 147, 148, 165, 166, 170, 177, 178, 190
- *melocana*, 171

Alepocursus myosuroides, 212
Alhizia spp, 172
Allelopathy, 65, 184

Alley cropping, 80, 167
Allium spp, 150
Alnus nepalensis, 131, 168, 169, 174
Alternate crop strategies, 42
Amomum subulantum, 131, 168
Amount of residue, 242
Anabaena azol'a, 254
Anacardium occidentale, 150, 171
Anaerobic decomposition, 232
Aneguissus latifolia, 169
Animal based component, 80
Apiculture with trees, 150, 177
Apiculture, 73
Apis cerana indica, 73
Apis mellifera, 73
Approach to achieve sustainability, 12
Aquaculture (fish culture), 72
Aquaforestry, 136, 150, 177
Argemone mexicana, 224
Arid regions, 165
Artocarpus sp., 149, 165
 – *heterophyllus*, 172
Ashrams, 130
Aspergillus awamori, 255
Assessment of agricultural system, 9
Atmosphere composition, 297
Atmospheric humidity, 293
Azadirachta indica, 146, 147, 150, 165, 167, 170, 178, 229
Azolla biofertilizer, 254
Azolla microphylla, 85
Azospirillum, 30
 – biofertilizers, 252
Azotobactor, 30
 – biofertilizers, 251

—*B*—

Bacillus circulans, 255
Bacillus megatherium, 255
Bacillus sublilis, 255
Bambusa arundinacea, 165
Bambusa nutans, 168
Bauhinia, 169
 – *variegata*, 170
Bench terracing, 18
Bending and coppicing, 193
Berseem, 39
Biodiesel, 322
Biodiversity, 319
 – conservation, 326
Biofertilizer, 247
Biogas, 76
Biological determinants, 103
Biological diversity, 319
Biological properties of soil, 245
Biotic factors, 304
Block plantation, 136
Blue-green algae (BGA) biofertilizer, 253
Bombax ceiba, 181
Bombax malabaricum, 149
Borassus flabellifer, 166, 167
Boundary plantation, 136
Brachiaria mutica, 174
Branch pruning, 194
Brassica compestris, 147
Brassica napus, 90
Brassica oleracea, 150
Breaks' in the monsoon rains, 41
Bulk density, 242
Butea monosperma, 172

—C—

C:N ratio of the organic materials, 240
Calotropis gigantea, 229
Canopy management, 193
Carbon compounds, 232, 233
Carbon dioxide, 322
Carbon-nitrogen ratio and decomposition, 233
Carcia papaya, 172

Cardinal points, 290
Cardinal temperatures, 290
Cassia auriculata, 166, 173, 228, 229
Cassia siamea, 146, 165, 170, 229
Cassia spp., 190
Casuarina, 161
- *equisetifolia*, 148, 165, 170, 171, 178, 179
- *glauca*, 173
Catch crop, 223
Cation exchange capacity, 244
Cattle / dairy management, 66
Ceiba pentandra, 166
Cenchrus ciliaris, 88, 166, 174, 176, 177
Cenchrus setigerus, 166
Ceop residues, 245
Chemical properties of soil, 243
Chloris gayana, 174
Chrysopogon fulvus, 174
Classification of green manures, 225
Climate change, 311
Climatic factors and crops, 282
Clusterbean, 227
Coastal (eastern and western) region, 99
Coastal and island regions, 172
Coastal zone agroforestry, 203
Cocos nucifera, 150, 171
Coefficient of sustainability, 11
Coffee arabica, 152
Coleus blumeri, 324
Commercial agroforestry system, 153
Community forestry, 135
Competitive ratio, 186
Conservation bench terraces,, 18
Conservation cropping, 78
Conservation of energy, 324
Conservation tillage, 24
Contingent crop production strategy, 41
Contour bunds, 17
Contour farming, 18
Conventional fossil fuel, 322
Conventional tillage, 27
Coppicing, 194
Cordia dichtotoma, 149
Cosmic rays, 283
Cotton, 292
Cover crops, 223
Crop components, 76
Crop diversification, 39
Crop geometry, 20
Crop life saving measures, 44
Crop management for rainfed condition, 40
Crop residue potential, 238
Crop residues, 237
Crop residues, 240
- management, 236
Crop residues, 242
Crop residues, 243
Crop rotation, 78
Cropping in acid soil, 45
Cropping in alkali and saline soils, 46
Cropping pattern, 78
Cropping system, 78
- concept, 37
Cropping with animal and biogas system, 86
Cropping with aquaculture, 84
Cropping with dairy, 86
Cropping with dairy, biogas and silviculture in irrigated areas, 89
Cropping with goat and silvi-pasture in rainfed areas, 88
Cropping with goat rearing, 86
Crotalaria burhia, 177
Crotalaria juncea, 222, 227

Cryptomeria japonica, 169
Cucurbita, 150
Cyamopsis tetragonoloba, 222, 227
Cyanobacteria, 297
Cymbopogon fulvus, 172
Cynodon dactylon, 170, 174

—*D*—

Dalbergia sp., 148
- *jambolana*, 149
- *sissoo*, 145, 147, 149, 165, 169, 172, 176, 177, 179, 181

Day neutral plants, 287
Decennial rainfall, 294
Decomposition of green manure, 231
Degradation of hazardous chemical by microorganism, 323
Degradation of heavy metals by microbes and plants, 323
Degradation of pesticide residue, 211
Depletion of ozone layer, 307
Derris indica, 149, 229
Determinants of farming system, 101
Dhaincha, 227
Dicanthus annulatum, 170
Dichanthium annulatum, 165, 174
Diplachne fusca, 174
Disadvantages of GM, 224
Diversification of crop and cropping systems, 19
Dodonia viscosa, 228
Double cropping, 77
Drought proofing, 328

—*E*—

Eastern himalayan region, 90
Eco-friendly management of cropping system, 38
Eco-friendly practices, 38
Ecological basis of classification, 153
Ecological sustainability, 278
Economic viability, 10
Edaphic factors and crops, 298
Effect of precipitation, 294
Effect of residue, 240
Effective rainfall, 294
Efficient cropping systems, 37
Eichhornia crassipes, 48
Emblica officianalis, 149
Endogenous, 104
Energy environment, 281
Energy plantation, 136
Environment and plant growth, 278
Environment soundness, 10
Environmental issues in agriculture, 306
Enzymes, 245
Erianthus munja, 176
Erythrina spp, 149, 161
- *abyssinica*, 149
- *indica*, 178
- *variegata*, 165

Essential components of a sustainable agriculture, 5
Eucalyptus sp., 143, 148, 166
- *citriodora*, 181
- *globus*, 145
- *tereticomis*, 145, 170, 173

Eulaliopsis binata, 174
Evaluation of sustainability, 9
Exogenous, 104

—*F*—

Factors affecting interactions, 181
Faidherbia albida, 181, 182
Farm- based agrorestry practices, 164
Farm forestry, 135
Farm, 59

Farming situations, 83
Farming system research (FSR), 107
- and agricultural development, 114
- characteristics, 114
- classification, 112
- concept, definition and approach, 109
- versus conventional agricultural research, 110
Farming system, 59, 60
- adoptive research, 113
- analysis, 112
- bare-line data analysis, 114
- components, 66
- component research, 113
- interactions, 81
Farming systems approach, 324
Farming, 59
Farmyard manure (FYM), 245
Fenugreek, 46
Feronia limonia, 165
Fertilizers, 212
Fish-cum-pig-farming, 88
Fish-cum-poultry /duck farming, 88
Fish-cum-sericulture, 86
Flemingia macrophylla, 183
Fodder, 178
Food of the animal, 246
Forage crops, 223
Forest, 80
Forest-based agroforestry systems, 161
Frankia, 140
Fresnel lense, 258
Fuel, 178
Function of the animal, 246
Fungicides, 210

—G—

Gamelina arborea, 172
Gamma rays, 283
Gangaetic plains, 169
Gaseous emission, 32
Gliricidia spp, 147, 161
- *sepium*, 146, 149, 165, 177, 183
Global warming potential, 309
Global warming, 309
Glycine max., 27
Glyricidia, 80, 229
- *cebium*, 178
- *maculata*, 223, 229
Gmelix arborea, 174
Goal of FSR, 111
Graded bunds, 18
Green leaf manuring, 222, 228
Green manure as a main crop, 226
Green manure in cropping systems, 230
Green manuring (GM), 222
- *in situ*, 222
- value, 178
Greenhouse effect, 309
Grevillea robusta, 145
Grewia optiva, 169
Groundnut, 292
Guar, 227
Gullied and ravenous lands, 174

—*H*—

Hardwiakia binata, 178, 181
Harmonic farming system, 100
Hedgerow cropping or alley cropping, 146
Herbaspirillum, 253
Herbicides, 212
Heteropsylla cubana, 143
Hetropogan contortus, 165
Himalayan region and North-Eastern India, 168
Holoptelea integrifolia, 181
Home gardens, 149
Home–gardens/homesteads, 164

Horti-agri-silviculture, 203
Horticulture with fish farming, 85
Human environment, 103
Humid and sub-humid regions, 154, 171
Hydraulic properties, 243

—*I*—

Immobilization of nitrogen, 241
Impact of fertilizer use on environment, 31
Impermeable covering, 22
Improved soil fertility, 182
Increased water use efficiency (wue), 23
Increasing efficiency of rain water, 21
Index of sustainability, 12
Indigenous technology knowledge, 121
Infrared thermal radiation, 284
Insecticides, 210
Integrated agriculture and livestock components, 86
Integrated agriculture, animal and tree-components, 88
Integrated aquaculture - agriculture components, 84
Integrated aquaculture and animal components, 88
Integrated farming system (IFS), 81, 82, 216
Integrated green revolution farming, 216
Integrated nutrient management practices, 37
Integrated nutrient management system (inms), 34
Integrated nutrients management – key for sustainable production, 33
Integrated pest management and strategies, 47
Intercropping, 39, 77
Intermediate agroforestry system, 153
Ipomea batatas, 150
Irrigation and rainwater management, 328
Island region, 99

—*J*—

Jatropha, 322
Jute, 292

—*L*—

Land equivalent ratio, 184
Lasiurus sindicus, 176
Lasiurus sindicus, 177
Leptadenia pyrotechnica, 177
Leptochloa fusca, 174, 188
Leucaena, 80, 161
– *leucocephala*, 138, 145, 146, 148, 149, 165, 166, 167, 172, 176, 177, 178, 183, 193, 223
Ley farming, 79
Limitation of agroforestry, 142
Limitation of no-tillage system, 26
Litchi chinensis, 170
Live fence/biofence plantation, 136
Lolium rigidum, 212
Long day plant, 287
Lopping, 194
Loss of biodiversity, 319
Low input sustainable agriculture (LISA), 6
Lycopersicon esculentum, 150

—*M*—

Madhuca indica, 170, 172
Maize, 292
Management of agroforestry system, 187

Management of animal components, 66
Management of component species, 188
Management of trees, 192
Management practices in agroforestry, 188
Mangifera indica, 150, 165, 170, 172
Manipulation of micro-climate, 329
Manner of storage, 246
Measurement of solar energy use, 284
Measures to reduce air pollution, 324
Mentha arvensis, 100
Metal toxicities, 32
Methane, 322
Methods of growing GM crops *in situ*, 226
Michelia oblonga, 174
Micro-climate improvement, 183
Microclimate of crops, 282
Micro-nutrients, 250
Mid-season corrections, 43
Mineral nutrient supply, 303
Mineral salts, 233
Minespoils, 174
Mitigating pollution in crop environment, 321
Mixed cropping, 77
Mixed intercropping, 77
Monocropping / monoculture, 76
Moriculture, 74
Moringa oleifera, 147, 172, 177
Morrenia adorata, 50
Morus alba, 74
Morus indica, 74
Morus latifolia, 74
Morus serrata, 74
Mulch tillage, 27
Multiple cropping, 76
Multipurpose tree species, 146, 177
Multipurpose woodlots, 150
Multistory, 136
Multi–tier or multi–storied cropping, 78
Mungo, 222
Musa spp, 150
Mushroom cultivation, 75

—*N*—

Native phosphate, 247
Natural environment, 103
Nature of interactions, 180
Nature of the litter, 246
Near infrared radiation, 284
Need for organic farming, 213
Neem, 229
Neochitina bruchi, 48
New farming system development, 113
N-fixing tree species, 178
Nitrate pollution, 31
Nitrogen, 250
- compounds, 232
- fixation, 245
Nitrogenous compounds, 233
Nitrous oxide, 321
Non-chemical pest management, 217
Non-chemical weed control, 217
Non-competitive interaction, 64
No–tillage, 25
Nursery management, 190
Nutrient accumulation, 30
Nutrient cycling, 183
Nutrient imbalance, 32
Nutrient management for crops and cropping systems, 36
Nutrient management, 217
- in organic farming, 221

—O—

Objectives of FSR, 112
Objectives of GM, 223
Onset of monsoon, 41
Opportunity cropping, 78
Optimum sowing date and seed rate, 19
Opuntia ficus, 149
Organic carbon, 244
Organic farming
- principles and practices, 220
Organic farming, 208
- in India:, 263,
- in other countries:, 262
- in USA:, 261
Organic farming, 260
Organic fertilizer, 247
Organic matter, 250
Organic versus natural farming, 215
Organic *vs* conventional farming, 217
Ozone layer, 32

—P—

Panicum spp, 174
- *antidotale*, 176
- *maximum*, 172, 174
Panicum turgidum, 176
Parika javanica, 174
Parkinsonia aculeata, 173
Participatory rural appraisal, 120
Penicellium digitarium, 255
Pennisetum pureum, 174
Pennisetum purpureum, 172
Perserianthes falcataria, 174
Pest management strategies, 50
Pesticide residues, 318
Pesticides, 210
Phalaris minor, 39, 212
Phanerochaete chrysosporium, 323
Phaseolus spp, 150
- *trilobus*, 228
- *vulgaris*, 90, 324
Philippesara, 223
Phoenix dactylifera, 165
Phosphate solubilizing biofertilizers, 255
Photoperiodism, 287
Photosynthetically active radiation, 284
Phylanthus emblica, 173
Physical environment, 280
Physical properties of soil, 242
Phytophthora palmivora, 49, 184
Piggery, 71
Pinus caribea, 182
Pinus kesiya., 161
Pinus roxburghii, 148
Piper nigrum, 152
Plant suitable for GM, 225
Planting and after care, 191
Plearotus sp., 75
- *ostreatus*, 323
Political desirability, 10
Pollarding, 194
Pongamia glabra, 229
Pongamia pinnata, 149, 165, 172, 223
Populus sp., 145
- *deltoids*, 181
Porosity, 242
Poultry and duck rearing, 69
Practices of organic farming, 221
Precision agriculture, 40
Precision farming, 123
Principles involved in FSR, 111
Prnus cerasoides, 174
Productivity, 195
Prosopis chilensis, 176
Prosopis species, 146, 148, 149, 161

- *cineraria*, 88, 131, 149, 165, 176, 178, 181
- *juliflora*, 148, 165, 166, 170, 173, 178, 188, 190

Protected agriculture, 327
Pruning, 193
Prunus nepalensis, 168
Pseudomonas putida, 324
Pseudomonas striata, 255
Psidium guajava, 150, 170, 173
Pterocarpus santalinus, 179
Pungamia glabra, 178
Punica granatum, 165, 173
Pure organic farming, 216

—*Q*—

Quadruple cropping, 78
Qualitative evaluation of a sustainable system, 11
Querens lamellose, 168

—*R*—

Rabbit rearing, 72
Radiant energy, 283
Rate of decomposition, 240
Ratooning, 78
Recession farming, 78
Redox potential, 302
Reduced tillage, 27
Reducing greenhouse gases from agricultural source, 321
Reflected radiation, 284
Relative crowding co-efficient, 185
Relative humidity, 293
Relative sustainability
- classification, 5

Relay cropping, 78
Relay intercropping, 77
Residue management practices, 240
Response farming, 328
Revines, 174
Rhizobia, 224
Rhizobial culture, 30
Rhizobium, 30, 140, 190, 248
- application, 250

Rhizopus arrhizus, 323
Rhus sinuta, 149
Rhytocerus, 49
Rice, 291
Rice-fish farming, 84
Rice-fish-azolla farming, 85
Rice-fish-vegetable farming, 85
Ridge tillage, 27
Robiniapseudo acacia, 169
Role of organic matter and bio-fertilizers, 29
Role of temperature, 289
Root pruning, 194
Row intercropping, 77

—*S*—

Sacharum munga, 166
Salinity, 250
Salix spp, 193
Salt affected soils, 173
Salvadora aleoidea, 166
Sand dune management, 176
Schleichera oleosa, 172
Schwannimyces accidentalis, 255
Scope of FSR, 111
Seasmum, 293
Sehima nervosum, 168, 174
Semi-arid and arid lands, 154
Semi-arid tropics, 166
Senna siamea, 183
Sequential cropping, 77
Sericulture, 74
Sericulture, 74
Sesbania or wild dhaineha, 227
Sesbania spp, 80, 149, 178
- *aculeata*, 46, 222, 224, 227
- *aegyptica*, 145
- *grandiflora*, 146, 178
- *rostrata*, 226

– *sesban*, 178
– *speciosa*, 223, 227
Shade crops, 223
Sheep and goat rearing, 68
Shelter belt plantation, 136
Shelter belts and wind breaks, 176
Shelterbelt / windbreak and plantation, 147
Shifting cultivation, 79, 150, 161
Shorea robusta, 172
Short day plants, 287
Silivipastoral system, 201
Silvi-agriculture, 136
Silviculture, 136
Silvi-horticulture, 136
Silvi-horti-floticulture, 136
Silvi-horti-pasture, 136
Silvi-pastoral, 136
– system, 148
Silvi-sericulture, 136
Size of straw, 241
Social acceptability, 10
Social forestry, 135
Socioeconomic classification, 152
Socioeconomic determinants, 104
Soil air, 298
Soil and water resource management, 15
Soil conservation, 183
Soil conservative measures, 17
Soil degradation and pollution, 317
Soil degradation, 15
Soil fertility
– concept, 29
– management, 29
Soil modifications, 22
Soil moisture, 301
Soil pollution
– control, 326
Soil reaction, 243, 302
Soil structure, 242
Soil temperature, 243, 299
Solar energy, 283
Solar radiation effect on crops, 285
Solar radiation, 283
Source of manure, 246
South–West monsoon, 41
Soybean, 293
Sporobolus spp., 174
Strategy for agroforestry developments, 199
Strip cropping, 19
Strip intercropping, 77
Stylosanthes guianensis, 172
Stylosanthes hamata, 168
Subsistance agroforestry system, 153
Sugarcane, 292
Suitable areas for agroforestry, 188
Summer-sown catch crop, 226
Sunflower, 293
Sunlight and temperature, 250
Sunnhemp, 227
Sustainability, 195
– criteria, 8
– criteria for evaluating, 10
Sustainable agriculture, 1
Sustainable utilization of water resources, 23
Swarming, 73
Swidden cultivation, 161
Syzygium cuminii, 170, 173

—*T*—

Tamarindus indica, 149, 165, 166, 167, 224
Tamaris aphylla, 173
Tamarix auriculata, 166
Tamarix troupii, 173
Taungya cutivation, 161
Taungya system, 146

Taungya system, 79
Teaktona grandis, 165
Technique of green manuring *in situ*, 230
Tecomella undulata, 181
Temperature and plant growth, 288
Temperature requirement of certain crops, 291
Tephrosia purpurea, 222, 228
Terminalia arjuna, 172, 173, 190
Terminalia myriocarpa, 147, 152
Terrestrial radiation, 284
Thermal conductivity, 300
Thinning, 194
Thysanolaena agrostis, 168
Tictona grandis., 191
Topography modifications, 22
Total factor productivity, 11
Traditional agroforestory systems in India, 161
Traditional tillage system in India, 28
Traditional tillage, 28
Trametes hirsuta, 323
Tree arrangement, 195
Tree based components, 79
Tree- crop interactions, 179
Tree density / spacing, 192
Tree density, 195
Tree gardens, 146
Tree plantation, 324
Tree rectangularity, 195
Tree species for agroforestry, 187
Tree-crop interaction, 186, 197
Trees and shrubs in pastures, 149
Trees with plantation crop, 147
Trichoderma sp, 255
Trifolium alexandrinum, 222
Trigonella spp., 46
Trilobus, 222
Triple cropping, 78
Tripsacum laxym, 171
Tropical highland, 154
Tropical region, 170
Turnover of organic materials, 241
Types of GM, 222
Types of tillage systems, 25

—U—

Undersown crop, 226
Urban forestry, 136

—V—

Vapour pressure deficit, 293
Vasicular- arbuscular mycorrhiza (VAM), 255
Vicia faba, 262
Vigna radiata, 226
Vigna unguiculata, 222
Volvariella volvacea, 75

—W—

Waste recycling, 256
Water conservation in irrigated agriculture, 24
Water conservation, 183
Water crises and pollution, 315
Water harvesting and recycling, 21
Water pollution
- control of, 325
Water resource conservation, 20
Waterlogged and marshy lands, 175
Weather and climate, 283
Weed control, 183
Weed management, 257
Western dry region, 99
Western himalayan region, 90
Wheat, 292

Wind velocity, 295
Wodlots, 178

—X—

X-rays, 283

—Z—

Zea mays, 27

Zizyphus, 148
– *mauritiana*, 165, 170, 172, 173, 176
– *nummularia*, 165, 178
– *jujuba*, 149